NASA TECHNICAL NOTE

NASA TN D-7647

NASA TN D-7647

PARAMETER ESTIMATION TECHNIQUES AND APPLICATIONS IN AIRCRAFT FLIGHT TESTING

A Symposium held at
Flight Research Center
Edwards, Calif.
April 24-25, 1973

NATIONAL AERONAUTICS AND SPACE ADMINISTRATION • WASHINGTON, D. C. • APRIL 1974

1. Report No. NASA TN D-7647	2. Government Accession No.	3. Recipient's Catalog No.
4. Title and Subtitle PARAMETER ESTIMATION TECHNIQUES AND APPLICATIONS IN AIRCRAFT FLIGHT TESTING - A Symposium at the NASA Flight Research Center, April 24-25, 1973		5. Report Date April 1974
		6. Performing Organization Code
7. Author(s)		8. Performing Organization Report No. H-806
9. Performing Organization Name and Address NASA Flight Research Center P. O. Box 273 Edwards, California 93523		10. Work Unit No. 501-26-01
		11. Contract or Grant No.
12. Sponsoring Agency Name and Address National Aeronautics and Space Administration Washington, D.C. 20546		13. Type of Report and Period Covered Technical Note
		14. Sponsoring Agency Code

15. Supplementary Notes

16. Abstract

Technical papers were presented by selected representatives from industry, universities, and various Air Force, Navy, and NASA installations. The topics covered included the newest developments in identification techniques, the most recent flight-test experience, and the projected potential for the near future.

17. Key Words (Suggested by Author(s)) Parameter estimation techniques, systems identification, nonlinear identification, aircraft stability and control derivatives, flight test analysis methods, estimation of structural parameters	18. Distribution Statement Unclassified - Unlimited		
19. Security Classif. (of this report) Unclassified	20. Security Classif. (of this page) Unclassified	21. No. of Pages 390	22. Price* $8.00

FOREWORD

Over the past several years there has been a renewed interest in determining dynamic aircraft parameters, such as stability and control derivatives, from flight-test measurements. The need for these data has persisted for many years, but only recently have highly automated data acquisition systems and advanced estimation techniques been available that can extract such information efficiently. Most flight-test organizations now have experience with one or more parameter estimation techniques to determine aircraft stability and control derivatives. The technology stands at the threshold of applications for numerous other parameter estimation problems in aircraft flight testing.

To provide a forum for discussion of the status and future of this technology, the NASA Flight Research Center hosted a Symposium on Parameter Estimation Techniques and Applications in Aircraft Flight Testing. Technical papers were presented by selected representatives from industry, universities, and various Air Force, Navy, and NASA installations who are actively working in the field. The topics covered included the newest developments in identification techniques, the most recent flight-test experience, and the projected potential for the near future.

Both formal and informal papers were presented in the technical sessions. The formal papers were complete technical papers, and the informal papers were brief summaries of preliminary results not ready for final publication or recent results that were important to the symposium but had been published previously. This volume contains the complete formal papers and abstracts of the informal papers. The publication of these papers does not constitute approval of their technical content by the National Aeronautics and Space Administration, but rather provides for exchange of information and, hopefully, stimulation of new ideas.

Dr. A. V. Balakrishnan, Chairman, Systems Science Department, School of Engineering and Applied Science, University of California, Los Angeles, was the guest speaker at the symposium dinner on April 24. A transcription of his talk, which concentrated on the major activities in systems identification outside the United States, is included.

Herman A. Rediess
Symposium Technical Chairman

CONTENTS

General Chairman: Herman A. Rediess, NASA Flight Research Center

INTRODUCTORY PAPERS

FLIGHT TEST EXPERIENCE: PART 1

Session Chairman: D. T. Berry, NASA Flight Research Center

FLIGHT TEST EXPERIENCE: PART 2

Session Chairman: M. J. Queijo, NASA Langley Research Center

NONLINEAR MODEL IDENTIFICATION

Session Chairman: R. C. Wingrove, NASA Ames Research Center

NEW TECHNIQUES AND ALGORITHMS

Session Chairman: R. K. Mehra, Harvard University

IDENTIFIABILITY, SENSITIVITY, AND ACCURACY

Session Chairman: C. M. Fry, Southern Methodist University

SPECIAL APPLICATIONS

Session Chairman: E. E. Kordes, NASA Flight Research Center

CONCLUDING PAPER

GUEST SPEAKER

AN OVERVIEW OF PARAMETER ESTIMATION TECHNIQUES AND

APPLICATIONS IN AIRCRAFT FLIGHT TESTING

Herman A. Rediess
NASA Flight Research Center

SUMMARY

This paper introduces parameter estimation as it applies to aircraft flight testing and presents an overview of the symposium. The evolution of techniques used in flight testing is reviewed briefly, and it is pointed out how the changing character of the aircraft tested and the availability of advanced data systems have promoted this evolution. Recent advances in optimal estimation theory have stimulated widespread interest and activity in parameter estimation. The framework of these advanced techniques is outlined to set the stage for subsequent papers. The session topics are introduced and related to the requirements of flight-test research.

INTRODUCTION

Over the past several years there has been a renewed interest in determining dynamic aircraft parameters, such as stability and control derivatives, from flight-test measurements. The need for these data has long persisted, but only recently have highly automated data acquisition systems and advanced estimation techniques been available that can extract such information efficiently. Most flight-test organizations now have experience with one or more parameter estimation techniques to determine aircraft stability and control derivatives. The technology stands at the threshold of application to numerous other parameter estimation problems in aircraft flight testing.

This symposium was organized to provide a forum for discussion of the status and future of the technology. This paper discusses, first, the need for this technology in aircraft flight research and the evolution of techniques over many years; then the basic concepts and approach of contemporary parameter estimation techniques are outlined; and finally the session topics are introduced and related to the needs of flight-test research.

1

BACKGROUND

Flight test has always been important in advancing aircraft technology, not by merely showing that a new design flies, but by obtaining quantitative measures of what is good and bad about the design so that future designs can be improved. The application of parameter estimation techniques to aircraft flight testing is simply the process of obtaining quantitative measures of various aircraft characteristics. In general, the parameters may relate to aerodynamic, structural, performance, or other types of characteristics. Typically, the flight-determined characteristics are compared with predicted values to verify or point out deficiencies in the prediction techniques. They are used to substantiate design goals, to assess control system performance, to verify and improve piloted simulators, and to establish design criteria. Clearly, the need for such data has not diminished.

Unfortunately, most of the airplane characteristics of interest, such as the aerodynamic forces, cannot be measured directly in flight. It has been necessary to develop techniques that use measurable quantities, such as the motion resulting from a change in a force. There are many possible approaches to the problem, and the search for more accurate and efficient methods is the prime motivation for continued activity in this field.

Since the determination of aerodynamic characteristics in flight, mainly stability and control derivatives, has been the principal aircraft application of parameter estimation, this review concentrates on those activities. Only the key technology changes are traced; no attempt is made to provide a complete survey of the field.

The evolution of techniques has been motivated primarily by two factors: the changing nature of the dominant aircraft dynamics as higher performance was obtained, and the desire to have more effective techniques in terms of improved accuracy or improved efficiency in applying them, or both.

From the beginning of manned flight to about World War II, aircraft longitudinal dynamics were characterized by the long-period or phugoid motion and lateral-directional dynamics by the rolling and quasi-steady-state yawing characteristics. The most important aerodynamic characteristics affecting aircraft motion could be measured in flight with acceptable accuracy by analyzing steady-state conditions.

One of the first test programs to obtain quantitative measurements of aircraft aerodynamic characteristics in flight was reported on by Warner and Norton in 1919 (ref. 1). The tests were made on two Curtiss "Jenny" JN-4H-type biplanes at Langley Field, Va. Lift and drag coefficients were determined by means of a series of static maneuvers performed at different airspeeds. The lift and drag coefficients were estimated by equating the lift to the aircraft weight and the drag to the thrust, assuming certain engine thrust characteristics. The in-flight measurements were airspeed, inclination of the aircraft (pitch attitude), and engine speed. The instrumentation and analysis techniques were rudimentary, but it was an important beginning.

In 1922 Norton and Brown determined the roll control and damping coefficients of a biplane by analyzing the initial and steady-state portions of a rolling maneuver (ref. 2); and in 1923 Norton estimated the longitudinal static stability and damping coefficients by analyzing data from a combination of static maneuvers and phugoid oscillations (ref. 3). Soule and Wheatley (ref. 4) appear to be the first to have determined all the major longitudinal stability and control derivatives of an airplane from flight-test data and compared the results with theoretical predictions. The analysis in each of these studies used simplified equations representing one-degree-of-freedom motion and solved for one parameter at a time, assuming values for other parameters based on wind-tunnel tests or other flight tests. This basic approach was used with only minor changes up to the mid-1940's.

As higher performance aircraft were being developed during the 1940's, the nature of aircraft dynamics affecting longitudinal flying qualities was changing. A comparison of the discussions of Soule in 1940 (ref. 5) and Phillips in 1949 (ref. 6) shows that the dominant characteristic shifted from the phugoid mode in 1940 to the short-period mode in 1949. The quasi-steady-state techniques that had been used to define the phugoid parameter were inadequate for determining the short-period stability derivatives. Milliken also pointed out in 1947 (ref. 7) that the increasing use of automatic control systems required more accurate modeling of the aircraft dynamic characteristics. These factors, coupled with the research engineer's motivation to improve the accuracy of flight results, stimulated the development of several new techniques for determining stability and control characteristics from flight data.

In the late 1940's through the mid-1950's, servomechanism theory was expanded rapidly, and the frequency-domain techniques of Nyquist and Bode were popular. It was a natural extension to use frequency-response techniques for determining the dynamic characteristics of an airplane from flight-test data. The first approach (ref. 7) was to obtain a frequency response in flight by oscillating the airpane, by means of the autopilot, at discrete frequencies and measuring the steady-state amplitude ratio and phase angle between the control surface and a response variable, such as pitch rate. A disadvantage of this approach was the considerable flight time required to sweep through all the frequencies of interest at each flight condition. Seamans, Blasingame, and Clementson (ref. 8) used a method for determining the aircraft frequency response from a single transient response maneuver by Fourier analysis. This technique greatly reduced the flight-test time. Greenberg (ref. 9) discusses several frequency-response methods. If the aircraft frequency response or transfer function is the final result desired from a flight test, these methods are appropriate. However, if the stability and control derivatives are needed, another step must be taken to relate the derivatives to the measured frequency response (for example, see refs. 7 and 10).

Because the problem of determining stability and control derivatives is based on a linearized, small-perturbation model of the aircraft dynamics, it was natural to consider using a linear least-squares fit of flight data to the linearized equations of motion as Greenberg did in 1951 (ref. 9). In 1954 Shinbrot developed a generalized least-squares method which encompassed the earlier least-squares methods and had a greater potential (ref. 11). A real drawback to these methods at that time was that they involved extensive calculations which had to be done by hand, because digital computers were not yet available. Furthermore, it was desirable to "fit" the equations at many time points in order to obtain good accuracy; this meant that a large volume of data had to be processed manually from flight film or oscillograph recordings. A fundamental problem with linear least-squares methods is that noisy measurements result in biased estimates of the stability derivatives. However, the general lack of acceptance of the methods was attributed more to the difficulty of applying them than to concern over biased estimates.

A rather simple technique was often used in the 1950's for determining the longitudinal short-period parameters. This technique is still adequate for many situations. When the short-period-mode frequency is much greater than the phugoid frequency and the damping ratio is low (less than 0.3), the primary short-period-mode stability derivatives can be estimated directly from measurements of the frequency, damping, and amplitude ratio of normal acceleration to angle of attack (ref. 12). Similar approximate methods were not satisfactory for the highly coupled lateral-directional dynamics, but an effective graphical technique developed by Doetsch (refs. 13 and 14) was used extensively. These techniques were straightforward and not difficult to apply but required ideal, free-oscillation maneuvers. As aircraft performance reached progressively higher Mach numbers, the damping decreased to such low values that at times it was too risky to obtain test data without the damper systems turned on. Attempts were made to correct for the effect of the damper system (ref. 15), but the empirical approach used left considerable uncertainty in the results. Basically, these simple techniques were applicable only if there were no pilot or automatic control system inputs during the free oscillation.

A technique called "analog matching" was used to overcome the problem of poorly conditioned maneuvers (ref. 15). It is a manual curve-fitting technique in which an analog computer is used to compute the response of a model. This response is then "matched" to the flight-measured response by adjusting the stability and control derivatives of the model. This approach was not a spontaneous development for determining derivatives but was, rather, an outgrowth of the use of analog computers as flight simulators. Analog matching was used as early as 1951 to check aircraft parameters determined by other methods (ref. 16). Even though the technique of analog matching has been greatly improved (ref. 17), the accuracy of the results is highly dependent on the skill of the individual operator. Furthermore, it can take an excessive number of man-hours to obtain an acceptable solution if several parameters are to be determined.

The best techniques available up to 1966 are reviewed by Wolowicz in reference 18. The practical aspects of applying the techniques to flight data are covered particularly well.

Although several dedicated engineers were working throughout the 1950's and early 1960's to improve techniques, the effort was relatively small and was concentrated at two or three flight research installations. Two factors caused a revolution in parameter estimation techniques starting in the mid-1960's: (1) Highly automated data acquisition systems were becoming standard in flight testing, and (2) large-capacity, high-speed digital computers were available to solve complicated algorithms efficiently. The ability to transfer the flight data directly to the computer with no manual operations on the data and the availability of high-speed computation permitted techniques to be considered that were previously impractical. Though seldom mentioned in recent literature, in 1951 Shinbrot developed the concept (ref. 19) that is fundamental to many contemporary techniques. At that time, however, it was not practical to use his concept, which involved manually computing the numerical minimization of a nonlinear functional. An application to the simplest flight-test problem of determining only four longitudinal parameters took up to 24 hours.

Interest in parameter estimation was renewed in 1968. Larson applied the method of quasi-linearization at Cornell Aeronautical Laboratory (ref. 20), and Taylor and Iliff applied basically the same method, but referred to as the modified Newton-Raphson technique, at the NASA Flight Research Center (refs. 21 and 22). The latter technique was based on the theoretical works of Balakrishnan (refs. 23 to 25). There have been numerous parallel developments since then in universities, private research companies, and major aircraft companies, as well as at Air Force, Navy, and NASA installations (for example, refs. 26 to 31). The papers presented at this symposium will cover work performed since 1968, with projections to the future.

This review has considered only the activities that have been most specifically involved in applications to aircraft flight testing and that have taken place primarily within the United States. Much work has been done in developing the general systems identification, estimation, filtering, and control theories and in applying them specifically to other fields (ref. 32, for example). Often, systems identification or parameter estimation is included as an important aspect of adaptive control systems research (refs. 25, 28, and 33). A complete survey would obviously be much broader than this limited review.

SYSTEMS IDENTIFICATION AND PARAMETER ESTIMATION

The meaning of the term "parameter estimation techniques" is not always clear to the nonspecialist. In aircraft flight testing, the terminology that has been used traditionally is "determination of stability and control derivatives," but this does not cover the present scope of the technology. It is appropriate, therefore, to define clearly at the outset what is meant by "parameter estimation techniques."

Parameter estimation techniques are methods used in systems identification problems. The general problem of systems identification (fig. 1) is to determine certain characteristics of the physical system from experimental test data. Measurements are made of external inputs and resulting output responses that depend in some way on the system characteristics to be determined. There may also be external disturbances that cannot be measured directly. Systems identification is the process of estimating the characteristics from the input/output measurements. Usually something is known beforehand about the system, such as the set of equations that describes its dynamic responses and approximate values of the forces and moments on the system. However, systems identification theory also includes the situation in which nothing is known except the input/output measurements.

There are several approaches to solving systems identification problems, and all are strongly influenced by the amount and type of a priori knowledge available. Parameter estimation techniques are the most common approaches. The general concept is illustrated in figure 2 for a flight-test situation. As a specific example of this concept, consider the problem of determining the stability and control characteristics for small perturbations about a trim flight condition. The types of data used are shown in figure 3, which is from a flight test of a lifting-body vehicle (ref. 22). The control inputs are small-amplitude aileron and rudder pulses, and the measured responses are roll rate, yaw rate, sideslip angle, bank angle, and lateral acceleration. External random disturbances (turbulence) were negligible. These data were recorded as pulse code modulation signals on magnetic tape, then formatted, scaled, and restored on tape for reading into a digital computer. The recorded inputs were used as inputs to the mathematical model, and the recorded response was compared with the computed response. The model in this case was the set of linearized differential equations for lateral-directional motion, and the parameters to be estimated were the linear coefficients, which are referred to as stability and control derivatives.

Typically the techniques start with some a priori estimate of the derivatives, such as wind-tunnel data. Usually the wind-tunnel data do not provide a good match, as shown in figure 4. In applying parameter estimation techniques, some algorithm is devised to adjust the stability and control derivatives in the model until a set is obtained that minimizes the error between the computed and measured time histories. A typical match is shown in figure 5.

The conceptual diagram in figure 2 and this example point out five key aspects of parameter estimation techniques: (1) the mathematical model, (2) the estimation criterion, (3) the computational algorithm, (4) the total data acquisition system, and (5) the test input.

Mathematical Model

A model must be selected that adequately represents the aircraft characteristics to be measured. The aerodynamic forces and moments on an aircraft are nonlinear functions of several variables, such as Mach number, angle of attack, control surface deflection, and sideslip angle. There may be significant structural modes, aeroelastic effects, nonstationary aerodynamic effects, and flow separation. Yet in many instances it is adequate to use a stationary, linearized, rigid-body model. In other instances a more complicated model is necessary, such as at very high angles of attack for which a nonlinear model may be required. An inappropriate model, however, can degrade the accuracy of the parameter estimate and even prevent convergence of the computation algorithm.

Estimation Criterion

There must be some means of assessing the fit of the computed response to the flight-measured response. When the process is implemented manually, as in analog matching (ref. 17), the operator makes the assessment on the basis of his experience by visually comparing the time responses. In the automated techniques a "criterion function" is used as indicated in figure 2. It is usually some form of an integral square of the error between the computed and measured response. Under certain conditions, it corresponds to the maximum likelihood criterion (ref. 23), which is intended to produce the most probable values of the parameters. The "best" estimate of the parameters is the set of parameters that minimizes the criterion function.

Computational Algorithm

The criterion function is nonlinear with respect to the parameters to be estimated; therefore, it has to be minimized by an iterative computational algorithm. Several algorithms that have been used are steepest descent, Newton-Raphson, modified Newton-Raphson (also referred to as quasi-linearization or differential correction), various conjugate gradient methods, various direct and random search techniques, stochastic approximation, and an iterative Kalman filter method (refs. 32 and 34). The nature of nonlinear minimization is such that no one algorithm can be classified as the best for all problems. Important factors in selecting the minimization algorithm are startup routines, convergence, computational efficiency, dimensionality, local minima, and whether the data processing will be on-line or batch.

Data Acquisition System

Parameter estimation is highly dependent on the quality of the flight-measured data. In figure 2 the uncertainties in the measured flight response are represented by additive noise. In reality the "noise" comes from several sources, and it is necessary to consider the entire data acquisition process. Bias and random errors can arise from imprecise location or orientation of sensors, calibration of the measurement and recording system, and data drop out. Other errors can be introduced from electrical noise, engine vibration pickup, sensor dynamics, inappropriate signal filters, and quantization. A comprehensive discussion of flight-test instrumentation for aircraft parameter estimation is given in reference 18. Any elimination of errors, noise, or uncertainties within the data acquisition process will improve the accuracy of the estimates. However, it is not always possible to optimize the instrumentation specifically for parameter estimation because of project schedule or manpower limitations. It is desirable to have techniques that are effective despite measurement noise and uncertainties.

Test Input

As a minimum requirement, the test input must excite the principal response modes that depend on the parameters to be determined. In the previous example a combination of rudder and aileron pulses adequately excited the lateral-directional motion. Thus the question arises whether one type of control input might be better than another, in the sense that it provides better estimates. Several papers have considered that question (ref. 35, for example) and have shown that in theory a test input can be found that will tend to minimize the variance of the estimated parameters. This concept has not been fully explored in a flight-test application.

Although there are variations from the basic concept depicted in figure 2, depending on the specific situation being considered or the mathematical approach used, the five key aspects listed on page 6 are always included in some form. For example, if there are nonmeasured external disturbances, such as air turbulence, it is necessary to add a state estimator to these five aspects.

It should be clear at this point that parameter estimation techniques are not restricted to determining linear stability and control derivatives. Virtually any flight test that can be modeled as an input/output dynamic system is amenable to this approach. The search for possible new applications to flight testing has just begun. Areas being considered include nonlinear aerodynamics, aircraft structural dynamics, estimates of external disturbances such as air turbulence and vortex wakes, and airframe/propulsion system interaction effects.

INTRODUCTION TO SYMPOSIUM SESSIONS

This symposium was organized to emphasize the practical and important aspects of applying parameter estimation techniques to aircraft flight-testing problems. Determining linear stability and control derivatives is emphasized because this is where most of the practical experience has been gained. The intent is to discuss the good and the bad features of the various methods used and to consider what improvements are needed. Also, several promising new areas of application are introduced.

Sessions I and II, Flight Test Experience, present recent experience of seven different organizations in determining stability and control derivatives. The contemporary techniques have been applied effectively to virtually every class of aircraft, including general aviation aircraft, jet transports, high-performance fighters, STOL and V/STOL aircraft, helicopters, lifting-body vehicles, and a host of special research aircraft. The papers in these sessions discuss the effectiveness and practical aspects of using various techniques including one on-line method. They point out special considerations that are needed for certain aircraft such as STOL and helicopters.

Two papers present new methods that can extract derivatives from data contaminated by turbulence. Previously such data had to be discarded for lack of an analysis method. In one lifting-body flight program at the NASA Flight Research Center, about 10 percent of the data was unusable because of turbulence. A 10-percent loss of data was significant in a program as expensive to conduct as the lifting-body program or as the initial shuttle flight-test program may be.

Session III, Nonlinear Model Identification, presents the early efforts to determine nonlinear aerodynamic characteristics at high angles of attack. The goal is to estimate the force and moment coefficients directly as nonlinear functions of angle of attack and angle of sideslip rather than as the usual linearized derivatives. This topic is timely because of the great interest within DOD and NASA in the stall/spin problem. Considerable effort is being directed toward achieving a better understanding of the causes of spin departure and how to prevent it. Large- or full-scale flight test is important because of Reynolds number effects in this nearly separated flow region at high angles of attack.

Session IV, New Techniques and Algorithms, includes three new approaches to the parameter estimation problem. As mentioned previously, there is no clear-cut best method for all problems, and so the search for more efficient techniques continues. Computational efficiency increases in importance as more complex models are used. Nonlinear models increase the dimensionality of the estimation problem. To consider turbulence effects on flexible aircraft, it will be necessary to include structural modes in the model as well as a turbulence model. The only practical way to analyze these higher order systems is to develop more effective algorithms.

Session V, Identifiability, Sensitivity, and Accuracy, considers in detail several aspects of parameter estimation that are important to the specialist. Knowing the conditions under which a system can be identified has not been of concern to the practicing engineer in determining stability and control derivatives because the problem had been exercised for so long that he had a good "feel" for it. As new applications are being considered, identifiability and the closely related problem of modeling will become increasingly important.

Other topics covered in this session include effects of instrumentation errors on the accuracy of estimated parameters and optimizing the test input. Both of these areas have been recognized as important for many years, but only recently have they been treated in this fairly rigorous manner.

Session VI, Special Applications, is a collection of papers treating some of the intriguing new developments and applications of parameter estimation technology.

Systems identification is an integral part of adaptive control from a theoretical viewpoint. To develop an optimal or near optimal control law for an aircraft operating throughout a wide range of flight environmental conditions that affect its static and dynamic response, it is necessary to identify its characteristics on-line. The so-called optimal adaptive systems (ref. 36) are complex and have not been considered practical to implement.

An adaptive control paper is presented for two reasons: It represents an important class of applications of parameter estimation; and it presents a technique for on-line parameter estimation that not only provides for a practical implementation of an adaptive system but also could be used for rapid on-line analysis of flight-test data.

Two papers consider the problem of determining aircraft structural mode parameters. This is going to become an increasingly important area of research and development because of the need for structural modal suppression systems on large supersonic aircraft. Three recent, independent, unpublished studies for NASA by Boeing, McDonnell Douglas, and Lockheed on advanced supersonic technology all stated that an advanced supersonic transport should be designed with a structural modal suppression system to provide for passenger comfort. The need to define the structural mode characteristics precisely is pointed out in reference 37. Small uncertainties in the structural mode characteristics can cause unacceptable system performance.

The last paper discusses another problem area that can affect supersonic aircraft with sophisticated propulsion systems. Forces induced on the airframe by operation of the engine inlets can cause unfavorable effects on the aircraft dynamics. Interaction among the propulsion system, flight control system, and airframe must be better understood so that the aircraft stability, control, and performance can be optimized simultaneously over the entire operational envelope of future advanced supersonic aircraft.

FUTURE POTENTIAL

The future effect of applying parameter estimation techniques to aircraft flight testing is anticipated to be twofold: It should reduce the overall schedule of a flight-test program and at the same time increase the amount of information extracted from the tests. Figure 6 depicts these benefits qualitatively. The quantitative improvements are yet to be determined.

The reduced schedule is expected to result from savings in analysis time, through more effective methods; flight time, through use of dynamic tests in place of the more time-consuming quasi-static tests, such as those used in measuring lift/drag polars; and preparation time, by not requiring special instrumentation for each research task. These savings should result in lower flight-test costs.

Several papers in the symposium show the trend toward increased information output. Data previously discarded because of turbulence contamination are now usable. It should be possible to use a single dynamic maneuver to obtain lift and drag as well as stability and control coefficients. With further improvements and refinements, other combinations may be possible, such as determining structural modes, turbulence, and stability and control characteristics simultaneously.

In view of the capabilities and potentials of parameter estimation techniques, flight-test engineers should reevaluate the test procedures and instrumentation that have been used for many years. It would be interesting to see what innovative approaches might result.

REFERENCES

1. Warner, Edward P.; and Norton, F. H.: Preliminary Report on Free Flight Tests. NACA Rep. No. 70, 1919.

2. Norton, F. H.; and Brown, W. G.: Controllability and Maneuverability of Airplanes. NACA Rep. No. 153, 1922.

3. Norton, F. H.: A Study of Longitudinal Dynamic Stability in Flight. NACA Rep. No. 170, 1923.

4. Soulé, Hartley A.; and Wheatley, John B.: A Comparison Between the Theoretical and Measured Longitudinal Stability Characteristics of an Airplane. NACA Rep. No. 442, 1933.

5. Soulé, H. A.: Preliminary Investigation of the Flying Qualities of Airplanes. NACA Rep. No. 700, 1940.

6. Phillips, William H.: Appreciation and Prediction of Flying Qualities. NACA Rep. 927, 1949.

7. Milliken, William F., Jr.: Progress in Dynamic Stability and Control Research. J. Aeron. Sci., vol. 14, no. 9, Sept. 1947, pp. 493-519.

8. Seamans, R. C., Jr.; Blasingame, B. P.; and Clementson, G. C.: The Pulse Method for the Determination of Aircraft Dynamic Performance. J. Aeron. Sci., vol. 17, no. 1, Jan. 1950, pp. 22-38.

9. Greenberg, Harry: A Survey of Methods for Determining Stability Parameters of an Airplane From Dynamic Flight Measurements. NACA TN 2340, 1951.

10. Donegan, James J.; Robinson, Samuel W., Jr.; and Gates, Ordway B., Jr.: Determination of Lateral-Stability Derivatives and Transfer-Function Coefficients From Frequency-Response Data for Lateral Motions. NACA Rep. 1225, 1955.

11. Shinbrot, Marvin: On the Analysis of Linear and Nonlinear Dynamical Systems From Transient-Response Data. NACA TN 3288, 1954.

12. Angle, Ellwyn E.; and Holleman, Euclid C.: Determination of Longitudinal Stability of the Bell X-1 Airplane From Transient Responses at Mach Numbers up to 1.12 at Lift Coefficients of 0.3 and 0.6. NACA RM L50I06a, 1950.

13. Doetsch, K. H.: The Time Vector Method for Stability Investigations. Rep. Aero 2495, Royal Aircraft Establishment, Aug. 1953.

14. Wolowicz, Chester H.: Time-Vector Determined Lateral Derivatives of a Swept-Wing Fighter-Type Airplane With Three Different Vertical Tails at Mach Numbers Between 0.70 and 1.48. NACA RM H56C20, 1956.

15. Yancey, Roxanah B.; Rediess, Herman A.; and Robinson, Glenn H.: Aerodynamic-Derivative Characteristics of the X-15 Research Airplane as Determined From Flight Tests for Mach Numbers From 0.6 to 3.4. NASA TN D-1060, 1962.

16. Triplett, William C.; and Smith, G. Allan: Longitudinal Frequency-Response Characteristics of a 35° Swept-Wing Airplane as Determined From Flight Measurements, Including a Method for the Evaluation of Transfer Functions. NACA RM A51G27, 1951.

17. Rampy, John M.; and Berry, Donald T.: Determination of Stability Derivatives From Flight Test Data by Means of High Speed Repetitive Operation Analog Matching. FTC-TDR-64-8, Air Force Flight Test Center, May 1964.

18. Wolowicz, Chester H.: Considerations in the Determination of Stability and Control Derivatives and Dynamic Characteristics From Flight Data. AGARD Rep. 549-Part 1, 1966.

19. Shinbrot, Marvin: A Least Squares Curve Fitting Method With Applications to the Calculation of Stability Coefficients From Transient-Response Data. NACA TN 2341, 1951.

20. Larson, Duane B.: Identification of Parameters by the Method of Quasi-linearization. CAL Rep. No. 164, Cornell Aero. Lab., Inc., May 14, 1968.

21. Taylor, Lawrence W., Jr.; and Iliff, Kenneth W.: A Modified Newton-Raphson Method for Determining Stability Derivatives From Flight Data. Computing Methods in Optimization Problems - 2, Lotfi A. Zadeh, Lucien W. Neustadt, and A. V. Balakrishnan, eds., Academic Press, 1969, pp. 353-364.

22. Iliff, Kenneth W.; and Taylor, Lawrence W., Jr.: Determination of Stability Derivatives From Flight Data Using a Newton-Raphson Minimization Technique. NASA TN D-6579, 1972.

23. Balakrishnan, A. V., ed.: Communication Theory. McGraw-Hill Book Co., c.1968.

24. Balakrishnan, A. V.: Techniques of System Identification. Facoltá di Ingegneria, Universitá di Roma, Corso di Specializzazione in Ingegneria dei Controlli Automatici, SISTEMA (Via Umberto Biancamano 23, Roma), Sept. 1968.

25. Balakrishnan, A. V.; and Peterka, V.: Identification in Automatic Control Systems. Survey Paper 9, presented at Fourth Congress of the International Federation of Automatic Control, Warsaw, Poland, June 16-21, 1969.

26. Mehra, Raman K.: Maximum Likelihood Identification of Aircraft Parameters. Preprints of Technical Papers presented at 1970 Joint Automatic Control Conference of the American Automatic Control Council, Georgia Institute of Technology, Atlanta, Ga., June 22-26, 1970, pp. 442-444.

27. Denery, Dallas G.: Identification of System Parameters From Input-Output Data With Application to Air Vehicles. NASA TN D-6468, 1971.

28. Balakrishnan, A. V.: Identification and Adaptive Control: An Application to Flight Control Systems. J. Optimization Theory and Applications, vol. 9, no. 3, Mar. 1972, pp. 187-213.

29. Mehra, Raman K.: Identification of Stochastic Linear Dynamic Systems Using Kalman Filter Representation. AIAA J., vol. 9, no. 1, Jan. 1971, pp. 28-31.

30. Chen, Robert T. N.; Eulrich, Bernard J.; and Lebacqz, J. Victor: Development of Advanced Techniques for the Identification of V/STOL Aircraft Stability and Control Parameters. CAL Rep. No. BM-2820-F-1, Cornell Aero. Lab., Inc., Aug. 1971.

31. Sage, Andrew P.; and Melsa, James L.: System Identification. Academic Press, N.Y., 1971.

32. Åstrom, K-J.; and Eykoff, P.: System Identification — a survey. Identification and Process Parameter Estimation, Part 1, Preprints of the 2nd Prague IFAC Symposium, Czechoslovakia, 15-20, June 1970.

33. Price, Charles F.; and Koenigsberg, William D.: Adaptive Control and Guidance for Tactical Missiles. Vol. I: Parts I and II - Introduction and Adaptive Control Theory, Vol. II: Parts III and IV - Adaptive Control Applications and Guidance, TR-170-1, The Analytic Sciences Corp., June 30, 1970.

34. Aoki, M.: Introduction to Optimization Techniques. The MacMillan Co., N.Y., 1971.

35. Mehra, Raman K.: Optimal Inputs for Linear System Identification. Preprints of Technical Papers presented at 1972 Joint Automatic Control Conference of the American Automatic Control Council, Stanford Univ., Stanford, Calif., Aug. 16-18, 1972, pp. 811-820.

36. Rediess, Herman A.: Theoretical Perspective of Adaptive Control Techniques and Modern Control Theory. Advanced Control System Concepts, AGARD C. P. No. 58, 1970, pp. 31-39.

37. Wykes, John H.; and Kordes, Eldon E.: Analytical Design and Flight Tests of a Modal Suppression System on the XB-70 Airplane. Part 1: Design Analysis. Part 2: Flight Tests. Aeroelastic Effects From a Flight Mechanics Standpoint, AGARD C. P. No. 46, Mar. 1970, pp. 23-1 - 23-18.

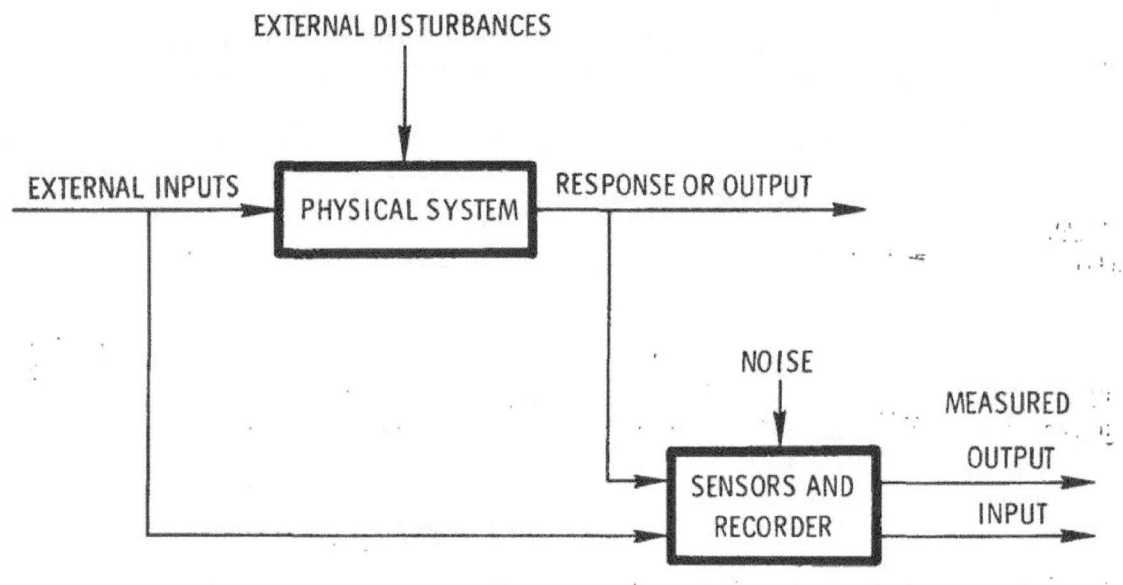

Figure 1.   General systems identification problem.

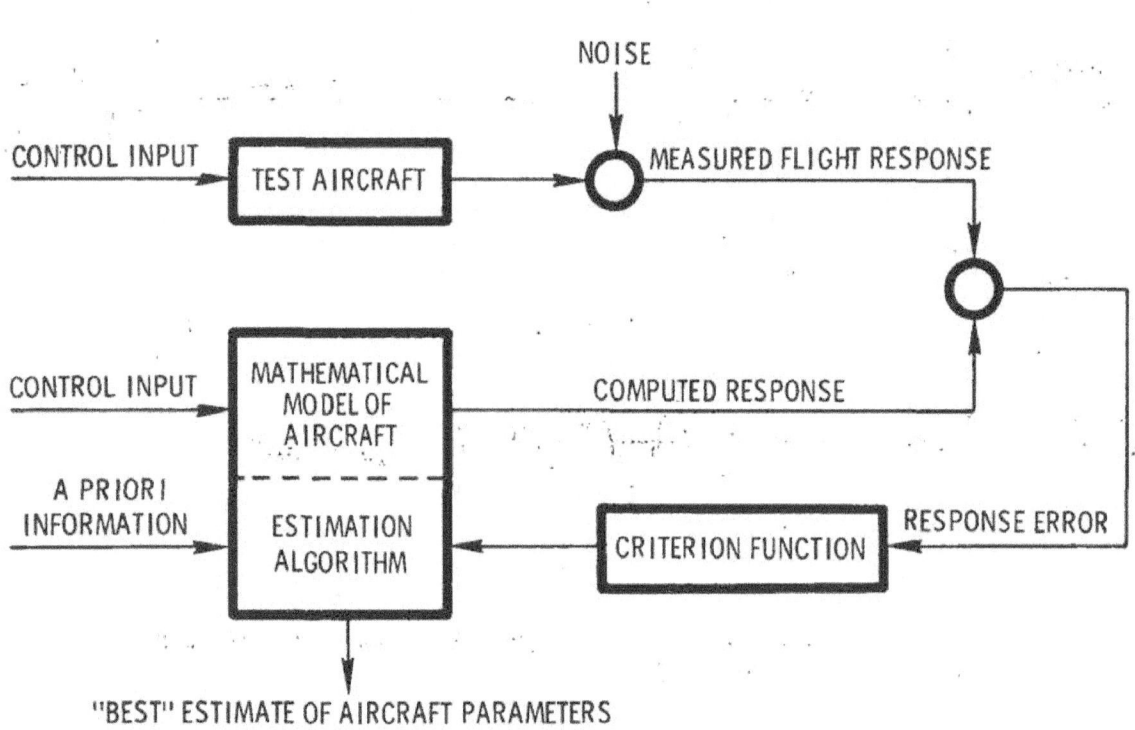

Figure 2.   Basic concept of contemporary parameter estimation techniques.

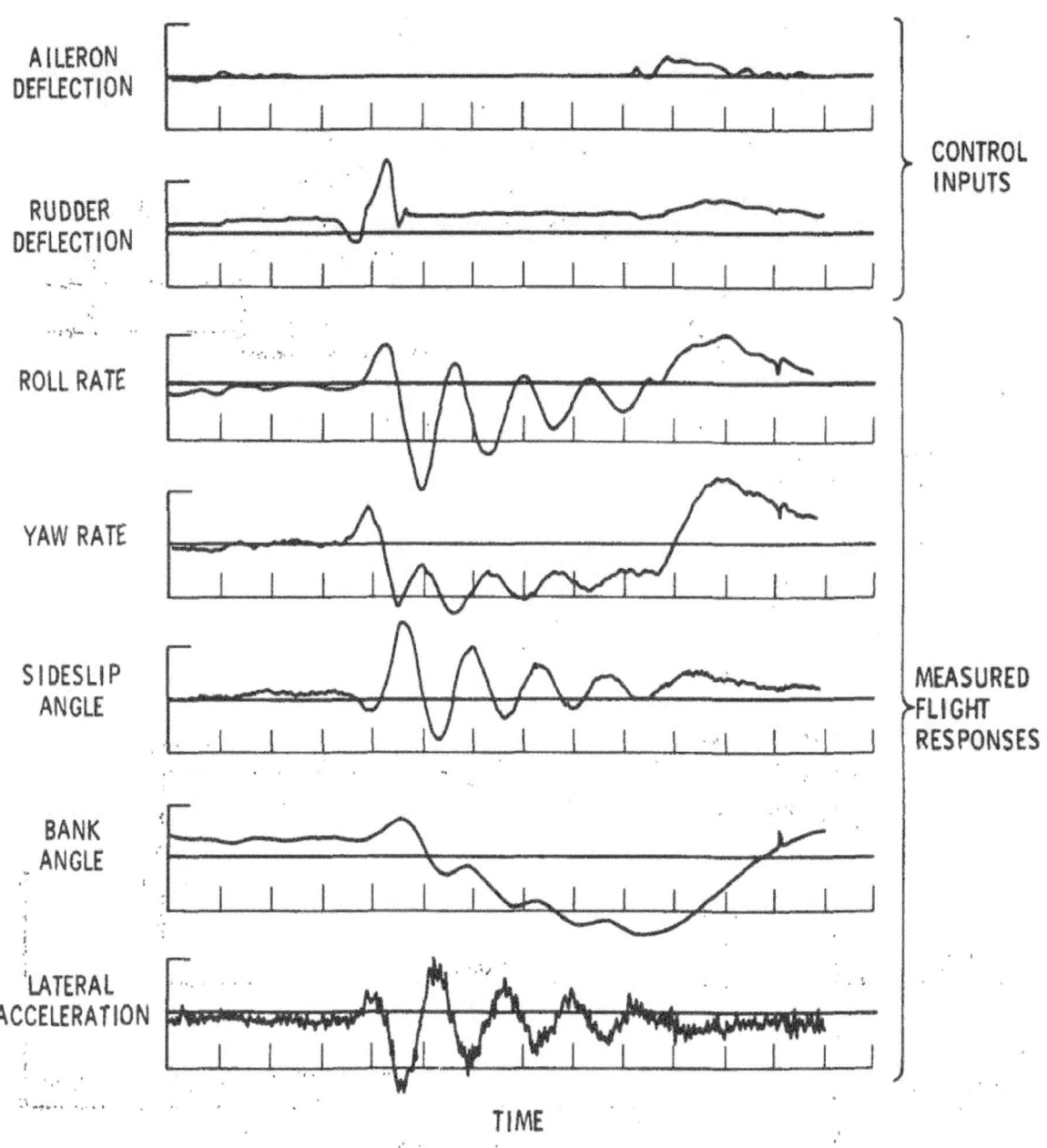

Figure 3. Typical lifting-body flight-test data for parameter estimation (from ref. 22).

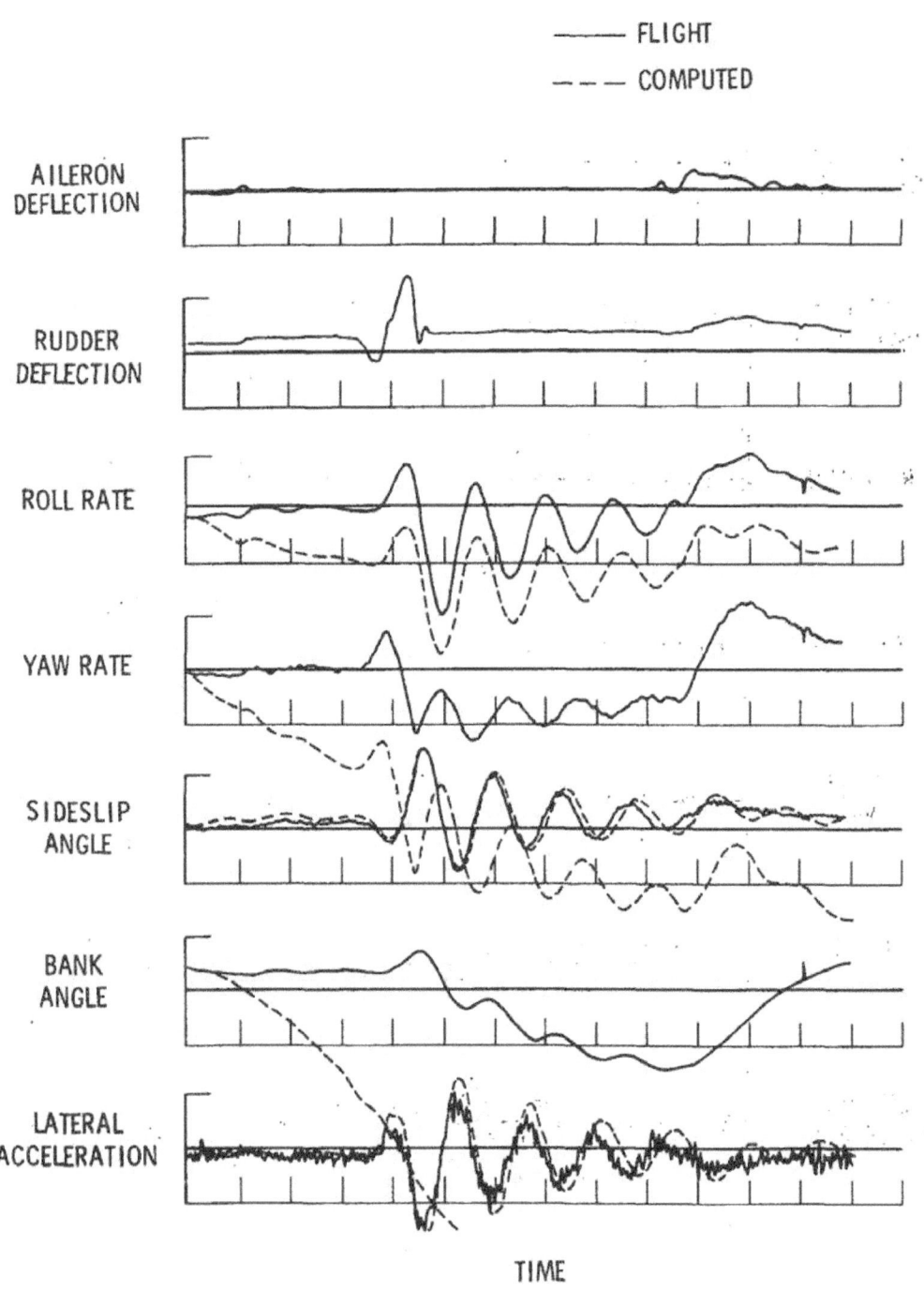

Figure 4. Comparison of computed response using wind-tunnel-parameter values with the flight-measured response (from ref. 22).

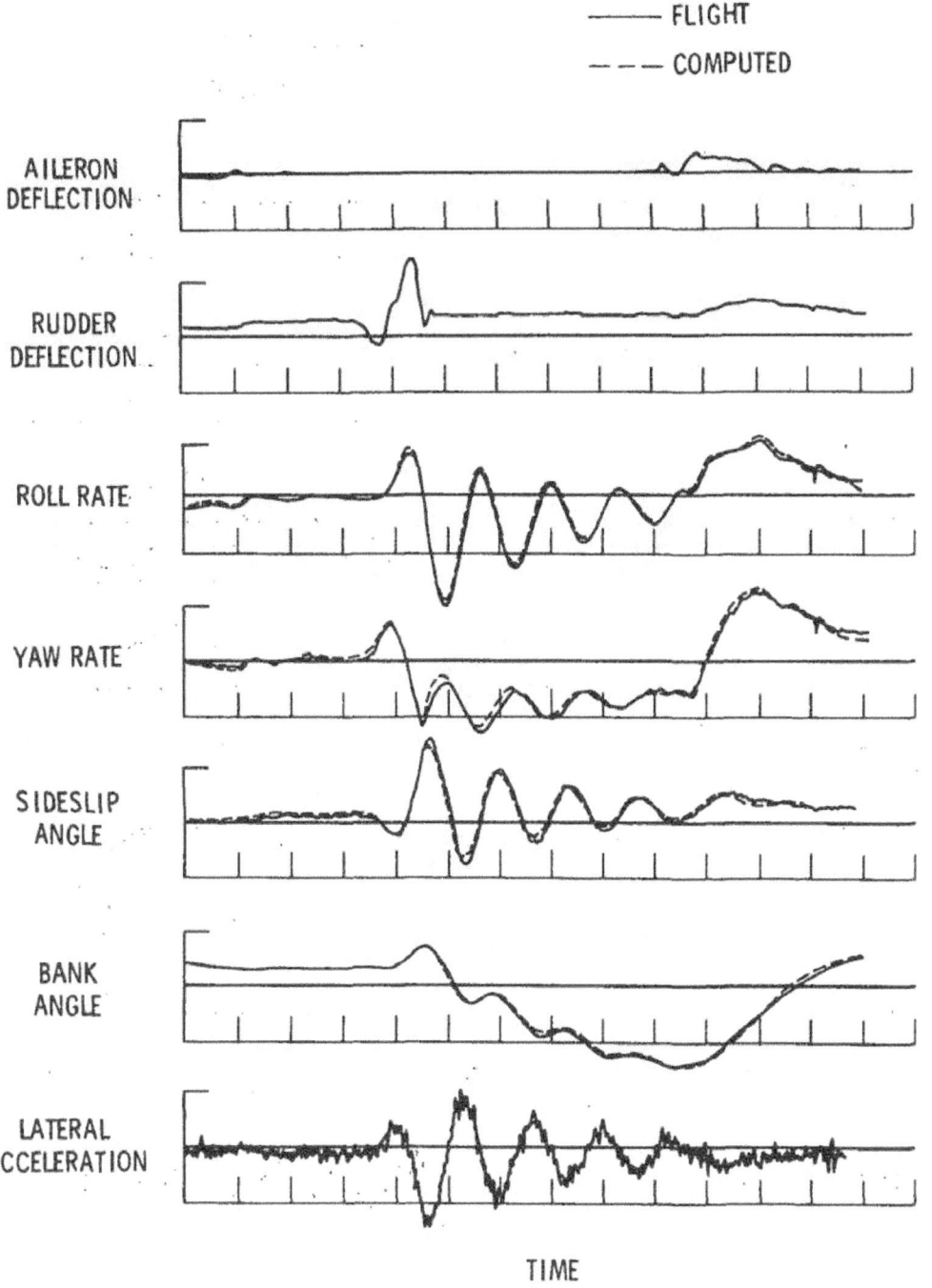

Figure 5. Typical match of computed response using estimated parameter values with the flight-measured response (from ref. 22).

17

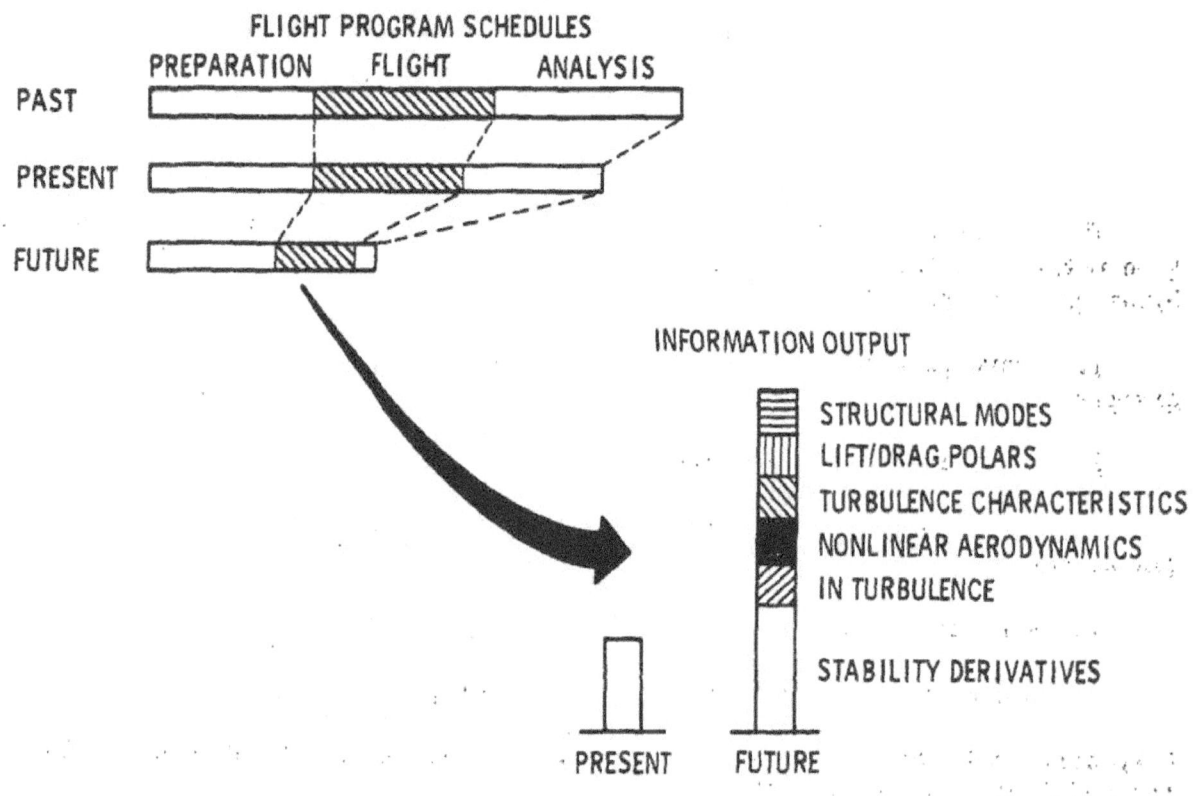

Figure 6. Potential of parameter estimation techniques.

A SURVEY OF AFFDL PARAMETER ESTIMATION

EFFORTS AND FUTURE PLANS

By Capt D. C. Eckholdt
Air Force Flight Dynamics Laboratory

And Dr. W. R. Wells
University of Cincinnati

SUMMARY

This paper presents an overview of the applications of parameter estimation methods to the following areas of interest at the Air Force Flight Dynamics Laboratory (AFFDL):

1. Conventional stability and control parameter estimation of "rigid" aircraft.

2. Extension to elastic aircraft.

3. Extension to stall/spin aerodynamics of "rigid" aircraft with a nonlinear model.

4. Application to the pilot model identification.

5. Correlation of wind tunnel, drop model and flight test data.

Currently, only well-documented algorithms are being used with modification to the model as required for the specific application.

The intent is to present the genesis of each problem and other background information which will enumerate the algorithms and explain how this information is used to improve existing operational aircraft characteristics as well as specify design criteria for future USAF aerospace vehicles.

INTRODUCTION

Over the past several years the Air Force Flight Dynamics Laboratory has renewed its efforts in the area of parameter identification and estimation. Several factors are involved in this renewal of effort. One is the fact that the evolution of analytical capability and computational facilities make possible what, in the past, would have been a most difficult, time consuming task or just a dream. Another factor involved is the increasing necessity, with the enormous cost of the development and flight test of modern weapon systems, for more complete and accurate analyses of new aircraft. Parameter identification, of course, plays a vital role in this more comprehensive analysis program, both in identifying the characteristics of aircraft to develop confidence in prediction procedures and in utilizing the information in development of the aircraft and its control system.

Several problem areas have accentuated the need for parameter identification and have also increased the difficulty of the solution. One of these problem areas is concerned with flight at high angles of attack and sideslip, the stall/spin problem, where a strong resurgence in research activity has resulted from an unacceptable level of accidents and incidents. Another is concerned with the increasing prevalence of highly elastic vehicles, coupled in many cases with complex flight control systems.

The object of this paper is to review past and present efforts of the AFFDL in the area of parameter identification and estimation, both contracted and inhouse. Some indication of trends of work that should be performed, and of our planned efforts will also be given.

More specifically, the body of the paper is divided into two main sections. The first section concerns in-house and contracted work of the WADC and the AFFDL Flight Control Divisions. This section includes a summary of the F-100A, XC-142, and NT-33A programs, as well as present parameter estimation work on large angle of attack and sideslip aerodynamics, on elastic aircraft, and on pilot modelling. The second section briefly describes the capabilities and some applications performed by the Vehicle Dynamics Division of the AFFDL. Their efforts are normally restricted to frequency domain analyses techniques as compared to the time domain techniques employed by the Flight Control Division.

FLIGHT CONTROL PARAMETER ESTIMATION METHODS

The F-100 A Program (References 1 and 2)

Looking back into history, it can be seen that modern aircraft and their unique problems arrived before digital computers and usable and reliable parameter estimation algorithms. In 1957 the former Cornell Aeronautical Laboratory (CAL) was asked by the Aircraft Laboratory of the Wright Air Development Center to conduct a research program on the F-100A in the transonic and lower supersonic speed ranges. Since 1944, CAL had been engaged in full-scale stability testing of aircraft under Air Force sponsorship and had successfully developed methods for obtaining and reducing response data to aerodynamic derivatives. The methods employed were based on those developed for conventional, low-speed World War II aircraft. However, they had successfully been applied to the unique flying wing configuration and had been extended for application on the F-80 which was an early high-subsonic turbojet fighter. The F-100A program had to again extend the existing methods or develop a new technique in order to test for longitudinal and lateral characteristics. The flight envelope and speed regimes for this aircraft were much larger than any other in existence. It included the transonic and lower supersonic regimes which were virtually unexplored. The longitudinal program was planned to obtain data which could be utilized to evaluate all the derivatives characterizing the short period mode. In order to obtain the necessary data, the aircraft was instrumented to measure the trim angle of attack and stabilizer deflection, as well as pitch rate and normal acceleration response to control input. A three-degree-of-freedom analog computer study was used to obtain the ranges of the dynamic-response instrumentation. Initially CAL intended to use static data to obtain frequency response data and then determine the lumped damping and stiffness parameter in order to solve for the damping derivatives. However, early in the static flight test it became apparent that considerable scatter was occurring in the measured angles of attack for trim and there was difficulty in determining the longitudinal static stability of the airplane at transonic and supersonic speeds. Thus, another method of obtaining and reducing the data was required. By considering the low short-period damping developed by the aircraft and the irreversible hydraulic boost control system for the all-movable horizontal stabilizer, it was found that the pilot could apply control pulses such that oscillation data suitable for analysis by the time vector method could be obtained. This resulted in the calculation of only an approximate value for the static longitudinal stability derivative and, hence, only a "lumped" value for the longitudinal dynamic derivatives. As a result, the predicted values and those reduced from flight test were not in complete agreement. However, the variations with Mach number and altitude were consistent. For the lateral tests, the time vector method was used to obtain solutions of the side force equation in order to obtain a position error calibration of the sideslip vanes. A five degree of freedom simulation was used to verify the form of the equation and to obtain the ranges of the dynamic response instrumentation. However, the main computations were accomplished from an equation of motion analysis method with a least squares averaging technique on a digital computer. The reason for this was the possibility of non-linear responses occurring in the transonic speed range. It was necessary that the form of non-linearity be known or assumed and the coefficients be constant. For the F-100A program the equations were originally written in a linear form with a least squares averaging performed to obtain the most probable values of the derivatives. Thus, the technique used consisted of substituting enough values of the variables in the assumed form of each of the equations to obtain a set of simultaneous equations which could be solved for all the unknown coefficients, namely the stability derivatives. Again complete agreement of the experimentally determined stability derivatives with predicted values was not obtained. Although the variations of the derivatives with Mach number and altitude were consistent. Time history calculations for the lateral-directional aircraft responses using the flight test stability derivatives showed excellent agreement with the reduced flight test responses. Moreover, the tests confirmed that it would be possible to extend the range of testing to higher speeds by analyzing data taken from a diving aircraft rather than one flying straight and level. It is interesting to note that aeroelastic effects and gyroscopic effects of the engine were encountered but not adequately modelled.

The previous example was used to define WADC's involvement in aircraft identification in the late 1950's. It should be emphasized that the old Flight Control Laboratory or the current Flight Dynamics Laboratory are not normally involved in flight testing or data reduction for stability and control derivatives. However, the Laboratory has always been involved in new and unique applications.

The XC-142 Program (Reference 3)

The next example considers the problem of obtaining the aerodynamic stability and control derivatives from the XC-142 airplane and the correlation of wind tunnel and flight test data. This study was performed in the late 1960's for the Air Force by LTV Aerospace Corporation. The results of the study were somewhat restrictive due to the flight data package used. The data were obtained from the Category I test program which is primarily a qualitative rather than quantitative evaluation of the aircraft. Not only was the quality of the data unsuitable, but it was impossible to estimate the control effectiveness in low speed flight. This hampered the accurate determination of the basic airframe characteristics. Thus, one of the major recommendations made as a result of this effort was that future programs include a critical review of the instrumentation system and flight maneuvers in order to obtain the best possible data for determining the low speed flight aerodynamic characteristics.

Another major obstruction was the inability to satisfactorily correlate the wind tunnel data obtained from tests of three models of the XC-142A airplane in four different size test sections. Again, the flight evaluation program was mainly aimed at pilot evaluation of the handling qualities of the aircraft. However, the instrumentation in airplanes No. 1 and 2 was sufficient in terms of the necessary variables measured for the purposes of parameter identification. The airborne data recorded was accomplished with a pulse duration modulation (PDM), tape system, camera coverage of cockpit control panels, and pilot and co-pilot verbal reports. The sampling rate of the PDM system was 10/sec and 20/sec. On certain flights, multiple gyro instrumentation, with different sensitivities, was available for comparison checks. In general, the flight data obtained during the program exhibited high frequency noise and/or vibration content, thus it was necessary to process the flight data through a numerical or digital filtering routine.

A least squares method was used to perform the data reduction. Basically, the mathematical model was fitted to the flight data where the criteria for obtaining the fit was the minimization of the squared error between the flight data and the model response. The model represented the uncoupled longitudinal and lateral-directional motions and was made up of only those variables which would be excited independently during a maneuver. The variables that were not excited during a particular maneuver, and those that were considered proportional to another variable, were eliminated from this general set-up by input specifications to the computer for each set of data. These normal equations also contained a constant coefficient which represented the total bias that may have resulted from errors in selecting trim or null values of the flight variables from which perturbations are taken for a particular maneuver.

The digital computer routine used for the least squares solutions was checked by determining solutions for "exact" time histories of assumed linear models obtained from established inverse Laplace routines. A comparison of the original or assumed aerodynamic coefficients with the least squares solutions was satisfactory to three or more significant digits for the important derivatives, the less significant derivatives being slightly less accurate.

The fact that the computed time histories were determined by single precision arithmetic would account for this difference. Single precision arithmetic was utilized in the least squares routine used in this program since aerodynamic models with six or fewer coefficients were selected. It was found that the least squares method lends itself readily to the type of flight data which were generally available for this study; i.e., continuously controlled maneuvers.

This study also used analog matching techniques independent of the least squares analysis to check the results obtained for the longitudinal, cruise configuration flight case. The particular flight maneuvers were pilot-forced sinusoidal oscillations. The unit horizontal tail time history was generated for the computer input forcing function and the aerodynamic derivatives were adjusted until the output traces for the variables matched the measured flight time histories.

Results from this study clearly indicated the need to plan the flight test program in such a way as to gather usable data. Due to high noise content, the flight data available were not of suitable quality to allow a high degree of confidence in inverse solutions for aerodynamic derivatives, except perhaps for predominant effects such as control effectiveness in the transition flight regime. Furthermore, the flight data available did not include controls fixed maneuvers which would have eliminated the predominant effect of control effectiveness in low speed flight and would have enabled more accurate determination of the basic airframe characteristics (neglecting instrumentation noise or accuracy limitations).

This study suggested the following recommendations to improve future efforts aimed at determining the aerodynamic stability and control derivatives for V/STOL aircraft from flight data:

1. Instrumentation systems should be critically reviewed to assure the necessary accuracies required for the determination of low speed flight aerodynamic characteristics. If sampled data systems are to be used, sampling rates of at least two times the highest vibration frequency of the airframe should be obtained, or some combination of high frequency electrical filters and high sampling rate be employed.

2. The sensors in the instrumentation system, particularly accelerometers, should be isolated from high frequency vibrations to the extent possible, by consideration of both location and local mounting.

3. Specific flight tests should be defined for obtaining the best possible data for isolating the aerodynamic effects. Due to the inherent instability or very low stability of V/STOL airplanes in low speed flight, maneuvers such as pilot forced sinusoidal oscillations appear to be the best means for obtaining good amplitudes in control inputs, angular rates, and acceleration responses while allowing the pilot to maintain a higher confidence in his control of the aircraft. Controls fixed maneuvers, or responses; i.e., stabilization augmentation inactive as well as stick fixed, should be obtained in all flight conditions where safety permits in order to eliminate

control effectiveness in the response and thus better isolate the remaining aerodynamic effects.

Based on the above study's recommendation AFFDL contracted with LTV Aerospace Corporation to do a detailed analysis on V/STOL flight test instrumentation requirements which would be aimed directly at the problem of extraction of aerodynamic coefficients. The results are reported in reference 4. Briefly, curve fitting the equations of motion by multiple linear regression (least squares) was selected as the best method of analyzing flight test time histories in order to extract aerodynamic coefficients. This algorithm had the most established statistical credentials; is easily programmed on digital computers; can handle non-linear equations; and uses digitized data which can undergo extensive preprocessing for error elimination before analysis is begun. The following types of instrument introduced errors were investigated:

- Gaussian distributed random noise

- High frequency sinusoids, simulating elastic modes or faulty instrument demodulation

- Instrument calibration slope errors

- Constant bias increments

- A time delay simulating a high order Butterworth filter or end instrument "stiction"

- A time lag through a lag filter or servo.

The following results were indicated:

- Gaussian distributed random noise of a known distribution.

- The numerical technique developed was effective.

- High frequency sinusoids. Digital filtering removed all the adverse effects.

- Instrument calibration slope. Error could be corrected accurately as soon as discovered because of the linear relation between calibration and error on associated coefficients.

- Constant bias removed automatically by the least squares method.

- The time delay could be removed exactly for one frequency input and compensated adequately for multiple frequency inputs by a combination of Fourier analysis and Laplace transormation.

- The time lag could be compensated adequately by use of Fourier analysis combined with Laplace transformation.

24

Thus, before one can state categorically that an instrument is inadequate, he should first ascertain to what extent he can compensate the output when he knows the instrument characteristics.

Further, the study showed that the effect of most of these errors, for instrument accuracies attainable at the present estimated state of the art, was to degrade the data unacceptably. By judicious preprocessing, useful data can be retrieved and good coefficients extracted. The preprocessing requires knowledge of the instrument characteristics as established by test.

Instrumentation requirements, which would satisfy flight test measurements and final reduced data "accuracies" such that the desired derivatives could be extracted, were established after a thorough investigation and appear in Table 1. Furthermore, the data of Table 2 illustrates the "target accuracies" for V/STOL instrumentation which must be achieved in order to optimize the instrument characteristics for derivative extraction purposes.

Finally, this analysis revealed that compensations for known instrumentation inaccuracies could be developed that would result in acceptably small errors in the aerodynamic coefficients. This indicates that the effort spent in testing to learn the nature of the instrument is as important as the continuing attempt to develop the more nearly perfect instrument.

The NT-33A Program

Although AFFDL has not sponsored all the applications work of applying different parameter identification methods to our variable stability NT-33A aircraft operated by Calspan, this extensive effort should be mentioned because of its significant contributions to the efforts of the AFFDL. Obviously the unaugmented, bare-airframe stability and control parameters must be known as accurately as possible, if the NT-33A is to be augmented using feedback techniques to simulate the characteristics of another aircraft or perform precision flying qualities experiments. As a result Calspan has employed every known computational technique to solve this problem, in fact, the NT-33A must be the most identified aircraft in the world. To date they have employed:

Frequency response   Kalman filtering

Analog matching ..   Quasi linearization (maximum likeli-
                                                       hood)

Least squares   Differential correction method.

Most of the results are published in the proceedings of the National Electronics Conference which was held in 1969 or in assorted Calspan and AFFDL technical reports.

For instance, reference 5 presents a method of matching the aircraft responses in terms of modal parameters instead of stability and control derivatives. The fundamental advantage of this formulation is the reduction of the number of parameters which must be matched and the guidance which is available on how to adjust the parameters. The results presented for this problem, roll rate response to an aileron step input, are very accurate.

In reference 6 the least-squares method or Equations-of-Motion Method is applied to the variable-stability T-33 airplane. Here it is reported that in order to properly extract the stability derivatives for either the basic airplane or simulated airplane, it was best to use a set of non-linear small perturbation equations. From the study it was determined that some of the non-linear kinematic and aerodynamic terms in the lateral-directional equations are as important as some of the small linear terms. These non-linear effects were modelled by retaining some second-order terms in the perturbations and their coefficients. The report also stressed that the extracted derivatives and response data may fit the equations well, yet the values for some or all of the extracted derivatives may be poor because of unknown errors in the calibration constants. Thus, one of the important results was a reasonable "goodness-of-fit criteria" which was established in the form of computed error coefficients. The application to the lateral-directional T-33 flight data was reported to have been accomplished with some success. It was suggested that the procedures and methods have general validity and could be applied to longitudinal motions, or a complete six-degree-of-freedom system.

In a separate report, reference 7, in-flight evaluation of certain lateral directional handling qualities of high performance aircraft is investigated. In particular, the landing approach task and the phenomena of decreasing directional stability during supersonic cruising flight is studied. An essential part of any handling qualities evaluation program is the identification of the stability parameters simulated in flight. Only with in-flight identification is there assurance of the configurations actually simulated. Pilot ratings can then be related to the measured stability derivatives. In this program, considerable effort was required in the analysis of flight test data to insure proper parameter identification. Rudder doublet responses were used to identify frequency, damping ratio and roll to sideslip ratio. These parameters and certain stability derivatives such as the dimensional yawing moment derivative with respect to the aileron control deflection were also identified by the Equations-of-Motion method and showed reasonable agreement with the results obtained by other approaches.

Reference 8 reports an effort to identify aircraft parameters using Kalman filtering. Basically raw T-33 flight test data involving sideslip angle, roll rate, yaw rate, and roll angle response to an aileron step deflection was operated on by a Kalman filter to estimate eleven parameters which define the lateral-directional equations of motion. This study indicated the feasibility of using Kalman filtering techniques as a means of both off-line and on-line parameter estimation. In particular, results indicated that the filter was capable of converging to a set of parameters which can be used to define an off-line model of the lateral motion; and that the filter can be used during on-line control to feedback the states and to update the model parameter. Based on the results of the study several operational suggestions were put forth:

1. Use the first 10-20% of the data to obtain an initial estimate and covariance matrix via a least square fit. Kalman filtering should be performed

on only the last 80-90% of the record.

2. At the end of each iteration try replacing R by the matrix of the averaged squared errors when convergence is close.

3. Investigate the effects of increasing the elements of the P matrix to see if there is an upper limit beyond which no benefit is attained.

4. Study the effects of assuming plant disturbances. This could possibly compensate for approximations made in the linearization procedures.

5. Analytically study the effects of incorrect values for Q(0) and R.

Reference 9 describes the combination of historical mathematics with the high-speed digital computer capability to produce an efficient device of searching for unknown parameters existing in a set of algebraic or differential equations, namely, the CAL computer program which uses the theory proposed by Bellman and Kalaba (ref. 10). This study describes a search technique to solve for a vector c (components of which are the set of unknown initial conditions and the unknown parameters of the given set of differential equations) that minimizes the equation

$$S(c) = \sum_{i=1}^{NP} \sum_{j=1}^{NK} \lambda_j (g_j(X(t_i)) - Y_{ji})^2$$

where λ_j are relative weights of NK differential functions $g_j(X(t))$ and $(g_j(X(t_i)) - Y_{ji})$ represents the error between the mathematical model and test data. The initial estimate of vector c was obtained by using spline functions. An example problem was applied to the T-33 flight test data for the lateral-directional case. The results indicated the feasibility of using this method of quasilinearization for solving the aircraft estimation problem. In fact, simplicity of application and speed (about 12 minutes to solve the example problem) appear to be some of the values of this technique. Further use and application of the Quasilinearization Method are reported in reference 11.

A final method applied to the NT-33 aircraft reviewed is that proposed in reference 12. Here a method is developed for the deterministic case in which the parameter estimates are obtained by minimizing a criteria function which is quadratic in the difference between the measurement vector and the model output vector. The differential correction method is best employed as an off-line parameter estimation method since it employs an iteration procedure. It applies to nonlinear differential equations, but for the example cases, Dolbin assumed linear, constant coefficient, differential equations. The data used in the example problems presented in the paper were measured in an airplane equipped with an automatic control system capable of altering the airplane parameters.

In particular, the algorithm was used to estimate the stability derivatives associated with the short period characteristics and the modal parameters associated with the lateral-directional transfer function of the roll rate response due to an aileron deflection. The results indicated that the differential correction method is entirely suitable to the estimation of airplane parameters from flight test data. The quality of the fit between the measured data and the outputs of the mathematical models was generally excellent. The parameter estimates were accurate whenever the measured responses contained sufficient information to provide adequate sensitivity of the criterion function to parameter changes.

Present and Future Programs

The previous section serves as a transition to that time period where AFFDL is applying modern estimation theory to obtain solutions to unique modelling and identification problems. Most of the activity for both inhouse and contracted efforts have been accomplished in the Control Criteria Branch of the Flight Control Division, in particular, the Aircraft Dynamics and Control Analysis Groups. The former group is charged with the mission of solving the stall/spin/loss-of-control problem exhibited by so many of the Air Force's high-performance fighters and developing a technical capability to handle aeroelastic stability and control problems associated with high performance vehicles. The latter group is involved in identification of stability and control parameters and pilot modelling. The following paragraph presents a review of their efforts.

Post Stall Characteristics

In order to eliminate an unacceptable number of flight accidents and incidents, the AFFDL has begun an Advanced Development Project concerned with solving stall/spin/loss-of-control problems. One portion of this project concerns correlating high angle of attack aerodynamic stability and control characteristics extracted from available full-scale flight test and free-flight drop model test data of current high performance aircraft.

As part of the effort, Calspan was awarded a contract in March of 1972 to correlate data available on existing high performance aircraft. The F-4, F-5, and F-111 aircraft were considered as possible study vehicles. Only the F-4 aircraft was selected however, since it has flight test, drop model, and wind tunnel, high angle-of-attack, post-stall data available.

To date, Calspan has formulated a mathematical model for the aerodynamic functions using a polynomial least squares fit in angle of attack and sideslip angle to approximate the F-4E wind tunnel data. Currently, time histories from a six degree of freedom simulation, developed from the wind tunnel data fit, are being matched with available flight test time histories. A raw data tape containing suitable flight test runs for the F-4E was acquired via a subcontract to McDonnel Douglas and has been segmented according to Mach number and angle of attack characteristics to aid in the identification of the aerodynamics. Instrument consistency checks have been made on the flight test data and the information is being used in conjunction with an initial identification of the flight data by a least squares technique. These results will provide the start up estimates for the Iterated Kalman Filter Program.

Correlation between the identified parameters from the drop model data and the wind tunnel data are nearly completed and the correlation appears very good. Calspan will present a more detailed paper later in the symposium (ref. 13).

During this same time period, an independent in-house effort was begun to study the feasibility of a maximum likelihood approach to the extraction of stability derivatives from an assumed non-linear aerodynamic model (ref. 14). This effort was supplemented by an Air Force Institute of Technology (AFIT) Master's Thesis project (ref. 15) in which a NASA/LRC computer program (ref. 16) was modified to handle the non-linear identification problem associated with high angle-of-attack flight. The program has successfully been applied to the F-4 low angle of attack data and is currently being checked out with the non-linear model of the F-4 aircraft. Their results will be presented later as an alternate paper (ref. 17) if time permits.

Finally, a new contract through the AFFDL Stall/Spin Advanced Development Program office with Atkins and Merrill, Inc to build and test several 3/10 dynamically scaled drop models of the light weight fighters have been awarded. This experimental program is designed specifically to develop a proper instrument package, flight test program, and data reduction procedure in order to extract the high angle of attack stability and control characteristics of the model.

Elastic Aircraft

Previous efforts to identify the parameters of elastic aircraft have concentrated on the frequency domain analysis techniques. These techniques are required if the frequency-dependent, unsteady aerodynamic methods employed in flutter analyses are to be improved. With the development of more highly aeroelastic aircraft, the coupling of the structural dynamics with the control system dynamics has necessitated a less complicated approach to the unsteady aerodynamics than the frequency domain approach. These approximate techniques, developed primarily for stability and control and flight control applications, employ a "stability derivative" representation of both the steady and unsteady aerodynamics that affect the body-fixed axis motions and the elastic deformations. The reduction of aerodynamic forces to "derivatives" leads immediately to the suggestion that the modern, time domain parameter estimation methods be used to analyze the flight test data of flexible aircraft.

A study has begun at the AFFDL to determine the feasibility of applying the modern estimation methods to elastic aircraft. The results of the study by Schwanz and Wells (ref. 18) will be reported in another paper of this symposium. Briefly, the study considered six different formulations of the dynamics of elastic vehicles and selected the MODAL TRUNCATION formulation as the most promising. A similar review of the "Equation Error", "Output Error", and "Advanced Methods" resulted in the selection of the advanced parameter estimation method referred to as maximum likelihood. Once the dynamics of the elastic aircraft and the sensor equations are defined, and the maximum likelihood method is selected, a computational algorithm may be developed. An examination of the algorithm, reported in reference 18, indicates the major computational problems are the inversion of large sized matrices and the time-wise integration of a large number of sensitivity equations. The study has now moved to the second phase in which a computer program called FLEXFLT is being developed in-house at the AFFDL. Analytical test cases as well as the B-52E CCV flight test data are to be used to demonstrate the method prior to applying it to other USAF flight test programs such as planned for the B-1 aircraft.

The AC-130 Gunship Program

Since February 1972, the Control Analysis Group has supported an in-house effort to develop computer programs designed specifically to find the stability and control derivatives of the AC-130 Gunship aircraft which is equipped with the Sight-Line Auto-Pilot (SLAP) to fly a trimmed 30° bank angle configuration (see reference 19). The dynamics of this configuration are represented by six degree of freedom, linear, coupled perturbation equations which contain a maximum of 31 unknown parameters. To solve this identification problem, three digital computer programs are utilized, the first two of which have been developed in-house:

Least squares program.

Maximum-Likelihood-Kalman Filter program.

Newton-Raphson program.

The Least Squares program is very effective when applied to cases where there is no measurement noise on the data and where there is an extended amount (i.e., derivatives of all states) of data available. However, for this purpose, Least Squares was written to be used as a start-up routine (i.e., it provides an initial estimate of the unknown parameters) for the Maximum Likelihood-Kalman Filter (MLKF) program.

The MLKF program is based on the algorithm presented in a paper by Mehra, Stepner, and Tyler (ref. 20). The program was written, using VASP (Variable Automatic Synthesis Program) subroutine, to handle the general, linear, fixed coefficient system with noise in the system and in the measurements. For the general case, it acts as both a state and parameter estimator.

The Newton-Raphson program was obtained from NASA Langley, It is to be used as an alternate method for identifying gunship derivatives. The results of this effort are not known as of yet since the aircraft has not been flight tested.

Pilot Modelling

Another in-house application of parameter identification techniques which appears to have a large pay-off is in the area of pilot modelling. In particular, Durrett and Pope have formulated a method of identifying pilot model parameters from experimental data. It is based on a state space representation of the pilot model which includes, as subcases, the various forms of the classical pilot models (ref. 21). Given measurements of the pilot output, time rate of change of output, and observation of an aircraft (or any linear system), the general computer program developed by Pope (ref. 19) for identifying linear system parameters can be utilized to identify such model parameters as pilot gain, lead, lag and time delay. Identification of these pilot parameters provides valuable information for the further development of some of the pilot-in-the-loop analysis techniques, for example, the Paper Pilot. Recently the AFFDL and AF Human Resources Laboratory have started a joint in-house program to identify the human operator in the closed loop situation for single or multi-loop feedback tasks.

Finally, two practical applications of this technique should be mentioned. The first is a joint program between AFFDL and the Aero Space Medical Division, Brooks AFB, Texas where pilot modelling identification methods are being used to aid in the evaluation of a pilot's capability to complete his mission once he has been exposed to a high level of nuclear radiation. The experimental portion of the study involves trained monkeys who fly a two-degree of freedom platform. Data is recorded before and after radiation exposure. Temporary loss of control is experienced, however, the effects change with time. Through parameter identification methods, it is hoped that the non-linear monkey operator model parameters can be extracted from the experimental data. These results will then be used in the "Paper Pilot" digital computer program in order to determine the degradation in closed loop flying qualities which has occurred due to radiation.

The second practical application involves data which were recorded during a handling qualities experiment being accomplished on the Air Force's Total In-Flight Simulator (TIFS) aircraft. In particular, an SST type configuration was flown to evaluate landing characteristics of the unaugmented aircraft. At the same time data was recorded to obtain a closed loop set of time histories of stick response, display error, pilot output, and plant output. The data is being processed in the pilot modelling identification program in order to obtain a pilot model for the SST type aircraft which can then be used to predict handling qualities.

VEHICLE DYNAMICS PARAMETER ESTIMATION METHODS

The Dynamics Technology Applications Branch is responsible for developing techniques and equipment for the measurement and analysis of experimental dynamics data. These capabilities are required to evaluate and assess the accuracy of analytical prediction methods and to accurately describe the operating environment for flight vehicles.

In the analysis area, a new high speed digital analysis system is now completing final acceptance testing. This system will provide the capability for performing the following analyses: amplitude spectra, power and cross power spectral density, auto and cross correlation, coherence, transfer and mechanical impedance functions, transient analysis and digital filtering.

A recent advancement in measurement technology by this organization is the development of an automatic gain control amplifier system for airborne flight test use. This system automatically provides the amplification necessary to maintain data signal levels in the optimum range for a tape recorder system. The gain status of each amplifier is sampled by a commutator system and recorded on a single tape recorder channel for recovery during data analysis.

Typical of system support efforts now being conducted is a test program in support of the Air Force Weapons Laboratory (AFWL) to determine the aerodynamic and acoustic environment of the F-111 bomb bay. In addition, a test program is also being conducted in support of the AFWL to determine the acoustic environment in the vicinity of high output laser devices. This program is being conducted to assess the potential safety hazards for personnel working near these lasers and to identify acoustic excitation levels which could cause misalignment of highly sensitive optical equipment.

Typically, an aircraft test program consists of instrumenting the vehicle with transducers at specific locations to measure vibration, sound, temperature, pressure, or strain. The test vehicle is then operated at various conditions and the output of the transducers are recorded on an airborne tape recorder. The tapes are then returned to the laboratory for playback and analysis. The data are analyzed with the digital analysis system and are presented in the form of engineering plots. To date, data from these test programs have been used to update test specifications, provide equipment design criteria, verify prediction methods for rotor powered V/STOL and high performance aircraft, refine the ground induced dynamic loads criteria, investigate methods of increasing damping and decreasing response, provide data for human factors studies, determine the effects of transients, e.g., gunfire, the response of external stores to vibration excitation and the aural detectability of limited-war aircraft.

Obviously, the expert instrumentation capability and high quality data processing equipment are a valuable resource to the Laboratory and will play an important part in any new major application of parameter estimation to identify stability and control parameters on rigid and elastic flight vehicles.

CLOSING COMMENTS

The main emphasis in this paper has been to relate the role of the AFFDL in the development of aircraft parameter estimation theory from the classical graphical approaches of the post World War II days to the modern theory of today which involves sophisticated mathematical analyses and liberal use of the digital computer. This paper no doubt has caused the reader to reflect back on his own experience and to see the progress and failures of past efforts. As has been noted, many of the basic problems associated with instrumentation packages, flight test plans, and data reduction have still not been completely solved.

Experience has proven that all aspects of the flight test program must be considered simultaneously using a systematic approach to integrate the different disciplines. Furthermore, the consolidated effort must be made very early in the initial design stage of an aircraft development program, if one intends to correlate wind tunnel, drop model, and flight test data. Trying to make things fit together after the fact becomes a very difficult task. The Navy is taking a step in the right direction on their T-2B program. It is specifically being designed and conducted to optimize the possibility of successfully extracting the stability derivatives. AFFDL is currently suggesting such an approach to the B-1 program and has already formulated such a plan for the AC-130 Gunship.

The Laboratory will continue to attack unique new applications of estimation theory when problems arise and constrain mission accomplishment. Our goal remains that of possessing the technical capability to solve the stability and control or handling qualities problems on any aerospace vehicle.

REFERENCES

1. Gowin, N. E.: Longitudinal Stability Investigations of an F-100A Airplane. WADC TR 57-562 or Cornell Aeronautical Laboratory Report No. TB-593-F-4, November 1957.

2. Gowin, N. E.: Lateral Stability Investigation of an F-100A Airplane. WADC TR 57-563 or Cornell Aeronautical Laboratory Report No. TB-593-F-5, November 1957.

3. Black, E. L.; and Booth, G.C.: Correlation of Aerodynamic Stability and Control Derivatives Obtained from Flight Tests and Wind Tunnel Tests on the XC-142A Airplane. AFFDL TR 68-167, 1968.

4. Hill, R.W.; Clinkenbeard, I.L.; and Bolling, N.F.: V/STOL Flight Test Instrumentation Requirements for Extraction of Aerodynamic Coefficients. AFFDL TR 68-154, Volume I, December 1968.

5. Hall, G. Warren: A Method for Matching Flight Test Records with the Output of an Analog Computer. Proceedings of the National Electronics Conference, Volume 25, 1969.

6. DiFranco, Dante A.: In-flight Parameter Identification by the Equations-of-Motion Technique-Application to the Variable Stability T-33 Airplane. CAL report No. TC-1921-F-3, December 1965.

7. Smith, Edward H.; and DiFranco, Dante A.: In-Flight Evaluation of Certain Lateral-Directional Handling Qualities of High-Performance

Aircraft. AFFDL TR 65-97, December 1965.

8. Kaufman, Howard: Aircraft Parameter Identification Using Kalman Filter-
    ing. Proceedings of the National Electronics Conference, Volume 25,
    1969.

9. Larson, Duane B. and Fleck, John T.: Quasilinearization Techniques.
    Proceedings of the National Electronics Conference, Colume 25, 1969.

10. Bellman, Richard and Kalaba, R.E.: Quasilinearization and Non-Linear
    Boundary-Value Problems. American Elsevier Publishing Company, Inc.
    New York, 1965.

11. Hall, G. Warren and Martino, Paul A.: Development of an Integrated
    Computer Program That Extracts Lateral-Directional Modal Parameters
    From Digitally Recorded Flight Test Data. AFFDL TR 70-72, November 1970.

12. Dolbin, Benjamin H., Jr.: A Differential Correction Method for the
    Identification of Airplane Parameters From Flight Test Data. Proceed-
    ings of the National Electronics Conference, Volume 25, 1969.

13. Eulrich, B.J. and Rynaski, E.G.: Identification of Non-linear Aerodyna-
    mic Stability and Control Parameters at High Angles of Attack.
    Symposium on Parameter Estimation Techniques and Applications In Air-
    craft Flight Testing, NASA/FRC, Edwards, CA. April 1973.

14. Wells, William R.: A Maximum Likelihood Method for the Extraction of
    Stability Derivatives from High-Angle-of-Attack Flight Data. AFFDL/FGC
    TM 72-16, September 1972.

15. Callahan, Jerry B.: Maximum Likelihood Estimation of Aircraft Stability
    and Control Coefficients for Low to Near Stall/Spin Angle of Attack
    Flight Regimes. Unpublished AFIT Master of Science Thesis. March 1973.

16. Grove, Randall D.; Bowles, Roland L.; and Mayhew, Standley C.: A Proce-
    dure for Estimating Stability and Control Parameters from Flight Test
    Data by Using Maximum Likelihood Methods Employing a Real-Time Digital
    System. NASA TN D-6735, May 1972.

17. Wells, W.R.; and Callahan, J.B.: Application of the Maximum Likelihood
    Method to the Identification of Aircraft at High Angles of Attack.
    Symposium on Parameter Estimation Techniques and Applications in Air-
    craft Flight Testing, NASA/FRC, Edwards, CA. April 1973.

18. Schwanz, R.C. and Wells, W.R.: Estimation of the Parameters of Elastic
    Aircraft Using the Maximum Likelihood Method. Symposium on Parameter
    Estimation Techniques and Applications in Aircraft Flight Testing,
    NASA/FRC, Edwards, CA. April 1973.

19. Pope, Rhall E.: A User's Manual for Program ID--A Parameter Identifica-
    tion Computer Program. AFFDL/FGC TM 73-34, January 1973.

20. Mehra, R.K.; Stepner, D.E. and Tyler, J.S.: A Generalized Method for the Identification of Aircraft Stability and Control Derivatives from Flight Test Data. 1972 Joint Automatic Control Conference of the American Automatic Control Council, paper number 16-4, Stanford University, CA. August 1972.

21. Durrett, John C.: On $\underset{\sim}{\dot{x}} = A\underset{\sim}{x} + B_u + M_\eta$ for the Airplane, Control System, Wind Gust. AFFDL/FGC TM 72-21, November 1972.

TABLE 1. RESUME OF MEASUREMENT REQUIREMENTS

		1 σ accuracy	fo (CPS)	SPS
1.	Roll Attitude	1/2°	6	256
2.	Pitch Attitude	1/2°	6	256
3.	Heading	1/2°	6	256
4.	Roll Rate	0.15°/s	8	256
5.	Pitch Rate	0.15°/s	8	256
6.	Yaw Rate	0.15°/s	8	256
7.	Normal Accel	0.02g	5	256
8.	Long Accel	0.02g	5	256
9.	Xvrse Accel	0.002g	5	256
10.	Pitch Accel	0.5%FS	5	128
11.	Roll Accel	0.5%FS	5	128
12.	Yaw Accel	0.5%FS	5	128
13.	Aileron Pos	0.4°	5	128
14.	Flap Pos	0.4°	1	64
15.	Wing Incidence	0.4°	1	64
16.	Horizontal Tail Incidence	0.4°	1	64
17.	Main Prop Blade Angle	0.4°	1	64
18.	Tail Prop Blade Angle	0.5°	1	64
19.	Propeller RPM	0.5%	1	64
20.	Rudder Position	0.4°	5	128
21.	Engine Shaft Torque	2%	1	32
22.	Airspeed	5K	1	64

TABLE 1. RESUME OF MEASUREMENT REQUIREMENTS (Concluded)

23.	Angle of Attack	5.0%	5	128
24.	Angle of Sideslip	5.0%	5	128
25.	Free Air Temp	2°F	1	64
26.	Pressure Alt.	10 feet	1	32
27.	Height Above Ground	5.0 feet	3	128
28.	Vert. Velocity	2 ft/sec	3	128
29.	Lateral Velocity	2 ft/sec	3	128
30.	Long Velocity	5 ft/sec	3	128
31.	Direct Thrust	3.0%	1	32
32.	True Airspeed (computed	3.0 Knot	1	32
33.	Time	N/A	N/A	1
34.	Camera Command	N/A	N/A	As Req'd
35.	Shutter Response	N/A	N/A	As Req'd

TABLE 2. STATE-OF-THE-ART ACCURACY TARGETS – V/STOL INSTRUMENTATION SYSTEM

FUNCTION	TARGET ACCURACY (% F.S.)	EST CH FREQ RESPONSE
ATTITUDES	0.2%	3 CPS
ANGULAR RATES	0.3%	3 CPS
ANGULAR ACCEL	0.5%	3 CPS
LINEAR VELOCITY	0.5%	3 CPS
LINEAR ACCEL	0.5%	10 CPS
ANGULAR POSITION	0.5%	2 CPS
ANGLE OF ATTACK	3.0%	3 CPS
ANGLE OF SIDESLIP	3.0%	3 CPS
AIRSPEED	1.0%	1 CPS
ALTITUDE	2.0%	1 CPS
FREE AIR TEMP	0.5%	1 CPS
DIRECT THRUST	2.0%	3 CPS

NAVY PARTICIPATION IN THE DEVELOPMENT OF AIRFRAME PARAMETER IDENTIFICATION TECHNIQUES

By

Roger A. Burton, Naval Air Test Center, and
Arthur J. Schuetz, Naval Air Development Center

ABSTRACT

The Navy is currently involved in the development of advanced parameter identification techniques for use in aircraft flight testing and refinement of aircraft dynamic systems modeling. This paper will present an overview of the Navy's research programs, capabilities, and facilities. The use of parameter identification techniques will be related to the flight testing, development, and simulation of aircraft and aircraft systems in the areas of flying qualities, automatic flight controls, flight dynamics, and advanced landing systems. Preliminary analytical and flight test results will be presented. The impact that new parameter identification technology will have on Navy flight test philosophy is discussed. Future plans will be outlined.

NAVAL AIR DEVELOPMENT CENTER (NADC)

An awareness of two substantial technological gaps has motivated the recent NADC interest in parameter identification. Extensive experience with computer-driven simulations of aircraft has shown that sufficiently complete and accurate stability and control data do not exist for many current aircraft, especially for high angle of attack flight conditions. In addition, no reliable method has been found for evaluating the in-flight simulation fidelity of a variable-stability aircraft.

Recognizing that parameter identification technique development has suffered from a lack of complete, high quality flight data for processing, NADC began a research program emphasizing the generation of such data by the NADC YT-2B. This program includes a flight data acquisition and reduction phase, a parameter identification phase utilizing the flight data, and an identified parameter evaluation and verification phase.

The flight test data were obtained from a specially instrumented YT-2B airplane operating at NADC. The aircraft has been equipped with twenty-four separate position and motion sensors. These include altitude, pitch and bank angles, angles of attack and sideslip, airspeed, three angular velocities, nine strategically located linear acceleration measurements, and five control deflections. Roll, pitch, and yaw angular accelerations are computed from linear acceleration data. For this project, aircraft response information was generated for three flight conditions, defined in Table I. A variety of carefully planned pilot control inputs were used. To aid in the estimation of measurement error statistics, one minute of information was recorded with the aircraft standing still on the ground with engines running. For many runs, every effort was made to minimize the external disturbances such as turbulence or gusts. In addition, moderate turbulence data are available to assess the effect of process noise on the parameter identification schemes.

For the current study, only the longitudinal data in the low angle of attack range are utilized for parameter identification. The identification schemes being employed include a modified Newton-Raphson, a specially-developed maximum-liklihood estimator, a combined Kalman filtering/smoothing technique, and a timesharing maximum-liklihood estimator. For uniformity, each identification technique is employing identical flight test data as well as the same aircraft equations of motion (which include fourteen unknown parameters).

In an effort to determine the relative effectiveness of the various identification techniques, the several sets of identified parameters will be compared with each other and with wind tunnel data. In addition to the direct comparisons of identified parameters and the examination of confidence intervals and/or standard deviations for the parameter estimates, an "untried inputs" procedure will be used to assess the validity of the identified parameters. A supporting effort, already in progress, will directly scrutinize the parameter identification process implementation.

A new study, generating and analyzing high angle of attack flight data, will soon begin. It will in many ways parallel the low angle of attack effort, but of necessity will also include instrumentation effectiveness analyses, identification technique development, and modelling studies. Extensive high angle of attack wind tunnel data for the YT-2B are being generated by NADC for use in this program.

NAVAL AIR TEST CENTER (NATC)

The steady growth in the complexities of aircraft specification requirements and automatic flight control systems has changed the complexion of the requirements set upon the flight test community for data accuracy and evaluation techniques. This change in flight test requirements is illustrated by the new flying qualities specification for piloted airplanes MIL-F-8785B(ASG) and the detail specifications for the F-14A and S-3A airplanes. For example,

MIL-F-8785B(ASG) sets new requirements for Dutch roll damping and frequency, spiral mode, roll mode, and roll rate specifications. The detail specifications for the F-14A and S-3A airplanes have a new requirement in the form of the Dutch roll coupling parameter and the Dutch roll excitation parameter. Recent requirements set upon the Navy by the Federal Aviation Administration (FAA) in connection with development of the microwave landing system have increased data accuracy requirements for determining aircraft frequency response. Existing flight test techniques are not able to fulfill these new data requirements; therefore, a program at NATC was established to develop airframe parameter identification technology to satisfy these new data accuracy and evaluation requirements.

Both classical and advanced airframe parameter identification methods are being investigated. This approach has been taken in order to develop a wide spectrum of experience in parameter estimation theory and applications. Classical methods (equation error and output error methods) that have been programmed are least squares regression, Prony's curve fit, Fourier transform, Z-transform, Newton-Raphson, and analog matching. These methods have been investigated using computer generated data and flight test data. An advanced method that is being developed utilizes the maximum likelihood approach.

The purpose of the flight test portion of the program was to obtain data to exercise these parameter identification methods and to determine the proper data gathering procedures. Of primary interest was the development of physically realizable pilot inputs that would give the best parameter estimates. Data were gathered using step, pulse, doublet, and random pilot control inputs and are presently being analyzed.

The successful application of parameter estimation theory to flight test data will have a significant impact on Navy flight testing philosophy. As data gathering and data reduction techniques have become more sophisticated during the last decade the emphasis in flight testing has been placed on the determination of specification compliance. Thus it has become easier and easier for the test pilot to slip away from qualitative flight testing and become merely an extension of the data recording system. This was illustrated by a survey of recent flight test programs showing that 70-90 percent of the flight tests were devoted to specification testing. The inordinate amount of time spent on specification testing is often irrelevant, since many specification requirements do not accurately describe the service suitability of an aircraft. For example, the static longitudinal stability of the F-4 aircraft in configuration PA at aft cg positions is essentially neutral and the stick force cues to airspeed are thus poor. These characteristics of the F-4 failed to meet specification requirements and using only this data it could be concluded that the airplane would have poor landing approach characteristics. However, pilots who have flown the F-4 are virtually unanimous in their opinion that it is comparatively easy to maintain the optimum approach angle-of-attack during visual approaches. The example presented simply illustrates the fact that there are factors which we do not measure quantitatively which affect the mission suitability of the aircraft. Thus the need for qualitative flight testing is clear. Therefore, NATC is developing parameter identification technology in order to reduce the amount of flight time spent on specification testing.

Future plans at NATC call for a continuation of the present effort in airframe parameter identification over the next three years. Analytical investigations will be extended to include:

a. Development of maximum likelihood and Kalman filtering algorithms.

b. Extension of existing algorithms to V/STOL and helicopter parameter identification.

c. Nonlinear model identification.

d. Instrumentation error effects on identification accuracy.

The flight test phase of the program will be continued to support the analytical investigations. A continuing effort will be made to develop flight test applications of parameter identification. In support of this effort it is planned to subcontract portions of the analytical investigations.

Table I
YT-2B FLIGHT CONDITIONS AND CONFIGURATIONS

Flight Condition	1	2	3
Mach No.	0.212	0.63	0.70
True Airspeed (ft/sec)	236	679	696
Altitude (ft)	sea level	10,000	30,000
Gear Position	down	up	up
Flap Position (deg)	16	0	0
Speed Brake Position	closed	closed	closed
Weight (lbs)	11,000	11,000	11,000
CG Position (% c)	20	20	20

A COMPARISON OF TWO METHODS OF EXTRACTING

STABILITY DERIVATIVES FROM FLIGHT TEST DATA

By Paul W. Kirsten and Captain Lawrence G. Ash

Air Force Flight Test Center
Edwards, California

ABSTRACT

Two methods for extracting stability derivatives from flight data are compared. A modified Newton-Raphson quasi-linearization minimization technique and a digital-analog (hybrid) matching technique were used to analyze the same data maneuvers obtained from two aircraft. About 70 maneuvers from an F-111E aircraft were analyzed over a Mach number range of 0.3 to 2.0 and an angle of attack range of 3 to 19 degrees. About 20 maneuvers were analyzed for the X-24A lifting body at Mach numbers of 0.5, 0.8, and 0.9, and an angle of attack range of 4 to 13 degrees. Stability derivatives were extracted from these maneuvers and the results from the two techniques, along with wind tunnel results, were compared.

The hybrid matching math model contained complete five-degree-of-freedom equations (no velocity derivatives) with variable dynamic pressure, whereas the Newton-Raphson model used three-degree-of-freedom equations with constant dynamic pressure. The hybrid matching technique required an experienced individual to supply the logic for the derivative extraction process. Although the Newton-Raphson technique did not use the human operator in the actual derivative extraction process, the effective use of the program required a comparable level of knowledge and experience in the selection of weighting functions and use of the a priori option.

Both techniques required that the source data tapes be edited and transferred to the proper format and computer language to be compatible with each of the analysis programs. This process, although required only once per test program, has proven to be both frustrating and time consuming.

Both programs were found to be capable of giving accurate results. Since Newton-Raphson tends to be less time consuming, it is better suited for processing large quantities of data maneuvers. Hybrid matching is more applicable for research vehicle type programs where a limited amount of data is processed for each flight; or for analyzing maneuvers which are highly coupled or transient in nature requiring complete five-degree-of-freedom equations.

43

BASIC PROGRAM STRUCTURE/DIFFERENCES

Newton-Raphson

1. 3-DEGREES-OF-FREEDOM EQUATIONS

2. DYNAMIC PRESSURE = CONSTANT

3. HUMAN OPERATOR NOT DIRECTLY IN LOOP, BUT REQUIRED FOR EFFECTIVE USE OF PROGRAM

4. TIME

Hybrid

1. 5-DEGREES-OF-FREEDOM EQUATIONS

2. DYNAMIC PRESSURE = VARIABLE

3. HUMAN OPERATOR OBTAINS RESULTS

4. TIME

HYBRID MATCHING PROGRAM

- STORES FLIGHT DATA DIGITALLY

- SOLVES EQUATIONS OF MOTION ON ANALOG IN REPETITIVE OPERATION MODE

- DISPLAYS FLIGHT DATA AND COMPUTED SOLUTION SIMULTAN- EOUSLY ON REPETITIVE-OPERATION SCOPE

- CAN UNCOUPLE EQUATIONS OF MOTION BY USING FLIGHT DATA IN EQUATIONS

- HUMAN OPERATOR MANUALLY CHANGES DERIVATIVES TO OBTAIN MATCH OF COMPUTED AND FLIGHT DATA

HYBRID EQUATIONS of MOTION - Body Axis

$$\dot{p} = \left(\frac{I_y - I_z}{I_x}\right)qr + \frac{I_{xz}}{I_x}\dot{r} + \frac{I_{xz}}{I_x}pq + \frac{\bar{q}Sb}{I_x}\left[C_{\ell_{\delta a}}\delta a + C_{\ell_{\delta r}}\delta r + C_{\ell_\beta}\beta + C_{\ell_p}\frac{pb}{2V} + C_{\ell_r}\frac{rb}{2V}\right]$$

$$\dot{q} = \left(\frac{I_z - I_x}{I_y}\right)pr + \frac{I_{xz}}{I_y}r^2 - \frac{I_{xz}}{I_y}p^2 + \frac{\bar{q}Sc}{I_y}\left[C_{m_o} + C_{m_\alpha}\alpha + C_{m_{\delta e}}\delta e + C_{m_q}\frac{qc}{2V}\right]$$

$$\dot{r} = \left(\frac{I_x - I_y}{I_z}\right)pq + \frac{I_{xz}}{I_z}\dot{p} - \frac{I_{xz}}{I_z}qr + \frac{\bar{q}Sb}{I_z}\left[C_{n_{\delta a}}\delta a + C_{n_{\delta r}}\delta r + C_{n_\beta}\beta + C_{n_r}\frac{rb}{2V} + C_{n_p}\frac{pb}{2V}\right]$$

$$\dot{\alpha} = q + \frac{1}{\cos\alpha\cos\beta}\left[\frac{g}{V}(\cos\theta\cos\phi) - p\sin\beta\right] - \frac{\bar{q}S}{mV\cos\alpha\cos\beta}\left[C_{N_o} + C_{N_\alpha}\alpha + C_{N_{\delta e}}\delta e\right]$$

$$\dot{\beta} = p\sin\alpha - r\cos\alpha + \frac{1}{\cos\beta}\left[\frac{g}{V}(\cos\theta\sin\phi) - \frac{g}{V}(\sin\theta\sin\beta)\right] + \frac{\bar{q}S}{mV\cos\beta}\left[C_{y_{\delta a}}\delta a + C_{y_{\delta r}}\delta r + C_{y_\beta}\beta\right]$$

RESULTS

● BOTH PROGRAMS ARE CAPABLE OF GIVING ACCURATE RESULTS FOR THE TYPES OF MANEUVERS ANALYZED.

● NEWTON-RAPHSON TENDS TO BE LESS TIME CONSUMING AND THEREFORE BETTER SUITED FOR PROCESSING LARGE QUANTITIES OF DATA MANEUVERS.

● HYBRID MATCHING IS MORE APPLICABLE FOR RESEARCH VEHICLE TYPE PRO-GRAMS WHERE A LIMITED AMOUNT OF DATA IS PROCESSED FOR EACH FLIGHT OR FOR ANALYZING MANEUVERS WHICH ARE HIGHLY COUPLED OR TRANSIENT IN NATURE REQUIRING COMPLETE FIVE-DEGREE-OF-FREEDOM EQUATIONS.

● A COMBINATION OF THE TWO METHODS WOULD BE IDEAL.

47

EXTRACTION OF DERIVATIVES FROM FLIGHT DATA FOR SEVERAL AIRCRAFT,

USING THE LRC INTERACTIVE COMPUTER SYSTEM

William T. Suit and James L. Williams
NASA Langley Research Center
Hampton, Virginia

ABSTRACT

Early in 1972 a procedure for estimating stability and control parameters from flight data, by using maximum likelihood methods employing an interactive computer system, was established at the NASA Langley Research Center. The system features a console where the flight data and computed aircraft motions are displayed as the computer performs iterations to determine the best fit to the flight data. The console operator can interact with the system in such matters as changing the cost function, biasing the flight data, selecting portions of the time history to be fitted, and so on.

Since installation of the system, it has been used to extract some aerodynamic derivatives for the Navion, XC-142, Kestrel, and F-8. The paper proposed for this confreence will review some of the results of these studies as well as some of the problems encountered.

INTRODUCTION

Numerous parameter-extraction techniques are available in the literature, and some have been used to extract aerodynamic coefficients from flight data. Most of the methods in use minimize a quadratic cost function through adjustment of the aerodynamic coefficients in the equations of motion. Published results have shown available techniques can be used to obtain good comparisons between computed and measured flight data (states), even in the presence of plant noise. However, published results often do not indicate the problems which are often encountered in obtaining the good comparisons. The purpose of this paper is to review some recent work done on parameter extraction at the Langley Research Center. The paper will show comparisons of computed and measured flight data for several aircraft, and will then indicate some of the problems encountered in the process.

SYMBOLS

a_x Acceleration along the x-body axis, m/sec^2 (ft/sec^2)

a_z Acceleration along the z-body axis, m/sec^2 (ft/sec^2)

49

θ   Pitch angle, radians

The following symbols are used on the computer-generated figures.

ALPHA   Angle of attack minus trim angle of attack, deg

AXI   Acceleration along the x-body axis, m/sec^2 (ft/sec^2)

AYI   Acceleration along the y-body axis, m/sec^2 (ft/sec^2)

AZI   Acceleration along the z-body axis, m/sec^2 (ft/sec^2)

DA   Aileron deflection minus trim aileron deflection, rad

DE   Elevator deflection minus trim elevator deflection, rad

DR   Rudder deflection minus trim rudder deflection, rad

P   Roll rate, rad/sec

PHI   Roll attitude, rad

Q   Pitch rate, rad/sec

R   Yaw rate, rad/sec

THETA   Pitch attitude, rad

U   Velocity along the x-body axis, m/sec (ft/sec)

V   Velocity along the y-body axis, m/sec (ft/sec)

W   Velocity along the z-body axis, m/sec (ft/sec)

The aerodynamic derivatives are given in the standard nondimensional notation.

The Langley Research Center Parameter-Extraction System

Much of the recent parameter-extraction work at LRC has been done on an iterative computer program using maximum likelihood techniques. The system displays on a cathode ray tube both the flight data and the computed states as the computer calculates the time histories of the states. The console operator can interact with the computer at any point during the iteration process. Figure 1 is a photograph of one of the consoles used in parameter-extraction work. The computer program and mechanization used are described in reference 1 and will not be discussed in this paper.

Results From Parameter-Extraction Process

The LRC interactive digital program has been used to extract some aerodynamic derivatives for the F-4, Navion, XC-142, Kestrel, F-8, and Helio Courier aircraft. Some results will be presented for the latter three aircraft. For the cases shown, 200 to 300 data points were used in most of the runs. Because of space limitations, less points are shown in figures.

Kestrel. The Kestrel is a vectored-thrust aircraft, and is the forerunner of the Harrier. The flight data analyzed were perturbations from equilibrium conditions for each of three thrust vector angles, and at three different airspeeds. Perturbations were generated by elevator deflections. Typical flight and final computer-generated time histories are shown in figure 2. The time histories compare quite well. The extracted aerodynamic derivatives and their standard deviations are shown in table I.

F-8. Time histories of the response of the F-8 to an elevator pulse are shown in figure 3. The calculated response compares quite well with the flight data, and the extracted parameters are listed in table II. The extracted parameters are compared with unpublished wind-tunnel data in figure 4. The agreement is quite good in all cases, except for the damping-in-pitch parameter $C_{m_q} + C_{m_{\dot{\alpha}}}$.

Helio Courier. Flight time histories for the Helio Courier are shown for various types of control inputs in figure 5 (elevator pulses), figure 6 (aileron and rudder pulses), and figure 7 (elevator, aileron, and rudder inputs). In all cases, the computed time histories matched the flight data quite well. The derivatives are arranged to permit easy comparison of the longitudinal and lateral derivatives for the decoupled (lateral or longitudinal) responses, and for combined lateral and longitudinal responses.

The results shown for the above three aircraft are typical of what is generally seen in the literature, and shows that it is usually possible to extract derivatives that result in computed time histories that match flight data quite well. What is not shown, however, is the fact that obtaining good results is not a straightforward, automatic process. Problem areas that the researcher must circumvent are left out of discussion, and only the successful results are reported. One area of parameter extraction which seems to have received too little attention, or at least too little documentation, is that of mechanizing a process for identifying problems and for circumventing the problem. This includes such problems as data bias, data incompatibility, linear dependence, and so forth. At present, our best parameter-extraction programs are a combination of an automated system with human judgment, which reminds one somewhat of analog-matching methods.

The remainder of this paper will discuss some of the problems encountered during use of the LRC interactive computer system, and techniques for working around the problems.

Problems Encountered in Parameter Extraction

Data Compatibility. One of the common problems encountered is that of instrument readings which contained biases, so that the measured states were not compatible at corresponding time points. For example, in equilibrium flight, there is a simple geometric relationship between the aircraft acceler- ometer readings and the aircraft pitch attitude.

$$a_x = g \sin \theta$$

$$a_z = -g \cos \theta$$

Figure 8 shows flight data for which these relationships are obviously violated; however, it is not clear whether the angle θ is incorrect, or if the longitu- dinal acceleration is incorrect. Figure 8 shows poor agreement between meas- ured and calculated longitudinal accelerations. However, the θ flight values were picked off of the cockpit eight-ball. The θ readings were felt to be less reliable than the accelerometer readings because of the basic accuracy of the eight-ball and because of a possible gyro erection problem, especially after successive maneuvers. Therefore, the flight θ was altered to be com- patible with the measured longitudinal acceleration. The results generally improved (fig. 9), but the match in the aircraft forward velocity deteriorated. The next step was to bias the flight longitudinal acceleration, and the results are shown in figure 10. Of course, as the various quantities are biased to achieve compatibility, extracted values of trim aerodynamic coefficients will automatically change. The point to be made is that where simple mathematical relationships exist which can indicate compatibility of data, they should be used, judiciously, to eliminate biases in the data. If such relationships are not apparent, then one must be very careful in adjusting data and making initial estimates of the trim coefficients in order to obtain acceptable histories and extract reasonable aerodynamic coefficients. Poor initial estimates of the trim coefficients will sometimes result in good convergence to the flight data, but incorrect aerodynamic derivatives will be extracted in the process. In other cases, we have found that poor initial estimates of the trim coefficients will result in divergent motion, and the computer will not iterate to a final solution.

Much of the parameter-extraction work done recently at LRC has required some adjustment of states and trim coefficients to initiate the iteration process on the computer. If the program can get through one iteration, it generally converges and exhibits repeatable behavior.

Validity of Extracted Derivatives. When flight data have been used and aerodynamic parameters extracted, it is necessary to evaluate the adequacy of the parameters. There are several criteria which can be used to establish some confidence level on the numerical values. These include such factors as how well the estimated motions fit the flight data, the incremental changes in the estimated parameters for successive iterations, the standard deviations, and knowledge gained from experience in estimating parameters or from wind- tunnel work. Additional confidence can be attached to the extracted parameters if they can be used to predict motions for other flights with similar control inputs. This approach is illustrated in the next two figures.

First of all, the flight data of figure 10 were divided into two segments, from 0 to 30 seconds and from 30 to 56 seconds. The parameter-extraction program was used to extract derivatives for the first 30 seconds of flight (fig. 11). The computed and experimental time histories compare quite well. The extracted derivatives are shown in the first column of table IV. These same derivatives were used to predict the motions for the flight (of fig. 10) from 30 to 57 seconds. This worked out fairly well except for the forward velocity u, which diverged (fig. 12). The problem was traced to the trim coefficient C_{X_O}, which was then computed to assure initial trim conditions. Reasonably good calculated time histories were then obtained (fig. 13); however, the forward velocity and acceleration were not as good as expected. The derivatives used to generate figure 13 are given in the second column of table IV. At this point the computer was allowed to iterate to match the flight data of figure 13. At convergence, the extracted derivatives were those listed in the third column of table IV, and the calculated time histories match the flight data quite well (fig. 14). Table IV shows that each section of the flight yielded still another set. In this particular case the differences are not great, but other cases have been found where differences are appreciable.

Another interesting situation arose in working with flight data for the Helio Courier aircraft. Several experiments were made using rudder and aileron inputs. In one series, the order of inputs was a rudder pulse followed by an aileron pulse. In another series the order was reversed. In one experiment, both controls were applied simultaneously. In working with these data, it was found that the extraction program provided good fits to the flight data for the rudder-aileron sequence (fig. 15), but not so good for the aileron-rudder sequence (fig. 16). In addition, there were some differences in extracted aerodynamics for various tests, even those with the same control sequence (see table V). It should be noted that the ranges in some of the important derivatives are quite large. The reasons for such variations have not been identified at this time.

The first column gives the results of combining two of the rudder-aileron runs. As previously mentioned, this longer run using more data seemed to give the most realistic set of numbers and the fit to the data was better than for the aileron-rudder run.

CONCLUDING REMARKS

The feeling of this investigator is that given perfect data and a perfect model, then unique results can be obtained. However, when using actual flight data this situation does not exist. The situation could probably be improved if better instrumentation were used, and if the analyst were fully familiar with the capabilities, limitations, position, and alinement of the instruments.

As pointed out, difficulty is still encountered in obtaining consistent results even with very similar control inputs. Areas which require some study are:

(1) Techniques for identifying and correcting incompatible data.

(2) Automating more of the preliminary data handling (before it enters the parameter-extraction program).

(3) Evaluation of model completeness.

REFERENCE

1. Grove, Randall D., Bowles, Roland L., and Mayhew, Stanley C.: A Procedure for Estimating Stability and Control Parameters From Flight Test Data by Using Maximum Likelihood Methods Employing a Real-Time Digital System. NASA TN D-6735, 1972.

TABLE I. EXTRACTED AERODYNAMIC PARAMETERS AND STANDARD DEVIATIONS FOR THE KESTREL AIRCRAFT

(Standard deviations are given in parentheses)

	$\theta_j = 0°$			$\theta_j = 15°$		$\theta_j = 30°$		
Approx. Mach No.	0.82	0.62	0.43	0.82	0.62	0.82	0.62	0.43
C_T	.121	.121	.155	.1195	.128	.139	.151	.196
α, deg	1.75	3.33	8.94	.722	4.70	.653	2.22	7.91
C_{z_o}	-.13 (.0012)	-.225 (.002)	-.46 (.0014)	-.068 (.0013)	-.21 (.0014)	-.042 (.0012)	-.127 (.0021)	-.37 (.0031)
C_{z_α}	-2.82 (.07)	-3.04 (.105)	-2.83 (.03)	-2.92 (.064)	-2.26 (.064)	-2.95 (.041)	-2.67 (.067)	-2.58 (.058)
$C_{z_{\delta_e}}$	-.42	-.50	-.63	-.405	-.47	-.42	-.33	-.12
C_{x_o}	-.029 (.0002)	-.01 (.0004)	.034 (.0003)	-.03 (.0001)	-.015 (.0002)	-.027 (.0002)	-.015 (.0004)	.02 (.0006)
C_{x_α}	.23 (.0067)	.32 (.013)	.535 (.0056)	.23 (.0049)	.26 (.0078)	.186 (.0052)	.29 (.0084)	.43 (.0095)
C_{m_α}	-.13 (.0035)	-.167 (.0046)	-.28 (.0035)	-.14 (.0034)	-.137 (.0016)	-.059 (.0008)	-.055 (.0011)	-.097 (.0014)
$C_{m_q} + C_{m_{\dot\alpha}}$	-10.44 (.70)	-10.79 (.75)	-14.90 (.39)	-11.61 -(.68)	-11.82 (.42)	-12.82 (.46)	-9.75 (.45)	-5.15 (.32)
$C_{m_{\delta_e}}$	-.91 (.034)	-1.07 (.04)	-1.37 (.022)	-.87 (.036)	-1.02 (.027)	-.901 (.027)	-.707 (.027)	-.257 (.016)

TABLE II. EXTRACTED AERODYNAMIC PARAMETERS AND

STANDARD DEVIATIONS FOR THE F-8 AIRCRAFT

(Standard deviations are given in parentheses)

Mach No. / Coeffs.	0.81	0.90	0.98
C_{x_o}	0.03 (0)	0.02 (0)	0.02 (0)
C_{x_α}	.58 (-)	.58 (-)	.58 (-)
C_{z_o}	-.44 (0)	-.44 (0)	-.41 (0)
C_{z_α}	-6.02 (.08)	-6.95 (.13)	-7.49 (.24)
$C_{z_{\alpha_e}}$	-.92 (-)	-1.01 (-)	-.97 (-)
C_{m_o}	.003 (0)	.003 (.003)	0 (0)
C_{m_α}	-1.62 (.01)	-2.04 (.10)	-3.31 (.02)
$C_{m_{\dot{\alpha}}}$	-7.00 (-)	-7.00 (-)	-7.00 (-)
C_{m_q}	-27.82 (1.04)	-29.17 (1.83)	-16.59 (2.6)
$C_{m_{\delta_e}}$	-2.34 (.02)	-2.58 (.73)	-2.46 (.07)

TABLE III. EXTRACTED DERIVATIVES FOR THE HELIO COURIER AIRCRAFT

FOR VARIOUS RESPONSE MODES

Longitudinal parameters			Lateral parameters		
Parameter	Elevator, aileron, and rudder inputs	Elevator inputs only	Parameter	Elevator, aileron, and rudder	Aileron and rudder inputs only
C_{x_o}	-0.026	-0.025	C_{y_o}	0.003	-0.016
C_{x_α}	-.056	-.33	C_{y_β}	-.81	-.92
C_{z_o}	-.60	-.45	C_{y_p}	.53	-.47
C_{z_α}	-4.72	-5.48	C_{y_r}	-.70	-1.0
C_{z_q}	-4.5	-4.5	$C_{y_{\delta_r}}$	.60	.09
$C_{z_{\delta_e}}$	-.44	-.47	C_{l_o}	.0013	-.0016
C_{m_o}	.0019	-.0045	C_{l_β}	-.014	-.060
C_{m_α}	-.75	-.35	C_{l_p}	-.13	-.26
$C_{m_{\dot\alpha}}$	-4.6	-4.6	C_{l_r}	.13	.07
C_{m_q}	-1.54	-2.65	$C_{l_{\delta_r}}$	.019	.011
$C_{m_{\delta_e}}$	-1.3	-1.4	$C_{l_{\delta_a}}$	.096	.081
			C_{n_o}	-.0014	-.0015
			C_{n_β}	.0805	.086
			C_{n_p}	.007	-.006
			C_{n_r}	-.078	-.105
			$C_{n_{\delta_r}}$	-.069	-.069
			$C_{n_{\delta_a}}$	-.015	-.005

TABLE IV.   DERIVATIVES USED TO PREDICT FUTURE RESPONSES AND THE

COMPARISON OF DERIVATIVES OBTAINED FROM SPLIT RUNS

Parameter	First half	Prediction of second half	Converged second half	Total run
C_{x_o}	-0.014	0.026	-0.010	-0.010
C_{x_α}	.3	.3	.33	.32
C_{z_o}	-.22	-.226	-.19	-.225
C_{z_α}	-3.16	-3.16	-3.25	-3.04
$C_{z_{\delta_e}}$	-.48	-.48	-.56	-.50
C_{m_o}	0	0	.0016	0
C_{m_α}	-.17	-.17	-.18	-.167
$C_{m_q} + C_{m_{\dot\alpha}}$	-10.17	-10.17	-12.00	-10.79
$C_{m_{\delta_e}}$	-1.03	-1.03	-1.22	-1.07

TABLE V. COMPARISON OF A NUMBER OF LATERAL RUNS FOR THE HELIO COURIER AIRCRAFT

Control input order \\ Parameters	$\delta_r - \delta_a$, $\delta_a - \delta_r$	$\delta_a - \delta_r$		$\delta_r - \delta_a$		
		1	2	1	2	3
c_{y_o}	-0.0158	-0.0017	0	0	0	0
c_{y_β}	-.918	-.706	-1.17	.89	.88	.98
c_{y_p}	-.465	-.126	-.68	-.55	-.51	-.70
c_{y_r}	-1.0	-.306	-2.68	-.85	-1.1	-1.49
$c_{y_{\delta_r}}$	.088	.225	-.097	.071	-.017	-.12
c_{l_o}	-.0016	0	0	0	0	0
c_{l_β}	-.62	-.056	-.077	-.049	-.041	-.055
c_{l_p}	-.26	-.267	-.34	-.25	-.21	-.31
c_{l_r}	.07	.10	.14	.15	.15	.13
$c_{l_{\delta_r}}$	-.013	-.0023	.005	.011	.012	0
$c_{l_{\delta_a}}$	.0807	.085	.105	.076	.082	.096
c_{n_o}	-.0015	0	0	0	0	0
c_{n_β}	.086	.093	.093	.082	.083	.087
c_{n_p}	-.0063	-.013	-.019	.003	-.006	.014
c_{n_r}	-.105	-.116	-.18	-.1	-.12	-.103
$c_{n_{\delta_r}}$	-.069	-.076	-.070	-.09	-.08	-.072
$c_{n_{\delta_a}}$	-.0046	-.0071	.0054	-.014	-.008	-.0085

Figure 1.- Typical control console used during parameter-extraction work.

L-69-8762

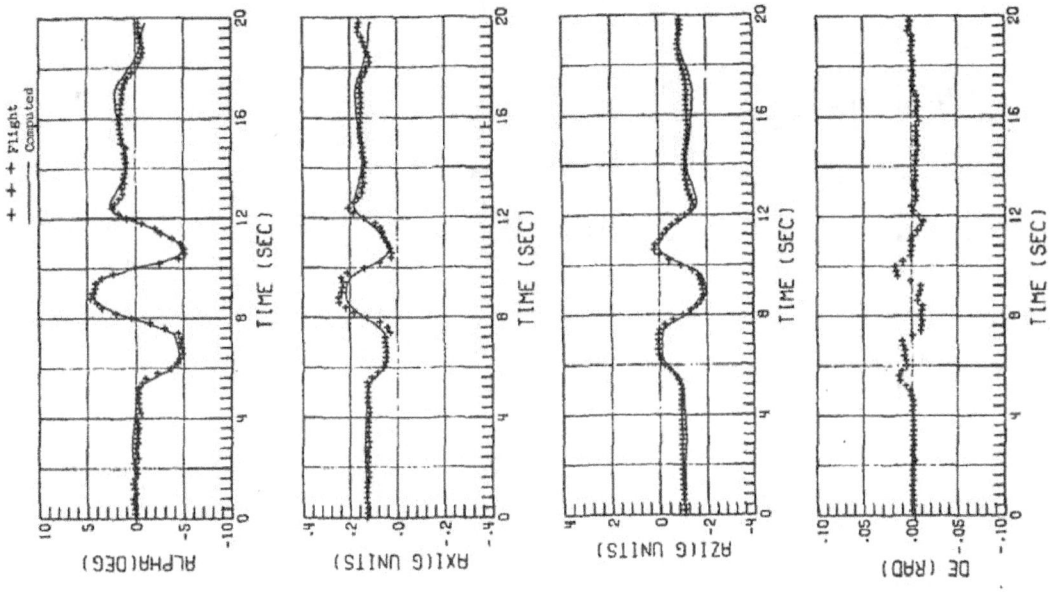

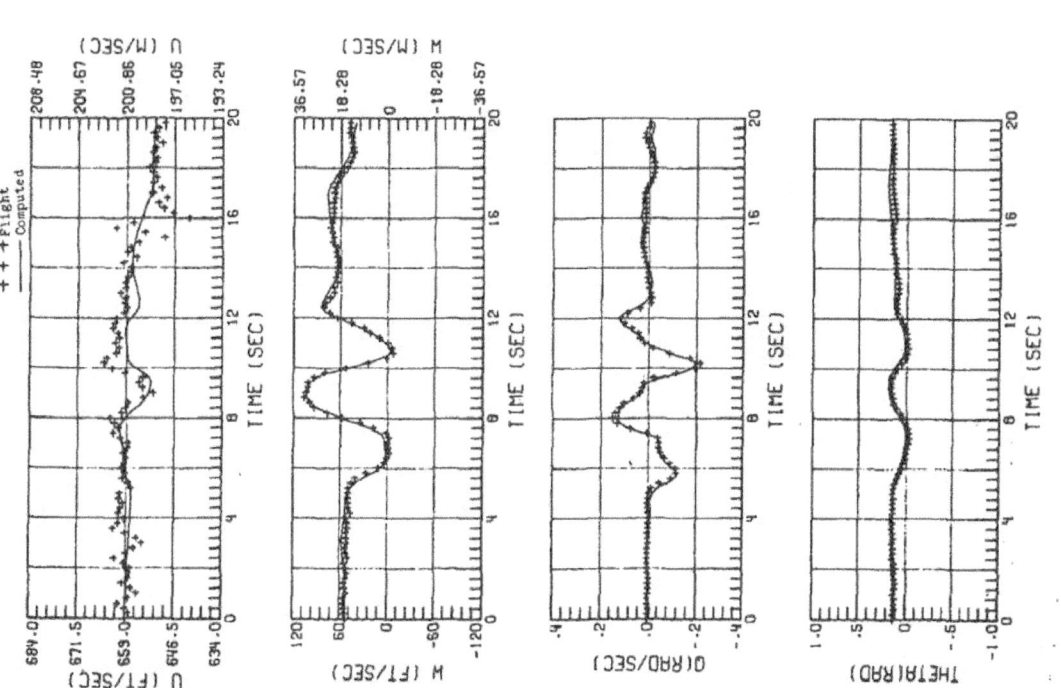

Figure 2.- Comparison of flight data with time histories for the Kestrel aircraft computed by using the parameters of table I for an elevator input. Mach number = 0.62, θ_j = 15 degrees.

61

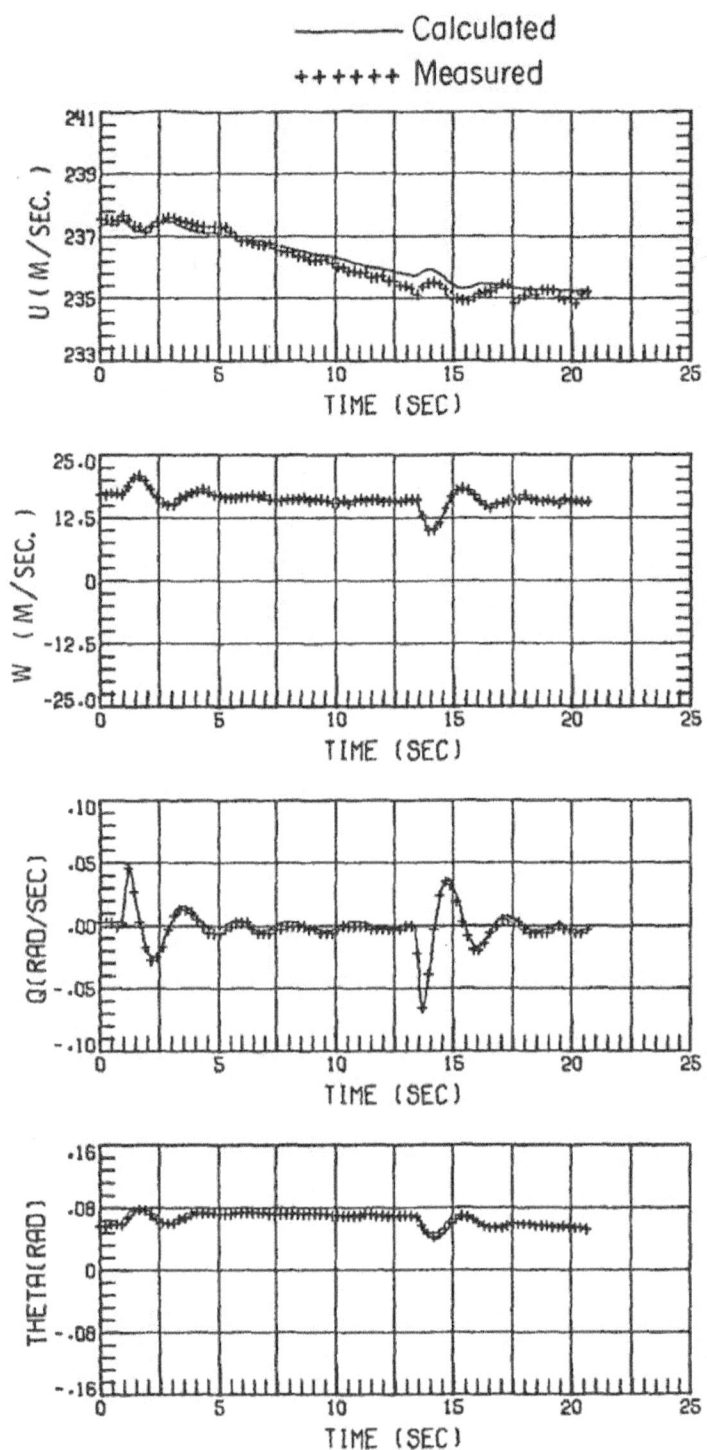

Figure 3.- Comparison of flight data with time histories for the F-8 aircraft computed using the parameters of table II. Mach number = 0.808.

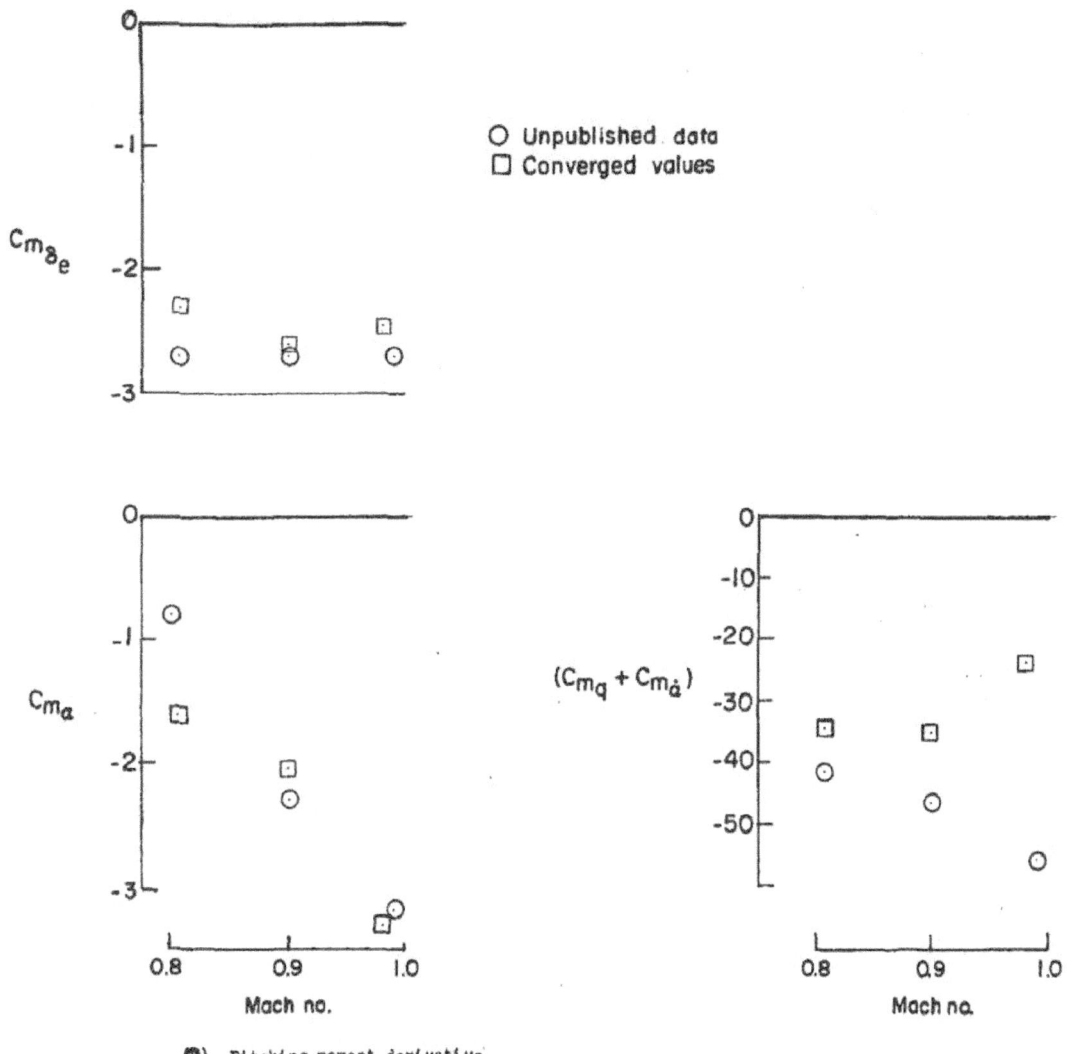

(a) Pitching moment derivative.

Figure 4.- Comparison of extracted derivatives with Mach number for the F-8 supercritical wing aircraft with unpublished derivatives.

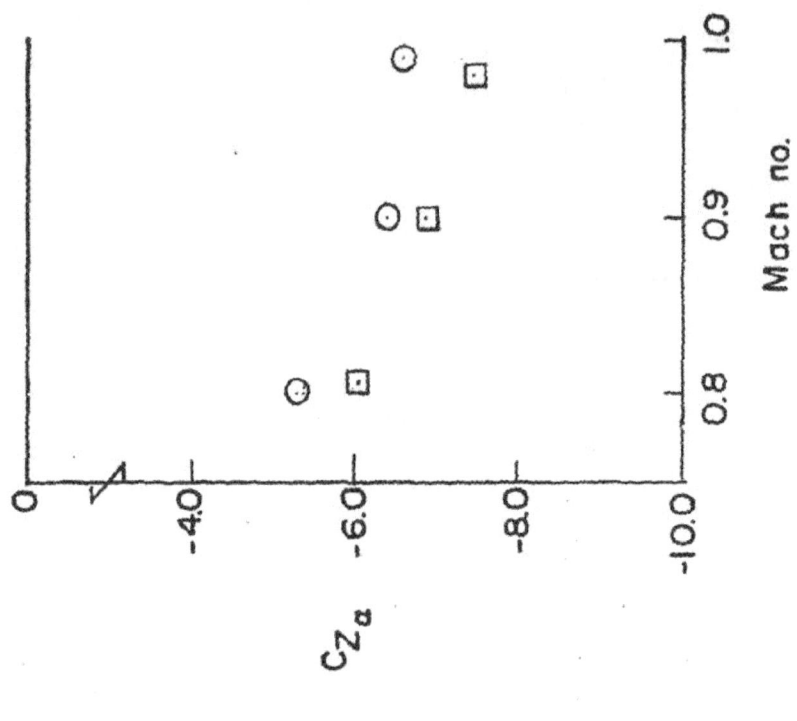

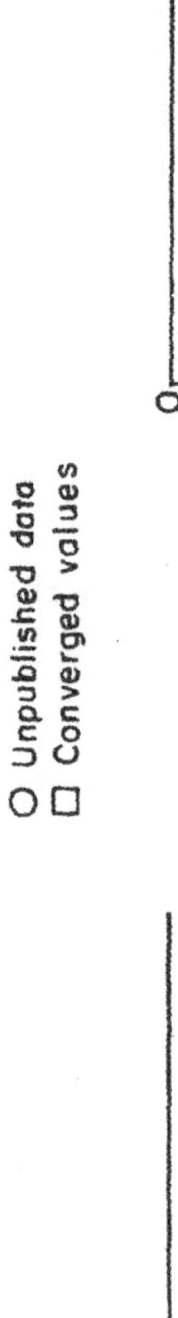

(b) Normal force derivatives.

Figure 4.- Concluded.

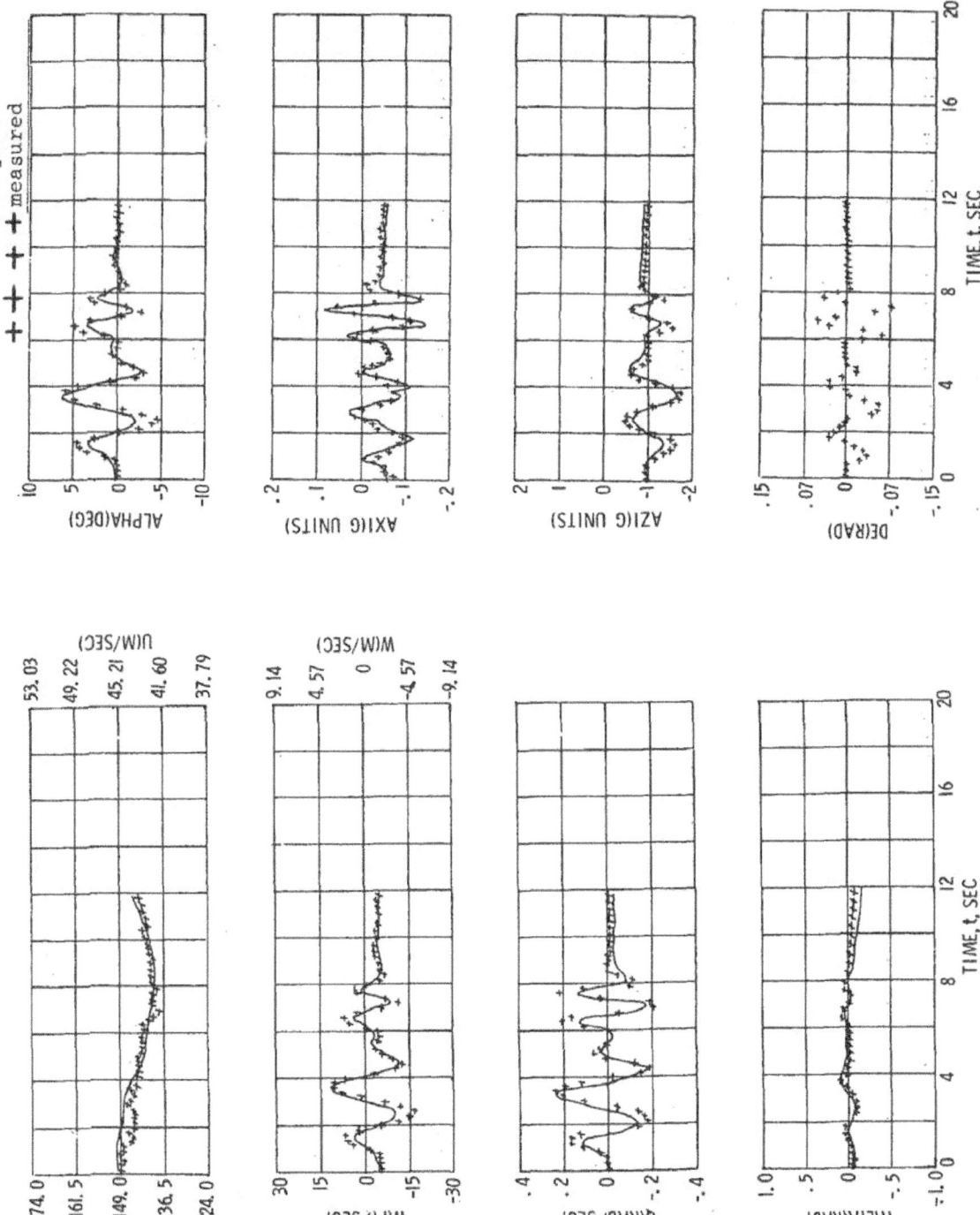

Figure 5.- Comparison of flight data with time histories for the Helio Courier aircraft computed using the parameters of table III and elevator control inputs.

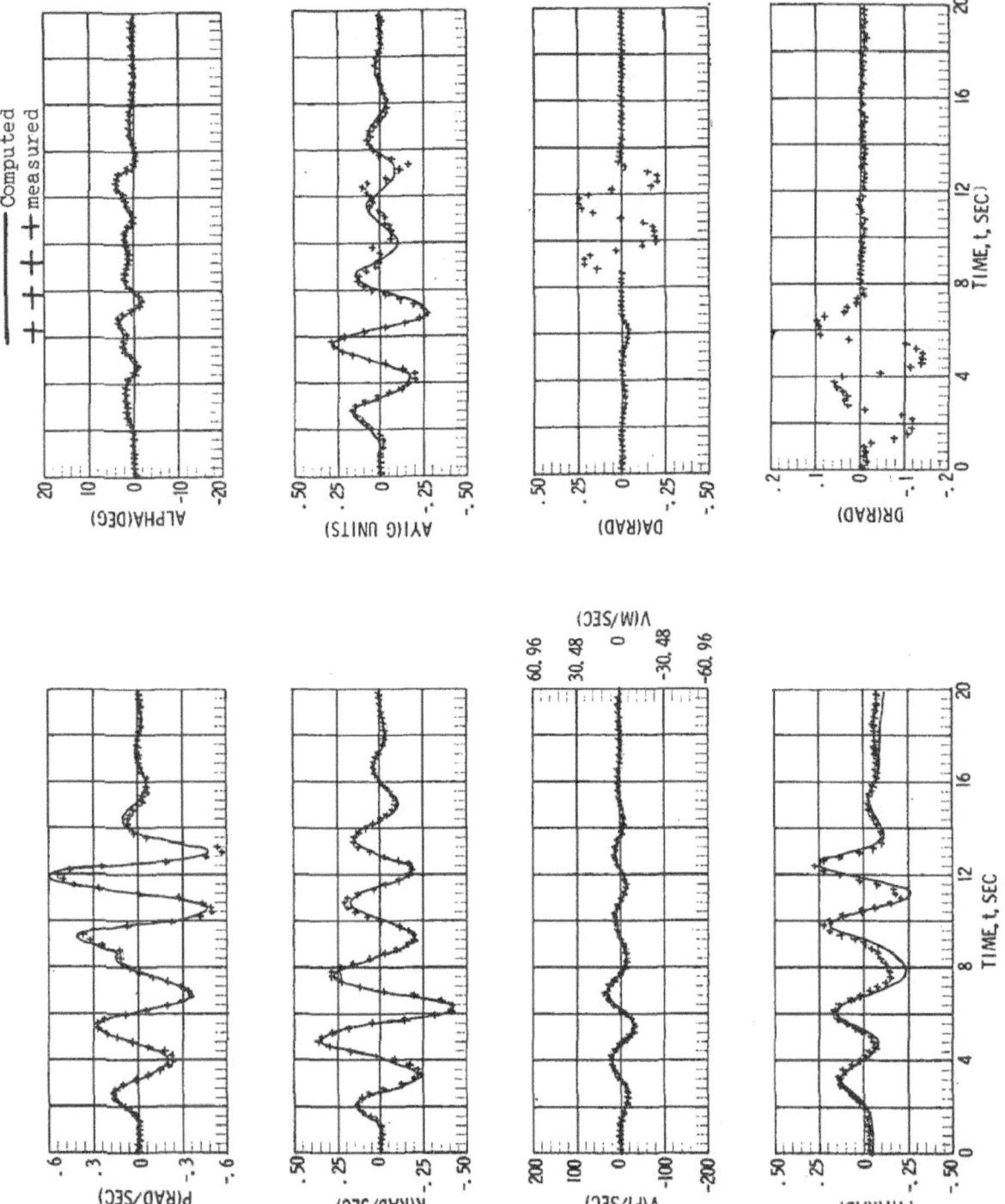

Figure 6.— Comparisons of flight data with time histories for the Helio Courier aircraft computed using the parameters of table III and rudder and aileron control inputs.

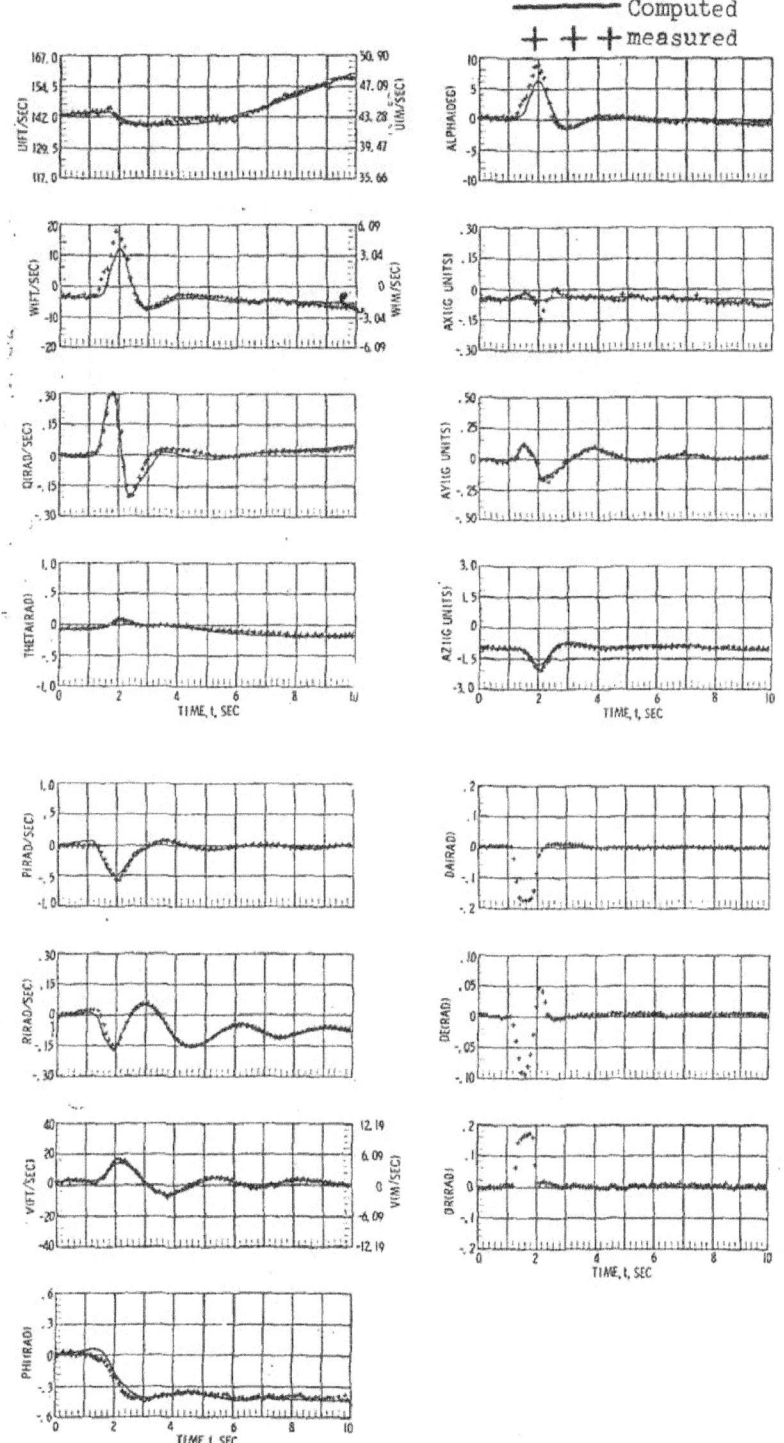

Figure 7.- Comparisons of flight data with time histories for the Helio
Courier aircraft computed using the parameters of table III and
elevator, rudder, and aileron inputs.

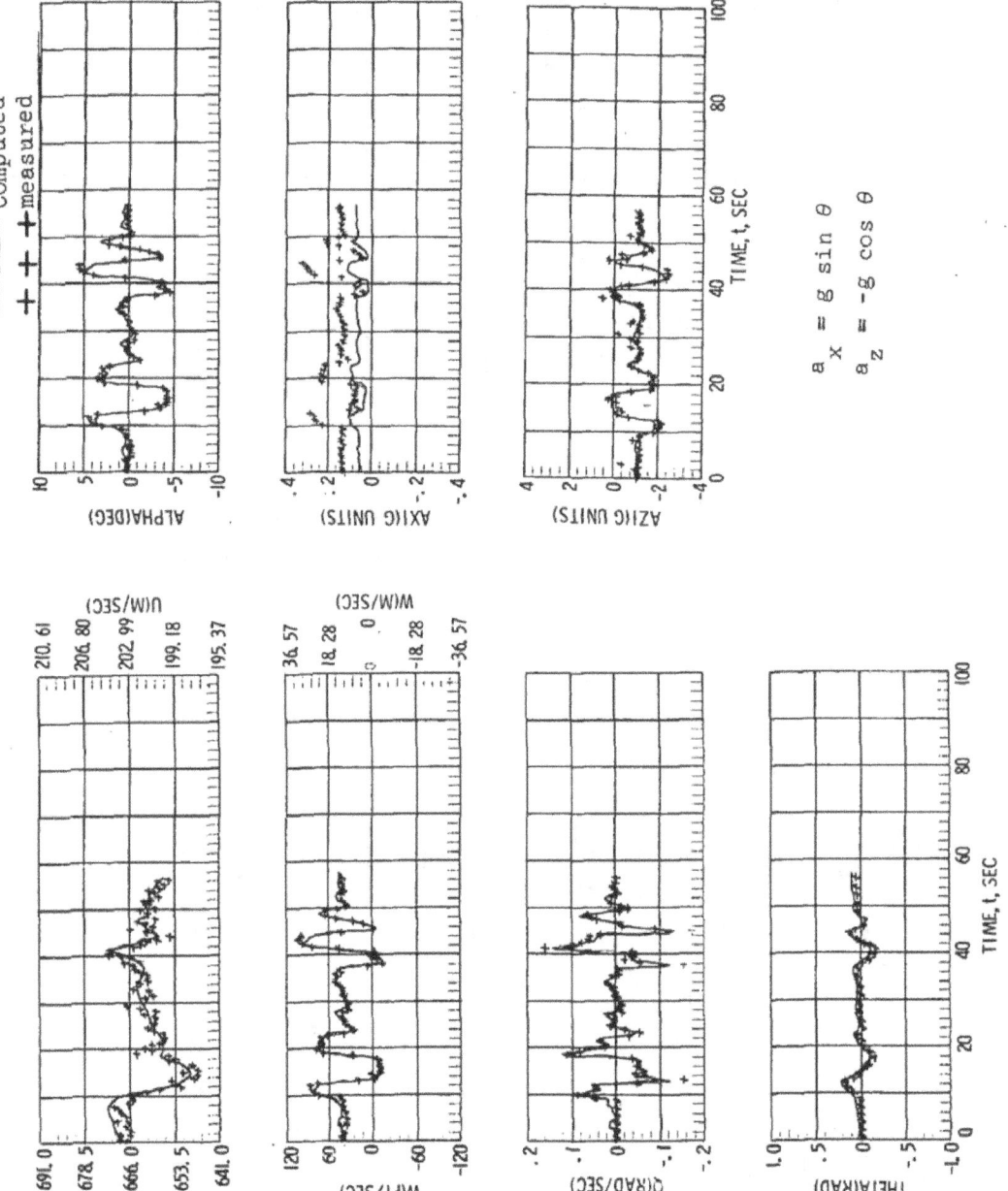

Figure 8.- Comparisons of flight data with computed time histories where some of the states are initially incompatible.

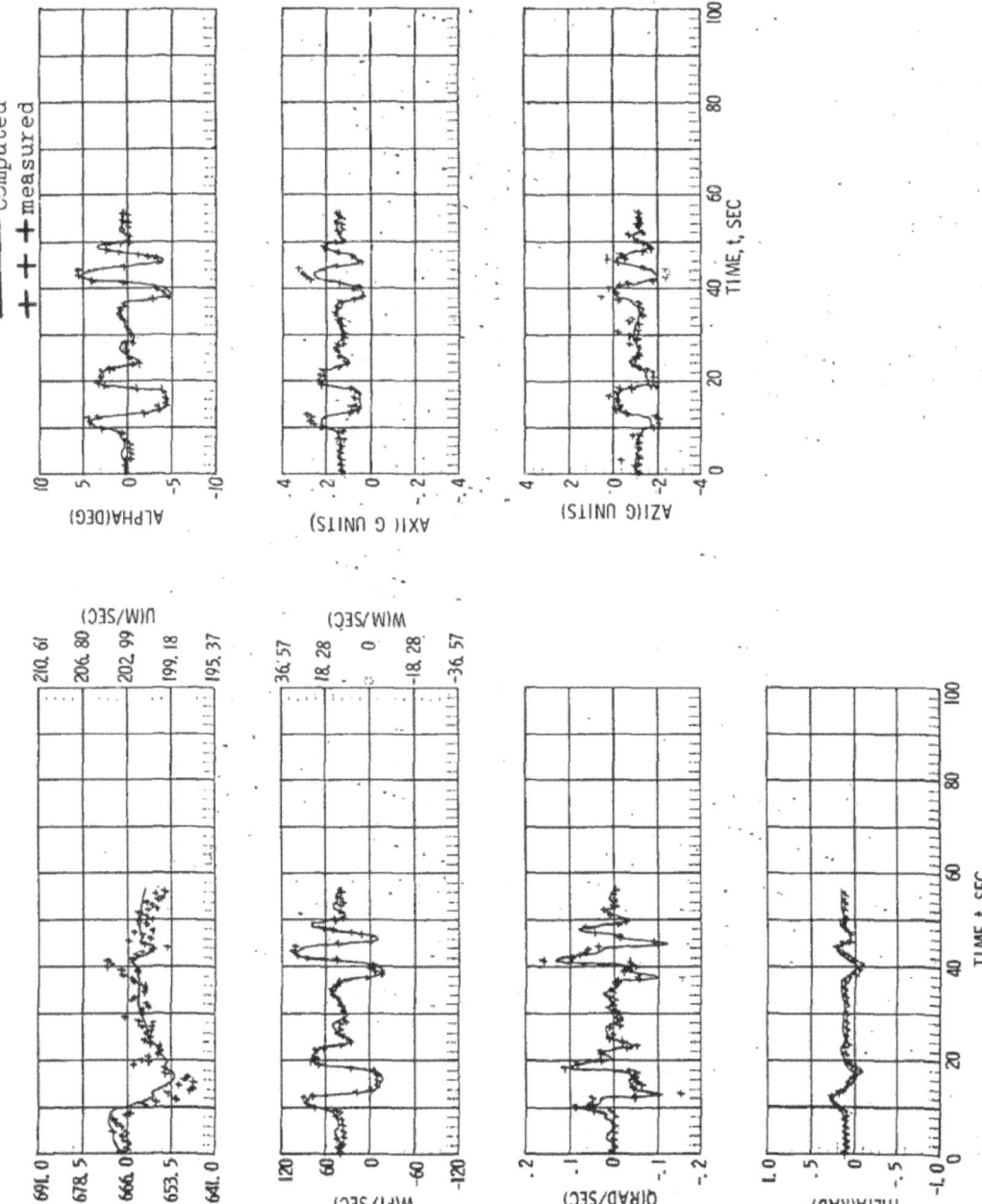

Figure 9.- Comparisons of flight data with computed time histories with the states biased to be initially compatible.

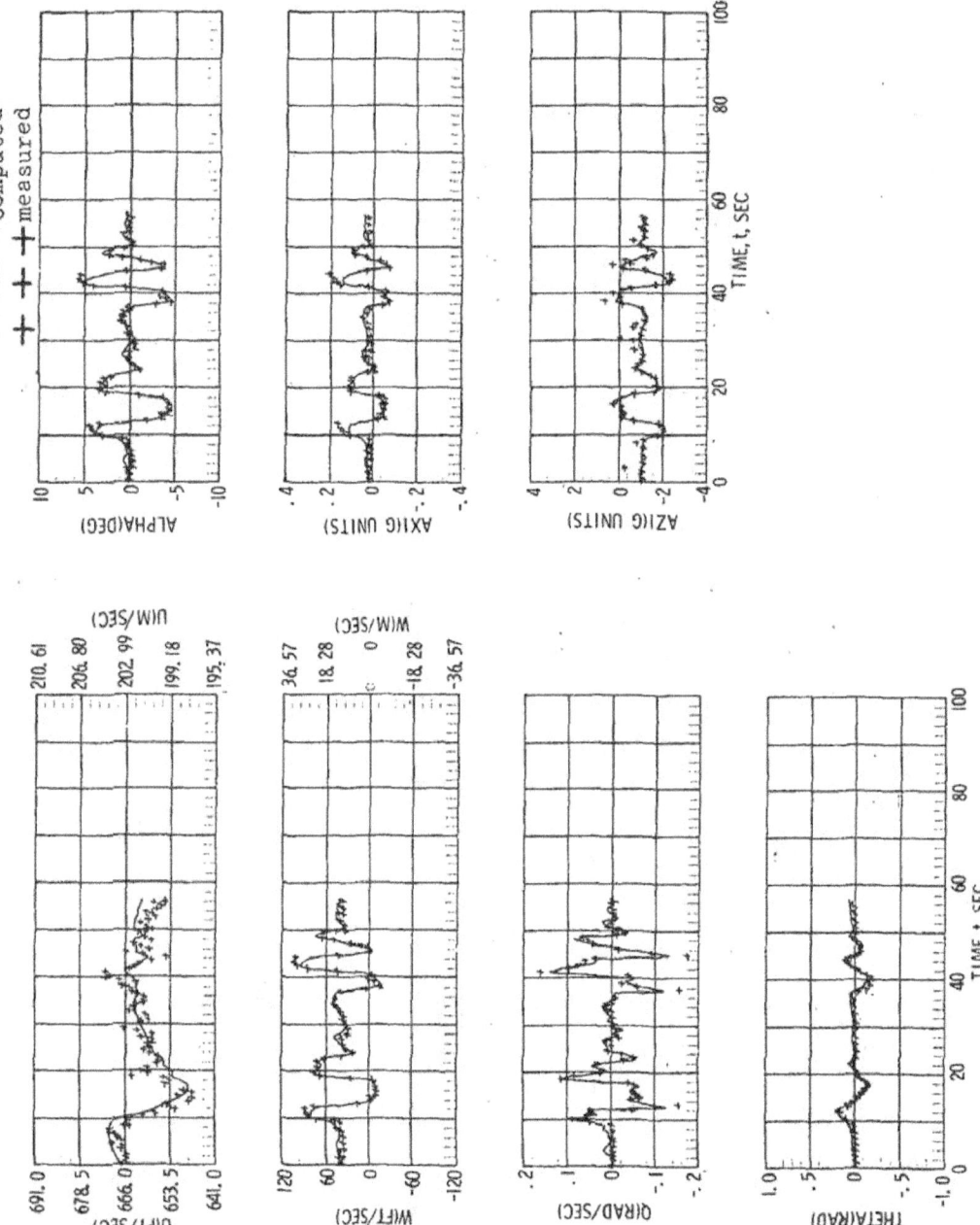

Figure 10.- Comparisons of flight data with computed time histories with the states biased to be initially compatible so that $a_x = \theta_0 = 0$ and $a_z = 1$.

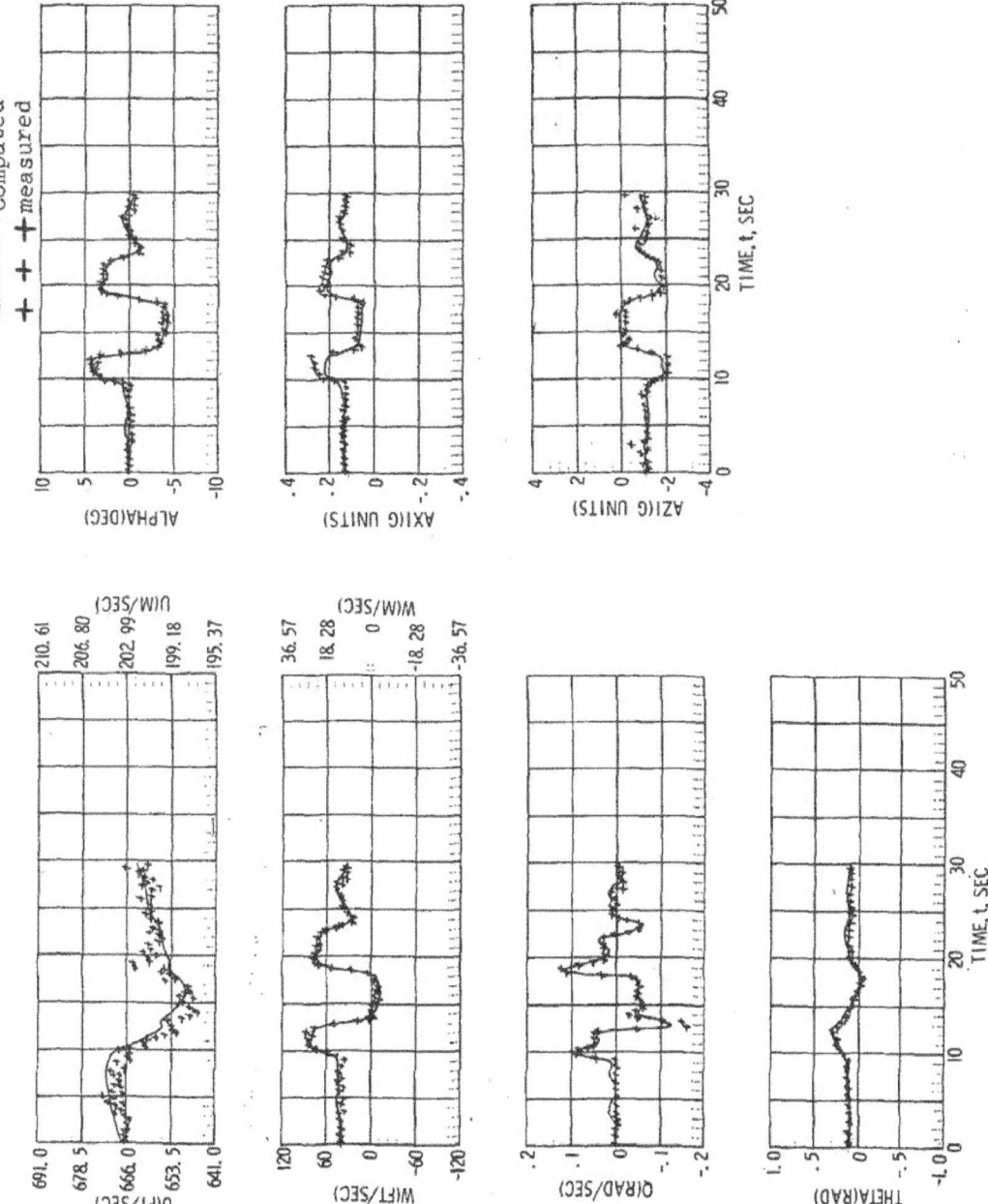

Figure 11.- Comparisons of flight data with computed time histories for the first half of the data of figure 10. The parameters used are shown in table IV.

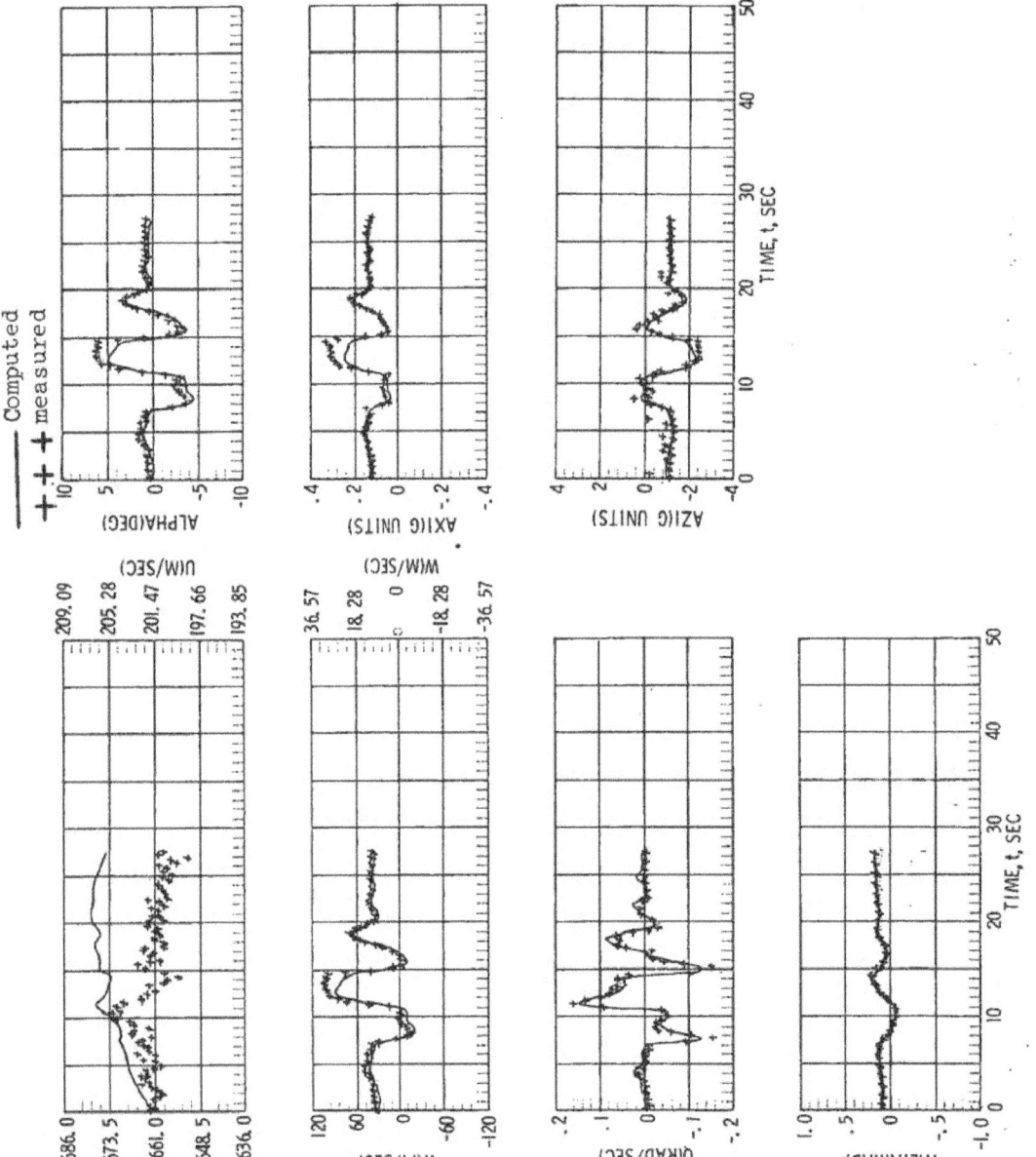

Figure 12.– Comparisons of flight data with computed time histories using the parameters in column 1 of table IV.

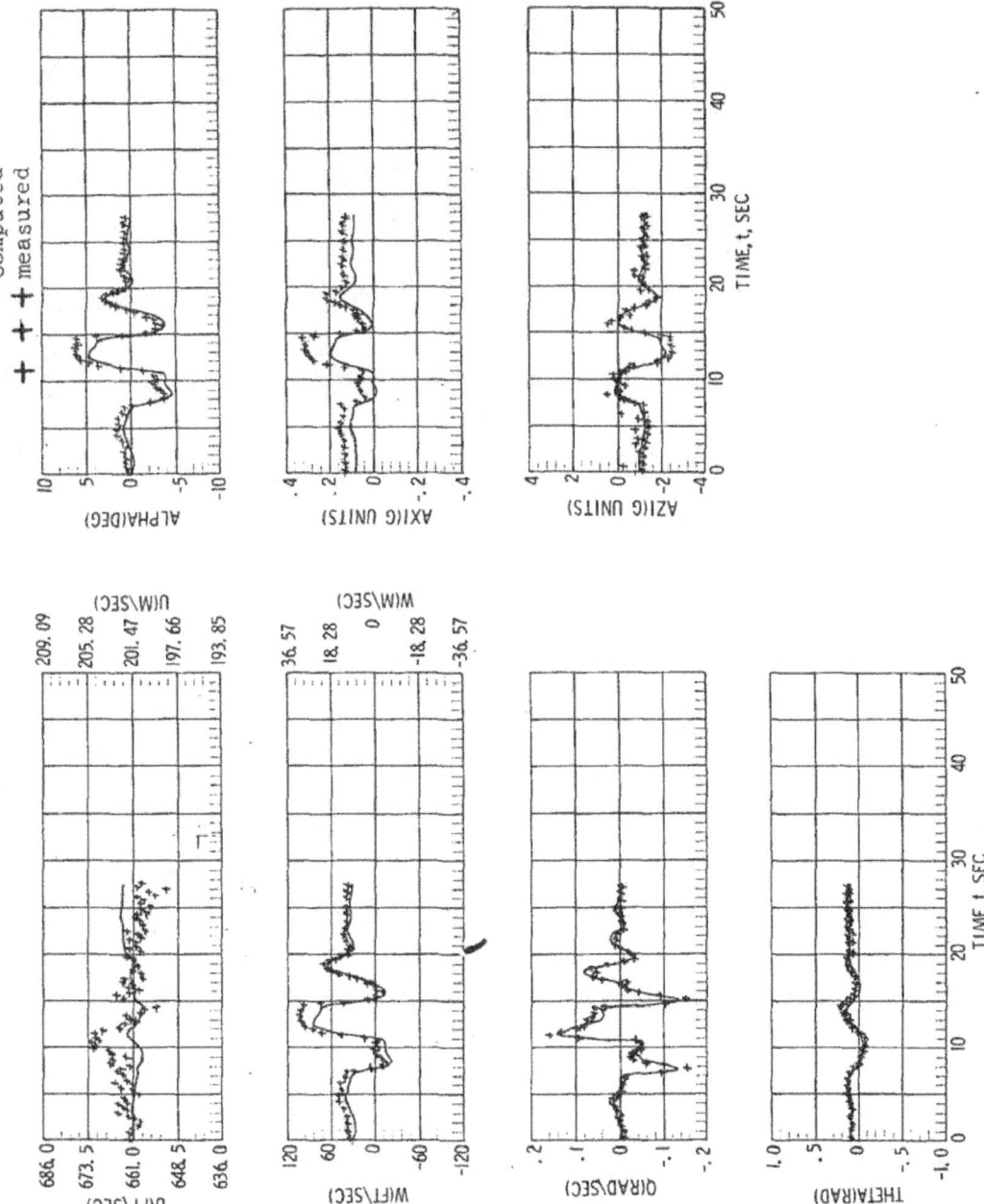

Figure 13.- Comparisons of flight data with computed time histories using the parameters in column 2 of table IV.

73

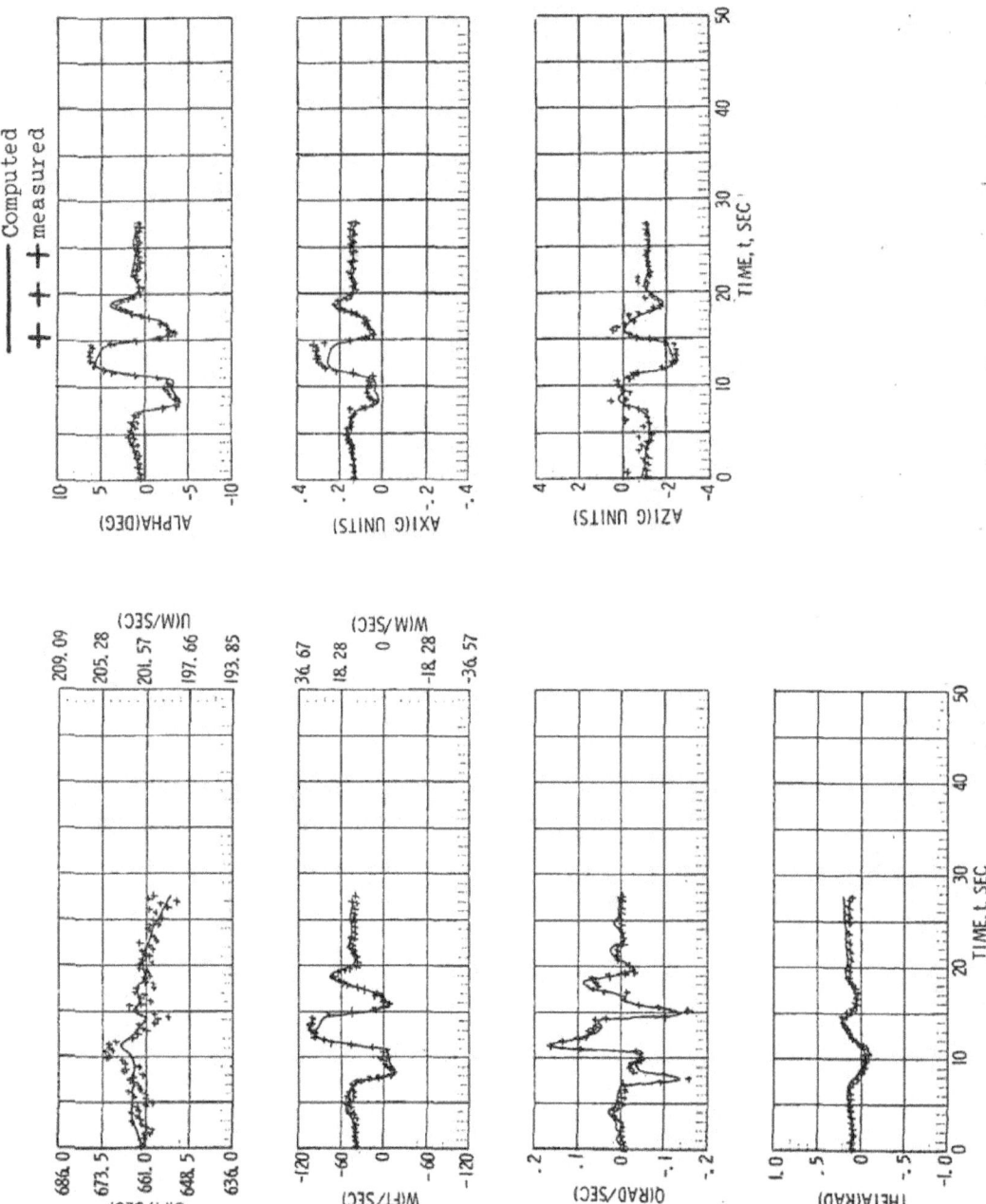

Figure 14.- Comparisons of flight data with computed time histories using the parameters in column 3 of table IV.

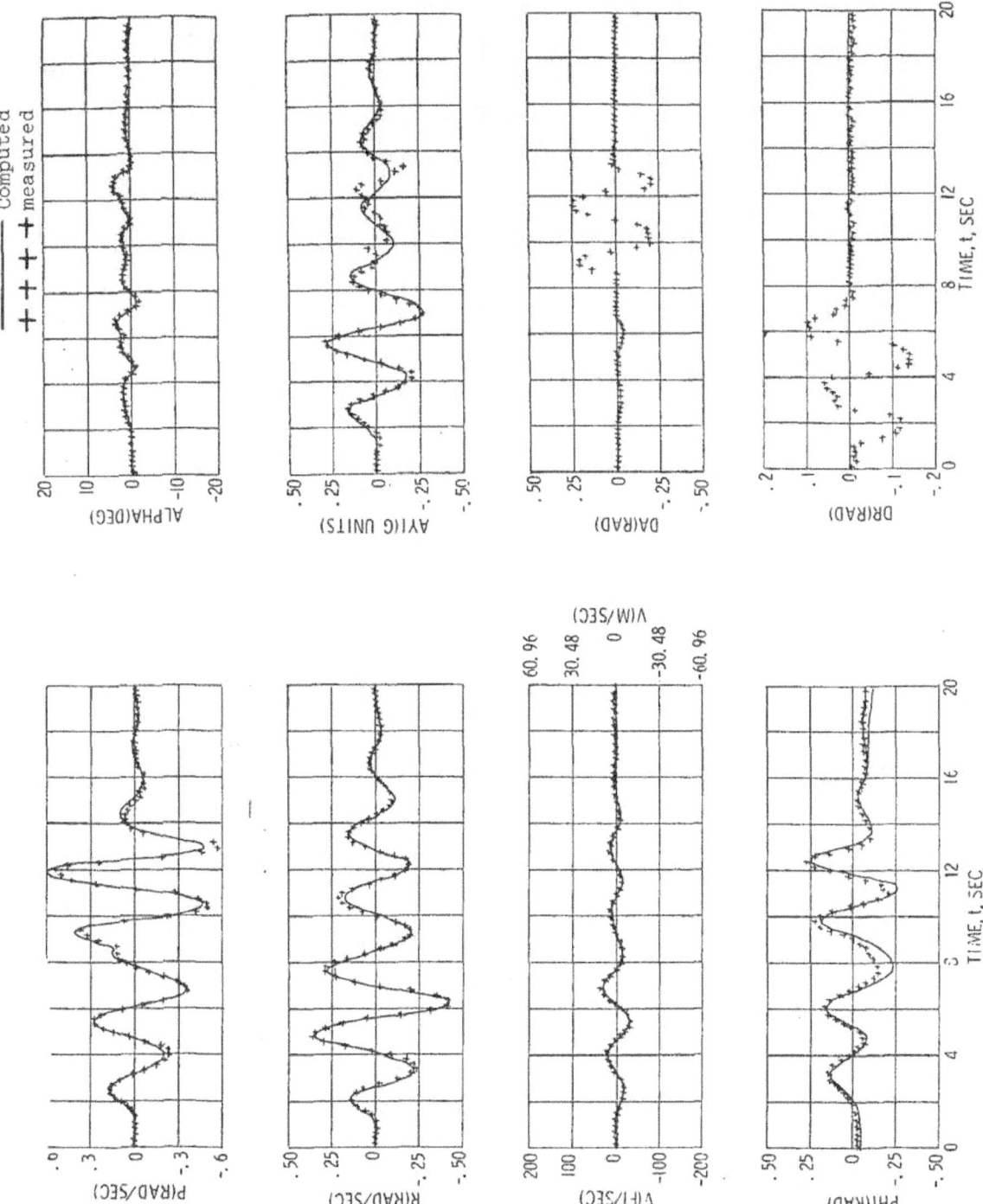

Figure 15.- Comparisons of flight data from the Helio Courier aircraft with computed time histories for rudder then aileron control inputs.

75

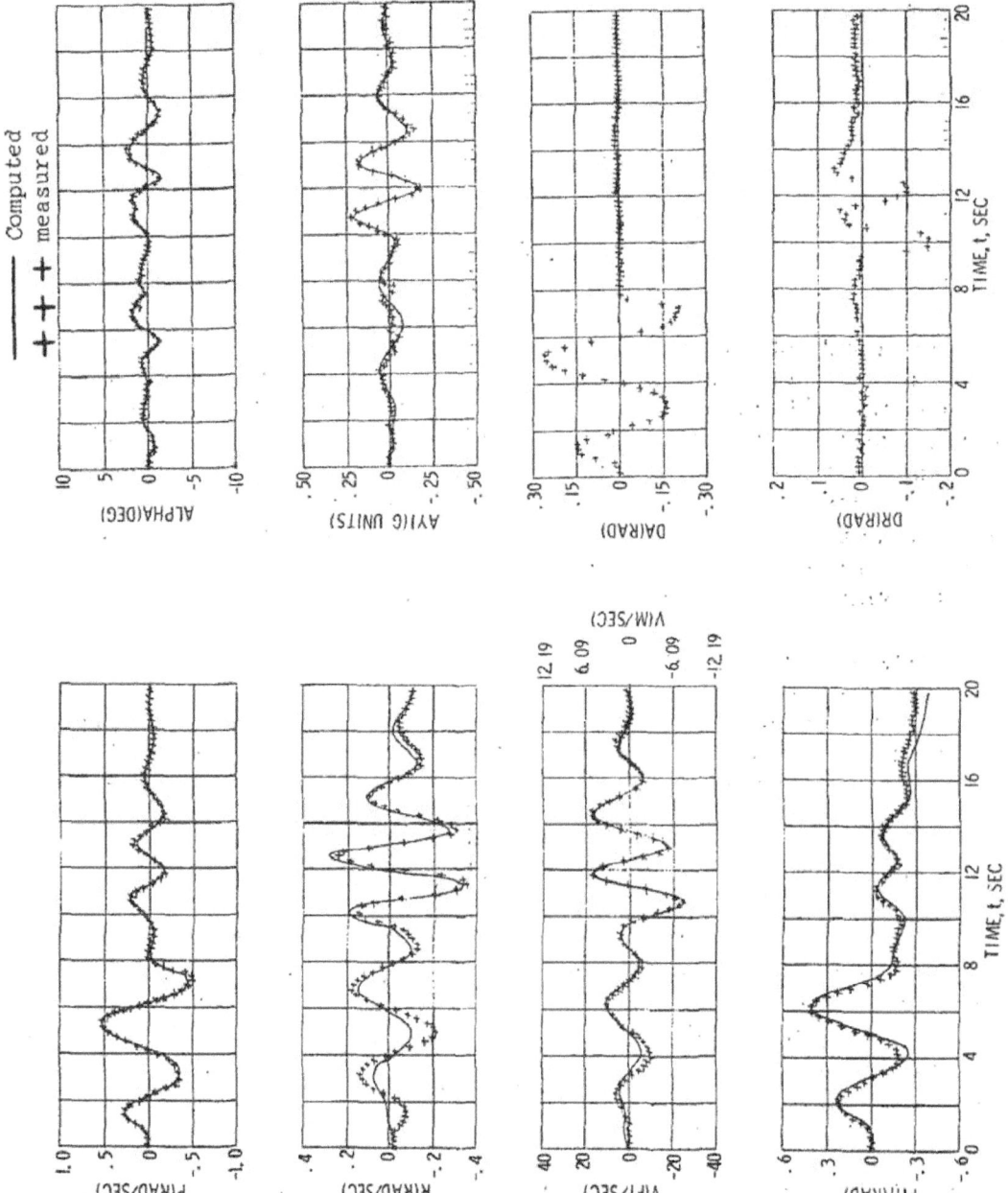

Figure 16.- Comparisons of flight data from the Helio Courier aircraft with computed time histories for aileron then rudder-control inputs.

76

IDENTIFICATION OF AIRCRAFT STABILITY AND CONTROL

DERIVATIVES IN THE PRESENCE OF TURBULENCE

Kenneth W. Iliff
NASA Flight Research Center

SUMMARY

A maximum likelihood estimator for a linear system with state and observation noise is developed to determine stability and control derivatives from flight data obtained in the presence of turbulence. The formulation for the longitudinal short-period mode is presented briefly, including a special case that greatly simplifies the problem if the measurement noise on one signal is negligible. The effectiveness and accuracy of the technique are assessed by applying it first to simulated flight data, in which the true parameter values and state noise are known, then to actual flight data obtained in turbulence. The results are compared with data obtained in smooth air and with wind-tunnel data.

The complete maximum likelihood estimator, which accounts for both state and observation noise, is shown to give the most accurate estimate of the stability and control derivatives from flight data obtained in turbulence. It is superior to the technique that ignores state noise and to the simplified method that neglects the measurement noise on the angle-of-attack signal.

INTRODUCTION

Although there are several well-established techniques for determining aircraft stability and control derivatives from flight data when the tests are performed in smooth air (ref. 1, for example), they are generally unable to treat data containing unknown external disturbances, such as turbulence. Turbulence may be of great concern in tests of future STOL aircraft because much of the flight-test time will be at low altitudes where turbulence is more prevalent.

Determining stability and control derivatives from flight-test data containing significant turbulence disturbances falls under the broad category of identifying a system with state and observation noise. This paper presents the results of applying the maximum likelihood technique to that class of problem specifically for longitudinal short-period aircraft dynamics. A simplified method that assumes there is no noise on the angle-of-attack measurement is considered in addition to the complete maximum likelihood estimator. These methods are applied first to simulated data in which the true values of system parameters and disturbance function are known, then to actual flight data recorded in strong turbulence. Comparisons are made of the fit of the measured and computed time histories. The coefficients estimated by using various methods on data obtained in turbulence are also compared with those obtained from smooth-air data.

In order to demonstrate the importance of accounting for state noise in the estimation technique, the results are compared with those obtained by a third method that ignores the state noise in its formulation.

SYMBOLS

A	stability matrix
a_n	normal acceleration, g
B	control matrix
C	observation matrix
D	observation control matrix
e	error quantity
F	state noise matrix
F_c	constant equating Z_{δ_e} to M_{δ_e}, sec
G	observation noise matrix
$G_{w_g}(\omega)$	turbulence power spectral density
g	acceleration due to gravity, m/sec^2 (ft/sec^2)
J	cost functional
K_α	scaler multiple of free-stream angle of attack to measured angle of attack
$\mathcal{L}$	Volterra operator given by Kalman filter
l_Z	distance from center of gravity to accelerometer, m (ft)
l_α	distance from center of gravity to angle-of-attack vane, m (ft)
M	pitching moment divided by moment of inertia, rad/sec^2

m	output vector due to control input only
N	number of data points
n_g	gaussian white noise
P	state estimator error covariance matrix
R	matrix used in J_{III}
T	total time interval, sec
Tr	trace of a matrix
t	time, sec
u	control vector
V	mean velocity, m/sec (ft/sec)
v	observation vector
W	cost functional weighting matrix
w_g	vertical turbulence component, m/sec (ft/sec)
x	state vector
Z	normal force divided by mass and velocity, rad/sec
α	angle of attack, rad or deg
Δ	increment
$\nabla_\kappa (\cdot)$	gradient of $(\cdot)$ with respect to κ
δ_e	elevator deflection, rad or deg
θ	pitch angle, rad or deg
κ	vector of unknown parameters
σ	standard deviation
σ_g^2	turbulence power, m^2/sec^2 (ft^2/sec^2)
$\sigma_{g_1}^2$	actual power in the turbulence time history, m^2/sec^2 (ft^2/sec^2)

$\hat{\sigma}^2_{g_1}$ indirect estimate of $\sigma^2_{g_1}$, m^2/sec^2 (ft^2/sec^2)

$\hat{\sigma}^2_{g_2}$ direct estimate of σ^2_{g}, m^2/sec^2 (ft^2/sec^2)

ω frequency, rad/sec

ω_c break frequency for turbulence spectrum, rad/sec

$\|\cdot\|$ norm of quantity enclosed

$\wedge$ denotes estimated quantity

$\cdot$ denotes time derivative of variable

$*$ transpose of a matrix

Subscripts:

b   bias

c   control

g   gust

m   measured

o   constant with time

s   state

$\alpha, \dot{\theta}, \delta_e$ partial derivatives with respect to subscripted variables

I, II, III   Estimators I, II, and III, respectively

DYNAMIC SYSTEM MODEL

The system to be identified is the longitudinal short-period dynamics of an aircraft subjected to external turbulence disturbances and specific test inputs by the pilot. The simplified longitudinal equations of motion for this system are given as

$$\dot{\alpha}_s = Z_\alpha \alpha_s + \dot{\theta} + Z_{\delta_e} \delta_e + Z_o + Z_\alpha \alpha_g \tag{1}$$

80

$$\ddot{\theta} = M_\alpha \alpha_s + M_{\dot{\theta}} \dot{\theta} + M_{\delta_e} \delta_e + M_o + M_\alpha \alpha_g \tag{2}$$

where

$$\alpha_g = w_g / V \tag{3}$$

and the random turbulence velocity w_g is considered to have zero mean value and a power spectral density given by

$$G_{w_g}(\omega) = \frac{2\sigma_g^2 \omega_c}{(\omega^2 + \omega_c^2)} \tag{4}$$

$$\omega_c = \frac{V}{1000} \tag{5}$$

The complete system state and observation equation (including the turbulence model) can be written as

$$\dot{x} = Ax + Bu + Fn_g \tag{6}$$

$$v = Cx + Du + Gn + v_b \tag{7}$$

where

$$x = \begin{bmatrix} \alpha_s \\ \theta \\ \dot{\theta} \\ \alpha_g \end{bmatrix}$$

$$u = \begin{bmatrix} \delta_e \\ 1 \end{bmatrix}$$

$$v = \begin{bmatrix} \dot{\theta}_m \\ \theta_m \\ a_{n_m} \\ \alpha_m \end{bmatrix}$$

$$A = \begin{bmatrix} Z_\alpha & 0 & 1 & Z_\alpha \\ 0 & 0 & 1 & 0 \\ M_\alpha & 0 & M_{\dot\theta} & M_\alpha \\ 0 & 0 & 0 & -\omega_c \end{bmatrix}$$

$$B = \begin{bmatrix} Z_{\delta_e} & Z_o \\ 0 & 0 \\ M_{\delta_e} & M_o \\ 0 & 0 \end{bmatrix}$$

$$C = \begin{bmatrix} 0 & 0 & 1 & 0 \\ 0 & 1 & 0 & 0 \\ \dfrac{l_Z M_\alpha - V Z_\alpha}{g} & 0 & \dfrac{l_Z M_{\dot\theta}}{g} & \dfrac{l_Z M_\alpha - V Z_\alpha}{g} \\ K_\alpha & 0 & -\dfrac{l_\alpha K_\alpha}{V} & K_\alpha \end{bmatrix}$$

$$D = \begin{bmatrix} 0 & 0 \\ 0 & 0 \\ \dfrac{l_Z M_{\delta_e} - V Z_{\delta_e}}{g} & \dfrac{l_Z M_o - V Z_o}{g} \\ 0 & 0 \end{bmatrix}$$

$$F = \begin{bmatrix} 0 \\ 0 \\ 0 \\ \frac{\sigma_g}{V}\sqrt{2\omega_c} \end{bmatrix}$$

$$G = \begin{bmatrix} g_{11} & 0 & 0 & 0 \\ 0 & g_{22} & 0 & 0 \\ 0 & 0 & g_{33} & 0 \\ 0 & 0 & 0 & g_{44} \end{bmatrix}$$

and n_g and n are, respectively, scalar- and vector-valued uncorrelated gaussian white noise processes with zero means and unity spectral densities. The unity element in u and the coefficients Z_o and M_o are to account for possible initial biases in the state equation. The vector v_b is the instrument bias. This formulation follows that of a challenge to design an automatic control system with unknown parameters based on external measurements which was presented by Lawrence Taylor, Jr., and Herman Rediess at the 1970 Joint Automatic Control Conference (JACC).

The unknown coefficients to be estimated are contained in the matrices A, B, C, D, and F and the vectors v_b and $x(o)$. For simplicity, the vector κ is defined to be the collection of the unknown coefficients to be estimated. Because different unknowns may be estimated at different times, κ does not always contain the same set of coefficients.

MAXIMUM LIKELIHOOD ESTIMATORS

The three methods considered in this paper are all based on a maximum likelihood criterion. They differ by virtue of the assumptions made about the state or observation noise. For convenience of presentation they are referred to as Estimators I, II, and III.

Estimator I - State Noise Neglected

If there is no state noise (i.e., $n_g = 0$), the method of reference 1, which is referred to as a Newton-Raphson or quasi-linearization technique, applies directly. It should be pointed out that no estimate of α_g is made by this Estimator; therefore, the estimate of α_g is identically zero. The maximum likelihood estimate is obtained by minimizing

$$J_I = \int_0^T \left\| v - Cx - Du - v_b \right\|_{W_I}^2 dt \tag{8}$$

where

$$W_I = \begin{bmatrix} \dfrac{1}{g_{11}^2} & 0 & 0 & 0 \\ 0 & \dfrac{1}{g_{22}^2} & 0 & 0 \\ 0 & 0 & \dfrac{1}{g_{33}^2} & 0 \\ 0 & 0 & 0 & \dfrac{1}{g_{44}^2} \end{bmatrix}$$

The details of this method and an algorithm to compute the parameter values that minimize the functional J_I are presented in reference 1.

Estimator II - Angle-of-Attack Measurement Noise Neglected

If there is a significant state noise, but the measurement noise on α_m can be neglected (i.e., $g_{44} \approx 0$), the state noise problem can be simplified to essentially that of Estimator I. This approach was first used by Balakrishnan at the 1971 Joint Automatic Control Conference in his treatment of Taylor and Rediess' design challenge. If $g_{44} = 0$, the equation for α_m (fourth row of eq. (7)) can be solved for α_g:

$$\alpha_g = -\alpha_s + \frac{l_\alpha}{V}\dot{\theta} + \frac{1}{K_\alpha}\alpha_m \tag{9}$$

Substituting equation (9) into equations (1) and (2),

$$\dot{\alpha}_s = \left(1 - \frac{Z_\alpha l_\alpha}{V}\right)\dot{\theta} + Z_{\delta_e}\delta_e + Z_o + \frac{Z_\alpha}{K_\alpha}\alpha_m \tag{10}$$

$$\ddot{\theta} = \left(M_{\dot{\theta}} - \frac{M_\alpha l_\alpha}{V}\right)\dot{\theta} + M_{\delta_e}\delta_e + M_o + \frac{M_\alpha}{K_\alpha}\alpha_m \tag{11}$$

the states α_s, θ, and $\dot{\theta}$ are determined by equations (10) and (11) in which the measurement α_m is treated as a forcing function in the same manner as δ_e. The state α_g is determined directly from equation (9). This reduces the problem to that of Estimator I. The cost functional is

$$J_{II} = \int_0^T \left\| v - Cx - Du - v_b \right\|^2_{W_{II}} dt \tag{12}$$

where

$$W_{II} = \begin{bmatrix} \dfrac{1}{g_{11}^2} & 0 & 0 & 0 \\ 0 & \dfrac{1}{g_{22}^2} & 0 & 0 \\ 0 & 0 & \dfrac{1}{g_{33}^2} & 0 \\ 0 & 0 & 0 & 0 \end{bmatrix}$$

Estimator III – State and Measurement Noise Included

The maximum likelihood method for the state noise problem in which the measurement noise on α_m cannot be neglected is more complicated. The derivation of this algorithm, which is beyond the scope of this paper, is presented in references 2 to 4. It is based on minimizing the functional

$$J_{III} = \frac{1}{T}\int_0^T \left[\left\| v - m - \mathcal{L}(v - m) \right\|^2_{(GG^*)^{-1}} - \left\| v \right\|^2_{(GG^*)^{-1}} \right] dt + \mathrm{Tr}(R) \tag{13}$$

where

$$\dot{x}_c = Ax_c + Bu$$

$$m = Cx_c + Du + v_b$$

$$R = C*(GG*)^{-1/2}P(GG*)^{-1/2}C$$

and $\mathcal{L}$ is a Volterra operator that is given by the Kalman filter, P is the state estimator error covariance matrix, and m is the response of the system due to the control input alone.

The basic concept of the algorithm is depicted in figure 1. The top sequence indicates the creation, either by computer simulation or flight test, of the data to be analyzed. The next sequence is the computation of the vehicle response and the gradient of the response due to the control input only. These values are put through a Kalman filter where the estimates of the states and gradients are computed, with all unknown coefficients held fixed. These quantities are then used with the computed responses of the control alone to perform the minimization of the likelihood functional. At each iteration a new vector κ is computed, and the process is repeated until the gradient of the likelihood is essentially zero. Thus a final set of estimated coefficients and estimates of the observations and states are obtained.

SIMULATION TEST RESULTS

The application of Estimator I to simulated data with turbulence is discussed in reference 2, and Estimators I, II, and III can be evaluated by applying them to simulated data. Estimators II and III are tested herein. The case chosen for the simulation is one from Taylor and Rediess' 1970 JACC presentation (also partially presented in ref. 2) inasmuch as probability distributions of the various coefficients to be estimated were included. The pertinent information for this case is presented in tables 1 and 2. For all the simulation results the startup values chosen were the 3σ (standard deviation) values of the corresponding coefficients. Thus the uncertainty usually associated with what initial estimates should be used is eliminated. This procedure should provide some information about the neighborhood needed to assure convergence.

The control input is a square-wave elevator command of 0.02 radian amplitude at 0.4 hertz. The conditions for the simulation are listed in table 1.

Estimator II

Figure 2 shows the fit of the simulated data with zero measurement noise on α with the estimated data obtained by Estimator II. The fit is virtually perfect. Even the estimated state variables α_s and α_g show an exact fit. This is important to note in the simulated results, inasmuch as this comparison cannot be made with flight data because the true values are not known. The unknown coefficients to be estimated are Z_α, M_α, $M_{\dot\theta}$, Z_{δ_e}, and M_{δ_e}. The instrument biases, state initial conditions, and aerodynamic bias terms Z_0 and M_0 need not be estimated for simulated data.

Table 2 shows the estimates of each coefficient at each iteration for 4.9 seconds of data as well as the correct results. Figure 3 is a plot of the same results. All coefficients have converged in four iterations, and only $M_{\dot\theta}$ and Z_{δ_e} fail to attain the correct value to four places.

Another item of interest for verification of the algorithm is how long an observation time is needed or, more importantly, how many data samples are needed. Inasmuch as $\Delta t = 0.01$ second, the number of data points is $N = T/\Delta t + 1$. Table 3 presents the values obtained as a function of the observation time, T. Figure 4 shows the same data. Fairly good estimates are obtained after only 0.3 second of data. After 2.5 seconds all the estimates are nearly at the true values except Z_{δ_e}. As in figure 3, all the estimates have converged to essentially the correct values by 4.9 seconds of observation time.

Estimator III

The procedure used to test Estimator III on simulated data is identical to that used with Estimator II, with two exceptions. Now α has measurement noise with a value of $g_{44} = 0.00005$ radian, and the turbulence power σ_g^2 is an additional unknown. Figure 5 is the fit obtained with Estimator III on 4.9 seconds of data. As before, the fit is essentially perfect. The estimate of the turbulence time history α_g is not as good as with the estimates of Estimator II, although it still appears exact.

Table 4 and figure 6 present the coefficients as a function of the iteration, again starting at the 3σ values. After six iterations, the solution converges to nearly the correct values. All are within less than 1 percent except Z_{δ_e} and the estimate of the turbulence power σ_g^2. The estimates of stability and control derivatives are not as good as with Estimator II, but this case is one of truly stochastic identification.

Table 5 and figure 7 present the estimates obtained by using Estimator III as a function of the observation time, T. The direct estimates in figure 7 refer to the estimates obtained directly as a consequence of minimizing the cost functional. The indirect estimate (applies only to the estimate of the turbulence power) is obtained by computing the turbulence power from the estimated turbulence α_g. After 0.5 second, fairly good estimates are obtained for Z_α and M_α. A full 12 seconds of data are needed before essentially exact values are obtained. Once again Z_{δ_e} is the last to converge. Inasmuch as the stochastic identification problem is much more complex than the deterministic input case, these results are encouraging.

The simulation of the turbulence needs some further explanation. Figure 7 shows the correct σ_g^2 varying as a function of observation time. The reason for this is that the power of the turbulence is a function of a random variable. It actually took about 25 seconds of data before the simulated turbulence reached the desired value. Therefore the correct value is taken as the power of the simulated turbulence for a given observation time. The estimation of turbulence factor may vary frequently by a factor of two (3 dB) from the actual power in a given time history. This would seem discouraging in that most coefficients can be estimated exactly. In this example the power estimates are much closer than a factor of two, but it would not be surprising if they deviated considerably more.

Because stochastic identification is complex, it would be advantageous to make the aircraft model as simple as possible. As was noted in the previous analysis, Z_{δ_e} was found to be the least reliable to estimate for both Estimator II and Estimator III. Fortunately, Z_{δ_e} can be constrained to be directly proportional to M_{δ_e} without significant compromise because the force imposed by the deflection of the control surface can be assumed to act at the elevator surface, i.e., the flow interference or interaction takes place only at the tail and the other effects can be ignored. Then $Z_{\delta_e} = F_c M_{\delta_e}$ can be substituted in the differential equations, where F_c is a known function of the flight condition and the geometry of the aircraft. There is a danger in this constraint in that it may degrade the estimation of M_{δ_e}; however, this did not occur. The preceding analysis was repeated on the simulation with Z_{δ_e} constrained to M_{δ_e}, and the results were essentially the same except that Z_{δ_e} was attained as easily as the other derivatives. All subsequent analysis with Estimators II and III was done with $Z_{\delta_e} = F_c M_{\delta_e}$, and wherever Z_{δ_e} appeared it was replaced with $F_c M_{\delta_e}$.

At this point it would seem that Estimator II is preferable to Estimator III, if there really is negligible noise on α. Although the simulation has shown that both methods can be applied successfully under ideal circumstances, it is by no means a complete assessment of the general validity of the methods. Both methods were also applied to actual flight data obtained in turbulence. Those results, which are presented in the next section, clarify the question of whether the simulated results are generally valid.

FLIGHT-TEST RESULTS

The flight data used were obtained from tests on a Lockheed JetStar airplane, which is described in reference 5. The airplane was specifically flown in turbulence while the maneuvers were performed to test these methods. The conditions of the test are listed in table 6. These data are analyzed more completely in reference 4. Figure 8 presents 65 seconds of continuous data containing five separate elevator pulse maneuvers, which are designated A, B, C, D, and E. These data were analyzed by Estimators I, II, and III. The coefficients to be estimated were Z_α, M_α, $M_{\dot\theta}$, Z_{δ_e}, M_{δ_e}, Z_o, M_o, $\alpha_s(o)$, $\dot\theta(o)$, $\alpha_g(o)$, v_{b_1}, and v_{b_3}. It was not always possible to obtain estimates for all the unkonwn coefficients; this is noted where applicable.

Estimator I

Figure 9 compares the flight data for maneuver E with those obtained by Estimator I, which neglects state noise. Considering the amount of turbulence, it is surprising that any stable solution can be obtained. On closer examination, it is evident that Estimator I finds that the only significant information in the maneuver is during the control input. Therefore, the fit is best during that portion, and the rest of the data are essentially ignored. The poor fit in this example shows that state noise must be accounted for in the analysis method.

Estimator II

Figure 10 shows the fit for maneuver C, which was the poorest comparison of flight to estimated data by Estimator II. Not only was the fit poor, but convergence was difficult to achieve. Also, with the initial conditions taken to be unknown on α_s and α_g, no stable solution could be obtained. The same situation existed on maneuver B. The other maneuvers, although sometimes slow to converge, all attained stable solutions with the full set of unknowns. The best fit obtained with Estimator II is shown in figure 11, in which the comparison for maneuver E is shown for flight data and the data estimated by Estimator II. This maneuver was performed during the portion of the flight with the greatest turbulence. This fit is an improvement over that obtained by Estimator I in figure 9.

Estimator III

Estimator III was then used to obtain estimates of the coefficients for the same five maneuvers. The same κ vector was used, with the addition of the turbulence power. Convergence was obtained routinely, and all five maneuvers converged in three iterations. All combinations of consecutive maneuvers also converged in three iterations. All the fits obtained by Estimator III were considered better than the best fit obtained by Estimator II. Figure 12 shows the fit of the flight data of maneuver C compared with the estimated data obtained by Estimator III. This is a significant improvement over that obtained by Estimator II, and the complete unknown vector was also obtained. Figure 13 compares the flight data of maneuver E with the estimated data obtained by Estimator III. This was the poorest fit obtained by Estimator III. It is still significantly better than that obtained by Estimator II, which was considered the best fit obtained by that method.

It should be pointed out that Estimator II and Estimator I would converge only for the basic maneuvers, and then sometimes with difficulty. As previously pointed out, Estimator III converged for the basic maneuvers as well as all combinations of consecutive maneuvers. Figure 14 compares the flight data of maneuver ABCD (shown in fig. 8) with the estimated data obtained by Estimator III. The fit is considered as good as is usually obtained using smooth-air data with some discrepancies in θ and $\dot{\theta}$.

Comparison of Estimated Coefficients

Another significant question regarding any identification technique is "How good are the estimates of the coefficients obtained by the technique?" For the simulated data the correct answers are known, so it is easy to decide how "good" the coefficients are. When the true values of the coefficients are not known, two ways to assess each method (in addition to the fit obtained) are (1) by their agreement with other methods of estimating coefficients and (2) by the repeatability of the estimates for different maneuvers for which the coefficients would be expected to be identical.

Figure 15 shows the estimates obtained for each method as a function of the maneuver analyzed. The raw data are given in table 7. The values obtained by Estimator III are also shown in the figure for some of the combinations of maneuvers. Only the major coefficients of interest that are common to all three estimators are shown. Although the plots are somewhat confusing because of the scatter in the results of the various methods, the Estimator III estimates have less scatter and the estimates obtained from the longer maneuvers show even less variation.

A more compact way of displaying the data would be to show the first- and second-order statistics of the estimates obtained for each of the basic maneuvers. Estimates obtained by other methods could then also be shown. It is of interest to compare data obtained from two other sources with the results of the preceding methods: the coefficients estimated by Estimator I from data obtained in smooth air; and the estimates obtained from wind-tunnel data. Data were analyzed from seven maneuvers performed in smooth air at essentially the same flight condition as the maneuvers performed in turbulence. The first- and second-order statistics of these maneuvers were then determined.

The mean and standard deviation for each of the methods except for the wind-tunnel estimates are presented in table 8 and figure 16. Unfortunately, the wind-tunnel estimates provide only a single data point, but that point can still be compared with the other data. The symbol in figure 16 indicates the mean for each method, and the vertical line represents ± 1 standard deviation for the five maneuvers analyzed. Estimator I shows the greatest standard deviation for all the coefficients except $M_{\dot{\theta}}$. The greatest standard deviation for $M_{\dot{\theta}}$ was obtained by Estimator II. Of the three methods of estimating coefficients from data obtained in turbulence, the standard deviation of the estimates of Estimator III were considerably smaller than those of the other two methods, with the exception of M_{α}. The standard deviation of M_{α} for Estimator II is about equal to that of Estimator III.

Of most significance in figure 16 is, perhaps, that the standard deviation of three of the five coefficients is better for Estimator III than for the estimates obtained from smooth-air data. The means of the smooth-air estimates and the estimates of Estimator III are in good agreement except for Z_{α}. The wind-tunnel estimates and those obtained by Estimator III are also in fairly good agreement. It should be pointed out how well the means obtained by Estimator III for the basic maneuvers agree with those obtained from the longer maneuvers. (Compare with table 7 or fig. 15.)

In summary, the estimates obtained by Estimator III were (1) in better agreement with those obtained from smooth-air data and (2) showed greater repeatability of the estimates for different maneuvers at the same flight condition. Therefore, on the basis of the analysis of flight data obtained in turbulence, Estimator III provided a superior fit, better agreement with estimates from other methods, and greater repeatability of estimates at the same flight condition.

Estimator III is more complex and therefore requires more computer time for each iteration. This would seem to be a drawback; however, Estimator III converged in three iterations for all the flight data analyzed, whereas Estimator II always required six iterations and usually more. For the existing computer programs, Estimator III always converged in less computer time than did Estimator II.

This study showed Estimator III to be preferable in every respect, including computation time, to the other methods used in analyzing flight data obtained in turbulence. Estimator II was found to be preferable to Estimator I.

This analysis of flight data points out that evaluating a method on the basis of simulated data alone may result in an erroneous conclusion. For example, the study using simulated data showed that Estimator II was preferable to Estimator III, but the study using flight data showed Estimator III to be much better than Estimator II.

CONCLUSIONS

Three maximum likelihood estimators for identifying aircraft stability and control derivatives were applied to simulated and flight-test data containing turbulence disturbances. Estimator I did not account for the turbulence (no state noise); Estimator II accounted for the turbulence but assumed the noise on the angle-of-attack measurement to be zero; and Estimator III accounted for both turbulence and measurement noise. The study showed that:

1. It was necessary to use an estimator that accounted for turbulence if the data contained significant turbulence disturbances.

2. Estimator III was superior to Estimator II with respect to the time history fit, convergence rate, and better consistency in the estimates.

3. The estimated coefficients for data obtained in turbulence using Estimator III were nearly as accurate as those for data obtained in smooth air using Estimator I.

4. The variance of the estimates obtained with Estimator III decreased as the length of observation increased. Estimator III was the only method that converged when two or more maneuvers were combined.

5. Estimator III required no more computer time than Estimator II because Estimator III converged more rapidly.

6. The final decision regarding the desirability of an algorithm should be based on flight data rather than on simulated data.

REFERENCES

1. Iliff, Kenneth W.; and Taylor, Lawrence W., Jr.: Determination of Stability Derivatives From Flight Data Using a Newton-Raphson Minimization Technique. NASA TN D-6579, 1972.

2. Balakrishnan, A. V.: Identification and Adaptive Control: An Application to Flight Control Systems. J. Optimization Theory and Applications, vol. 9, no. 3, Mar. 1972, pp. 187-213.

3. Balakrishnan, A. V.: Modelling and Identification Theory: A Flight Control Application. Theory and Applications of Variable Structure Systems, Academic Press, Inc., New York, Apr. 1972.

4. Iliff, K. W.: Identification and Stochastic Control With Application to Flight Control in Turbulence. UCLA-ENG-7340, UCLA School of Engineering and Applied Science, May 1973.

5. Clark, Daniel C.; and Kroll, John: General Purpose Airborne Simulator - Conceptual Design Report. NASA CR-544, 1966.

TABLE 1. SUPPLEMENTARY DATA FOR SIMULATION TESTS

$$V = 509 \text{ m/sec} \ (1670 \text{ ft/sec})$$

$$\sigma_g = 1.524 \text{ m/sec} \ (5 \text{ ft/sec})$$

$$\Delta t = 0.01 \text{ sec}$$

$$l_Z = 3.048 \text{ m} \ (10 \text{ ft})$$

$$l_\alpha = 9.76 \text{ m} \ (32 \text{ ft})$$

$$K_\alpha = 1.7$$

$$g_{11} = 0.0005 \text{ rad/sec}$$

$$g_{22} = 0.0001 \text{ rad}$$

$$g_{33} = 0.01g$$

$$g_{44} = 0.00005 \text{ rad}$$

TABLE 2. CONVERGENCE OF STABILITY AND CONTROL DERIVATIVES FOR 4.9 SECONDS OF DATA USING ESTIMATOR II

Iteration	Z_α, rad/sec	M_α, rad/sec^2	$M_{\dot\Theta}$, rad/sec	Z_{δ_e}, rad/sec	M_{δ_e}, rad/sec^2
0	-2.400	-39.00	-2.400	-0.6750	-36.00
1	-1.589	-53.48	-.857	-.4483	-50.14
2	-1.653	-52.75	-1.389	-.4489	-50.90
3	-1.650	-53.85	-1.638	-.4489	-52.35
4	-1.650	-54.00	-1.644	-.4498	-52.50
5	-1.650	-54.00	-1.644	-.4496	-52.50
Correct value	-1.650	-54.00	-1.650	-0.4500	-52.50

TABLE 3. EFFECT OF THE OBSERVATION TIME INTERVAL ON THE ESTIMATED STABILITY AND CONTROL DERIVATIVES USING ESTIMATOR II

T, sec	Z_α, rad/sec	M_α, rad/sec^2	$M_{\dot\Theta}$, rad/sec	Z_{δ_e}, rad/sec	M_{δ_e}, rad/sec^2
0.3	-1.504	-52.71	-3.627	-0.3933	-58.94
.5	-1.603	-51.61	-2.065	-.4293	-51.96
1.0	-1.649	-54.00	-1.760	-.4614	-52.27
2.5	-1.663	-53.94	-1.649	-.4652	-52.47
4.9	-1.650	-54.00	-1.644	-.4496	-52.50
11.9	-1.650	-54.00	-1.650	-.4500	-52.50
Correct value	-1.650	-54.00	-1.650	-0.4500	-52.50

TABLE 4. CONVERGENCE OF STABILITY AND CONTROL DERIVATIVES FOR 4.9 SECONDS OF DATA USING ESTIMATOR III

Iteration	Z_α, rad/sec	M_α, rad/sec²	$M_{\dot\theta}$, rad/sec	Z_{δ_e}, rad/sec	M_{δ_e}, rad/sec²	$\hat\sigma^2_{g_2}$, m²/sec² (ft²/sec²)
0	-2.400	-39.00	-2.400	-0.6750	-36.00	0.232 (2.500)
1	-1.709	-58.90	-3.289	-.5818	-57.44	.816 (8.775)
2	-1.666	-54.07	-1.288	-.4607	-52.47	.791 (8.500)
3	-1.662	-53.60	-1.612	-.4549	-52.06	.740 (7.950)
4	-1.664	-53.98	-1.663	-.4560	-52.49	.735 (7.900)
5	-1.664	-54.11	-1.641	-.4571	-52.55	.737 (7.925)
6	-1.664	-54.11	-1.640	-.4571	-52.55	.737 (7.925)
Correct value	-1.650	-54.00	-1.650	-0.4500	-52.50	0.915 (9.830)

TABLE 5. EFFECT OF THE OBSERVATION TIME INTERVAL ON THE ESTIMATED STABILITY AND CONTROL DERIVATIVES USING ESTIMATOR III

T, sec	Z_α, rad/sec	M_α, rad/sec²	M_θ, rad/sec	Z_{δ_e}, rad/sec	M_{δ_e}, rad/sec²	$\hat\sigma^2_{g_1}$, m²/sec² (ft²/sec²)	$\hat\sigma^2_{g_2}$, m²/sec² (ft²/sec²)	$\sigma^2_{g_1}$, m²/sec² (ft²/sec²)
0.3	-1.561	-70.28	0.9077	-0.3325	-41.10	0.327 (3.514)	0.614 (6.601)	0.520 (5.59)
.5	-1.599	-52.80	-1.530	-.3544	-46.69	.873 (9.387)	.847 (9.100)	1.326 (14.25)
1.0	-1.620	-53.94	-1.834	-.3797	-51.00	.907 (9.744)	.612 (6.581)	1.211 (13.01)
2.5	-1.663	-54.06	-1.662	-.4647	-52.58	.835 (8.975)	.677 (7.281)	.986 (10.60)
4.9	-1.664	-54.11	-1.640	-.4571	-52.55	.904 (9.714)	.734 (7.892)	.915 (9.83)
9.9	-1.658	-54.03	-1.648	-.4516	-52.52	1.568 (16.85)	1.145 (12.30)	1.632 (17.54)
11.9	-1.651	-54.00	-1.650	-.4505	-52.52	1.592 (17.11)	1.182 (12.70)	1.699 (18.25)
Correct value	-1.650	-54.00	-1.650	-0.4500	-52.50	----------	----------	----------

TABLE 6. SUPPLEMENTARY DATA FOR FLIGHT TESTS

$$V = 179.3 \text{ m/sec } (588 \text{ ft/sec})$$

$$\text{Weight} = 133,300 \text{ N } (30,000 \text{ lb})$$

$$\Delta t = 0.02 \text{ sec}$$

$$l_Z = 0 \text{ m } (0 \text{ ft})$$

$$l_\alpha = 7.94 \text{ m } (26 \text{ ft})$$

$$K_\alpha = 1.6$$

$$g_{11} = 0.00166 \text{ rad/sec}$$

$$g_{22} = 0.0007 \text{ rad}$$

$$g_{33} = 0.0126g$$

$$g_{44} = 0.00132 \text{ rad}$$

TABLE 7. STABILITY AND CONTROL DERIVATIVE ESTIMATES OBTAINED IN TURBULENCE
BY ESTIMATOR I, ESTIMATOR II, AND ESTIMATOR III

Estimator	Maneuver	Z_α, rad/sec	M_α, rad/sec^2	$M_{\dot\theta}$, rad/sec	Z_{δ_e}, rad/sec	M_{δ_e}, rad/sec^2	$\hat\sigma^2_{g_1}$, m^2/sec^2 (ft^2/sec^2)	$\hat\sigma^2_{g_2}$, m^2/sec^2 (ft^2/sec^2)
I	A	-1.68	-12.44	-.76	-.223	-8.55	---	---
I	B	-1.78	-19.20	-3.45	-.057	-14.46	---	---
I	C	-1.45	-9.46	-2.73	-.191	-9.63	---	---
I	D	-1.32	-11.95	-.91	.190	-7.03	---	---
I	E	-1.08	-4.79	-.77	-.037	-5.79	---	---
II	A	-1.39	-9.13	-3.26	-.088	-9.45	---	---
II	B	-1.35	-5.90	-4.24	-.082	-8.81	---	---
II	C	-1.32	-8.16	-5.82	-.107	-11.44	---	---
II	D	-1.43	-8.63	-1.51	-.054	-5.74	---	---
II	E	-1.25	-5.70	-2.90	-.073	-7.87	---	---
III	A	-1.54	-10.49	-1.35	-.074	-7.95	1.029 (11.06)	2.492 (26.77)
III	B	-1.67	-13.07	-.58	-.066	-7.09	3.534 (37.96)	8.188 (87.95)
III	C	-1.53	-10.32	-2.20	-.084	-9.00	3.958 (42.52)	15.687 (168.50)
III	D	-1.56	-10.94	-1.22	-.068	-7.30	2.513 (27.00)	10.566 (113.50)
III	E	-1.55	-8.02	-2.30	-.079	-8.46	13.615 (146.25)	42.593 (457.50)
III	AB	-1.64	-11.55	-1.45	-.079	-8.67	6.035 (64.83)	11.725 (125.95)
III	CD	-1.56	-11.05	-1.53	-.073	-7.86	3.798 (40.80)	15.477 (166.25)
III	ABCD	-1.58	-11.17	-1.52	-.075	-8.06	7.297 (78.38)	15.661 (168.22)

TABLE 8. MEANS AND STANDARD DEVIATIONS FOR FIVE METHODS OF
ESTIMATING STABILITY AND CONTROL DERIVATIVES

Method	Z_α, rad/sec		M_α, rad/sec^2		$M_{\dot\theta}$, rad/sec		Z_{δ_e}, rad/sec		M_{δ_e}, rad/sec^2	
	Mean	Standard deviation	Mean	Standard deviation	Mean	Standard deviation	Mean	Standard deviation	Mean	Standard deviation
Estimator I	-1.46	.250	-11.57	4.681	-1.72	1.140	-.049	.152	-9.09	2.984
Estimator II	-1.35	.061	-7.50	1.427	-3.54	1.435	-.081	.017	-8.66	1.874
Estimator III	-1.57	.049	-10.56	1.611	-1.53	.641	-.074	.007	-7.96	.710
Smooth air	-1.94	0.074	-11.31	0.341	-1.75	0.092	-0.093	0.014	-9.87	0.887
Wind tunnel	-1.60	---	-7.79	---	-1.31	---	-.123	---	-9.75	---

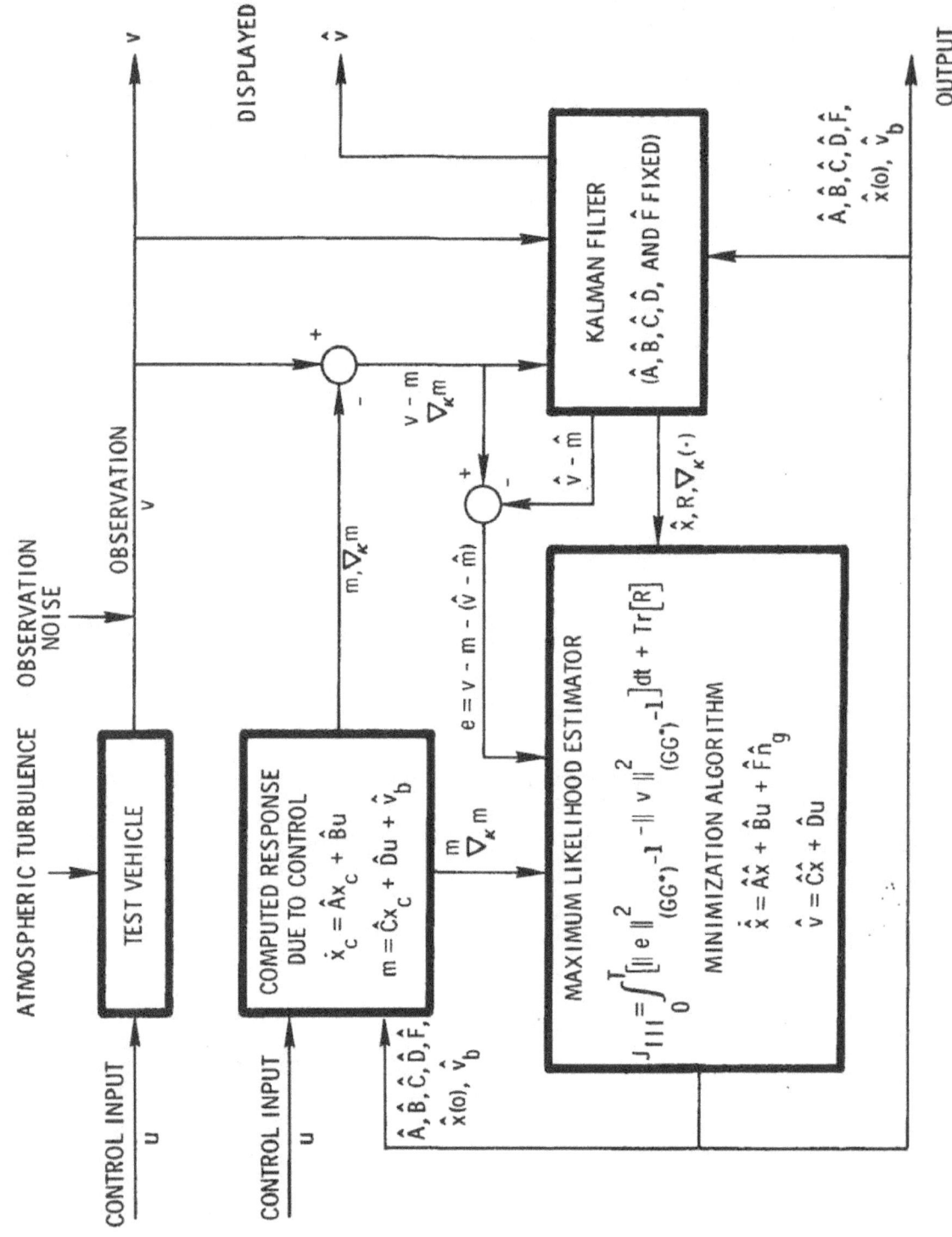

Figure 1. Block diagram of maximum likelihood estimates for stochastic identification.

98

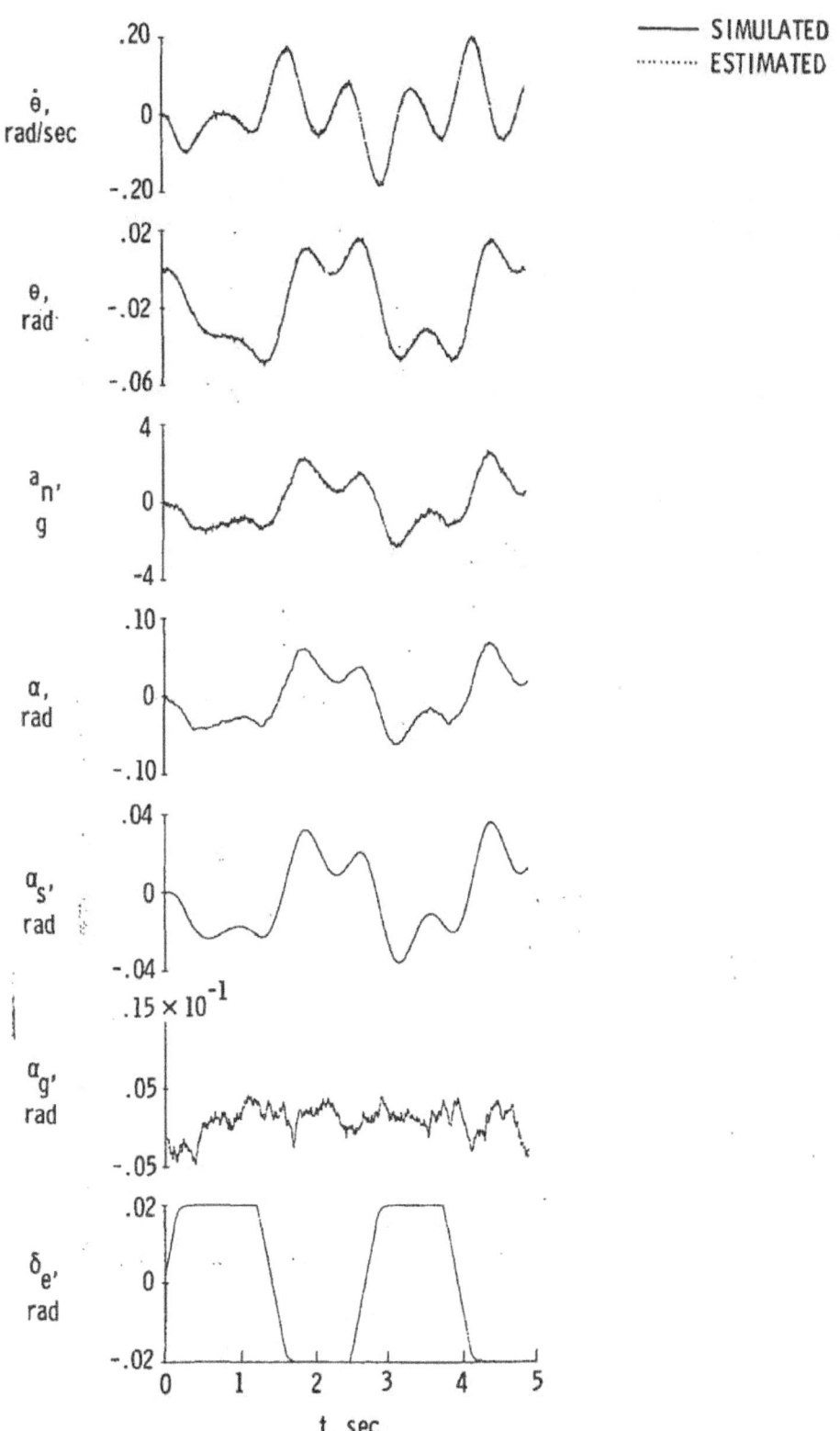

Figure 2. Comparison of computed data with no measurement noise on α with estimated data obtained by Estimator II.

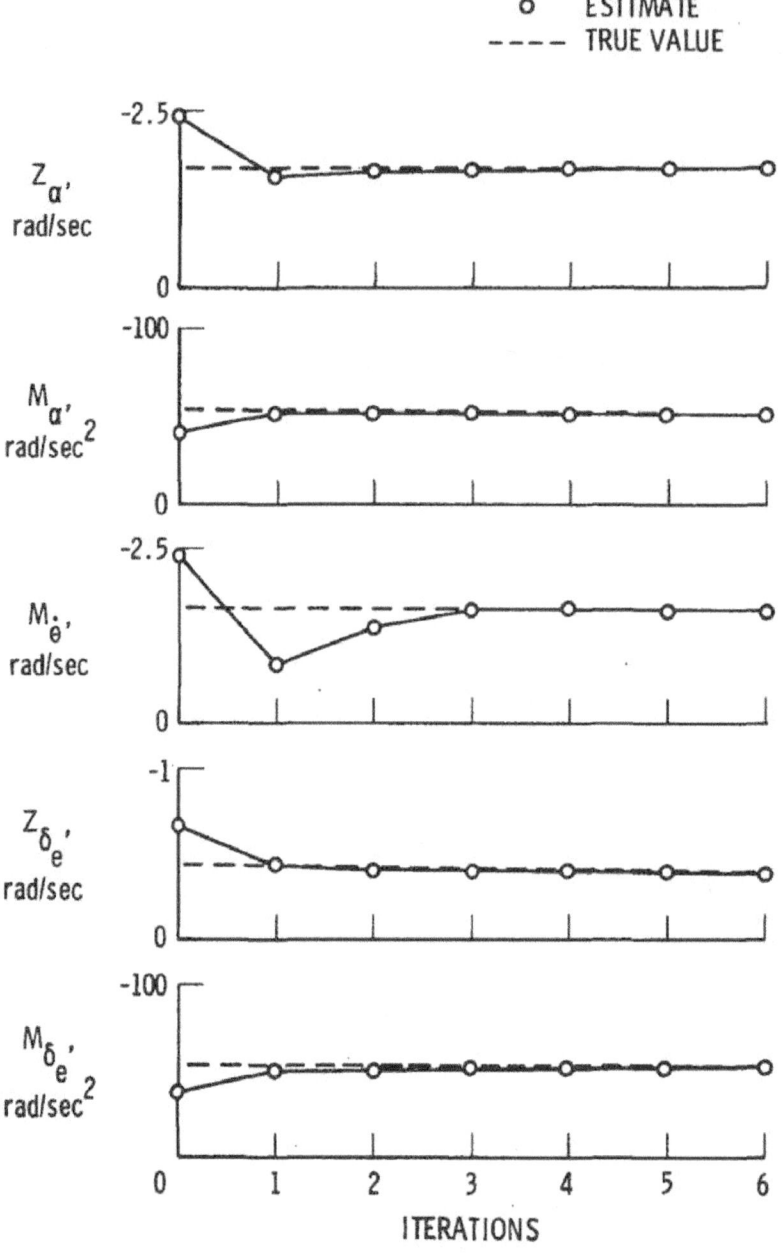

Figure 3. Convergence of stability and control derivatives for 4.9 seconds of data using Estimator II.

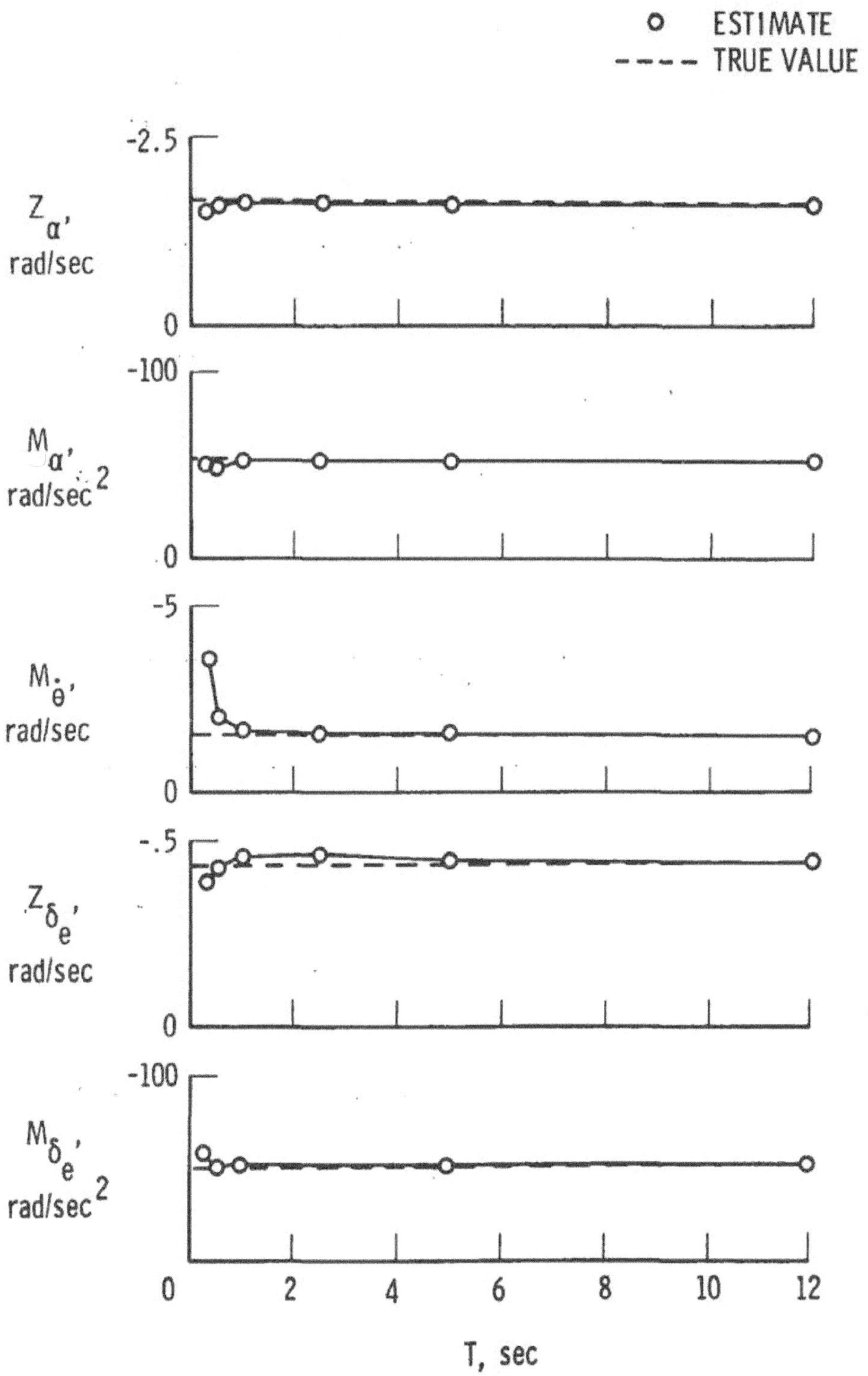

Figure 4. Effect of observation time interval on estimated stability and control derivatives using Estimator II.

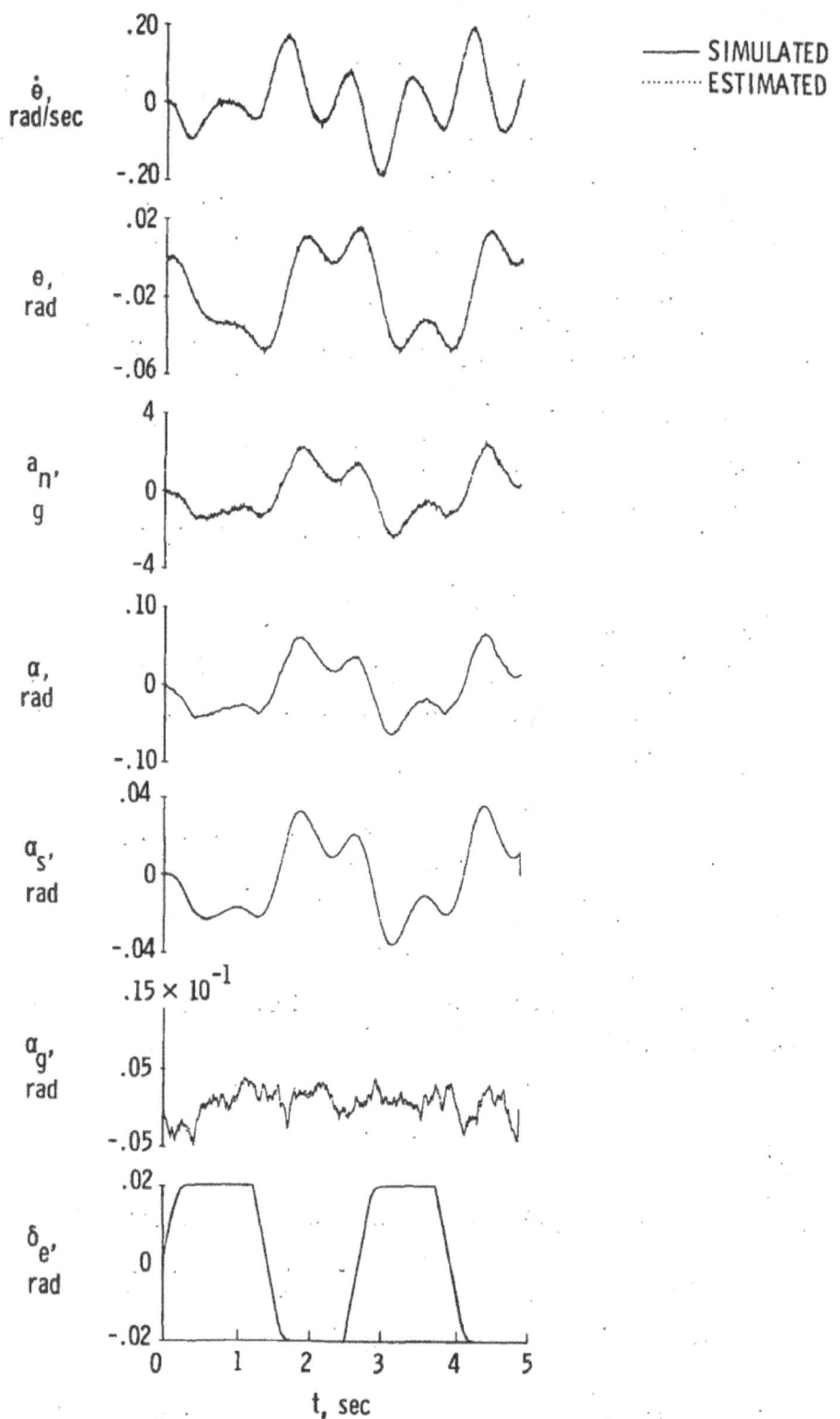

Figure 5. Comparison of computed data with measurement noise on α with estimated data obtained by Estimator III.

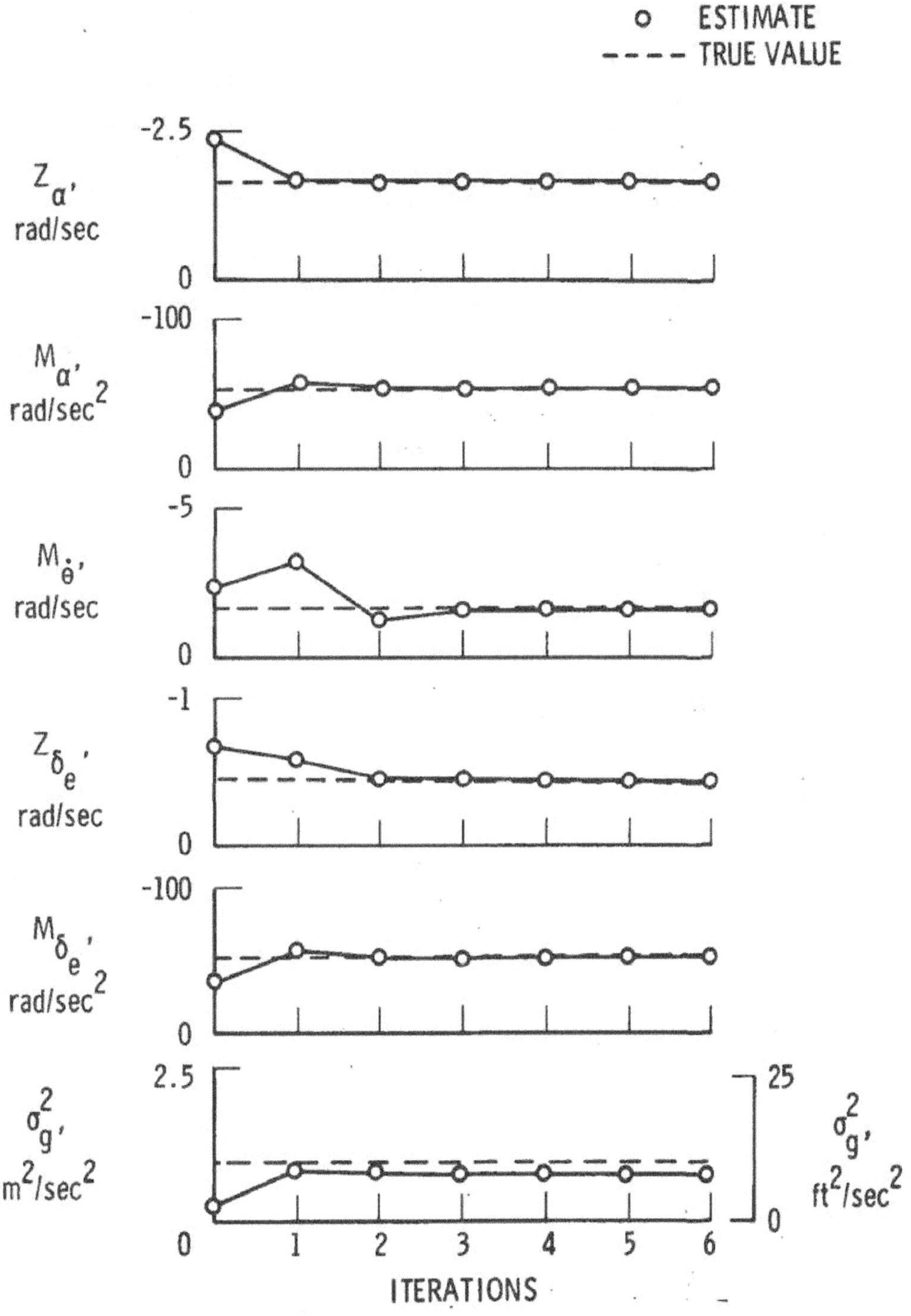

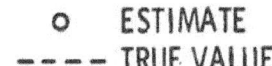

Figure 6. Convergence of coefficients for 4.9 seconds of data using Estimator III.

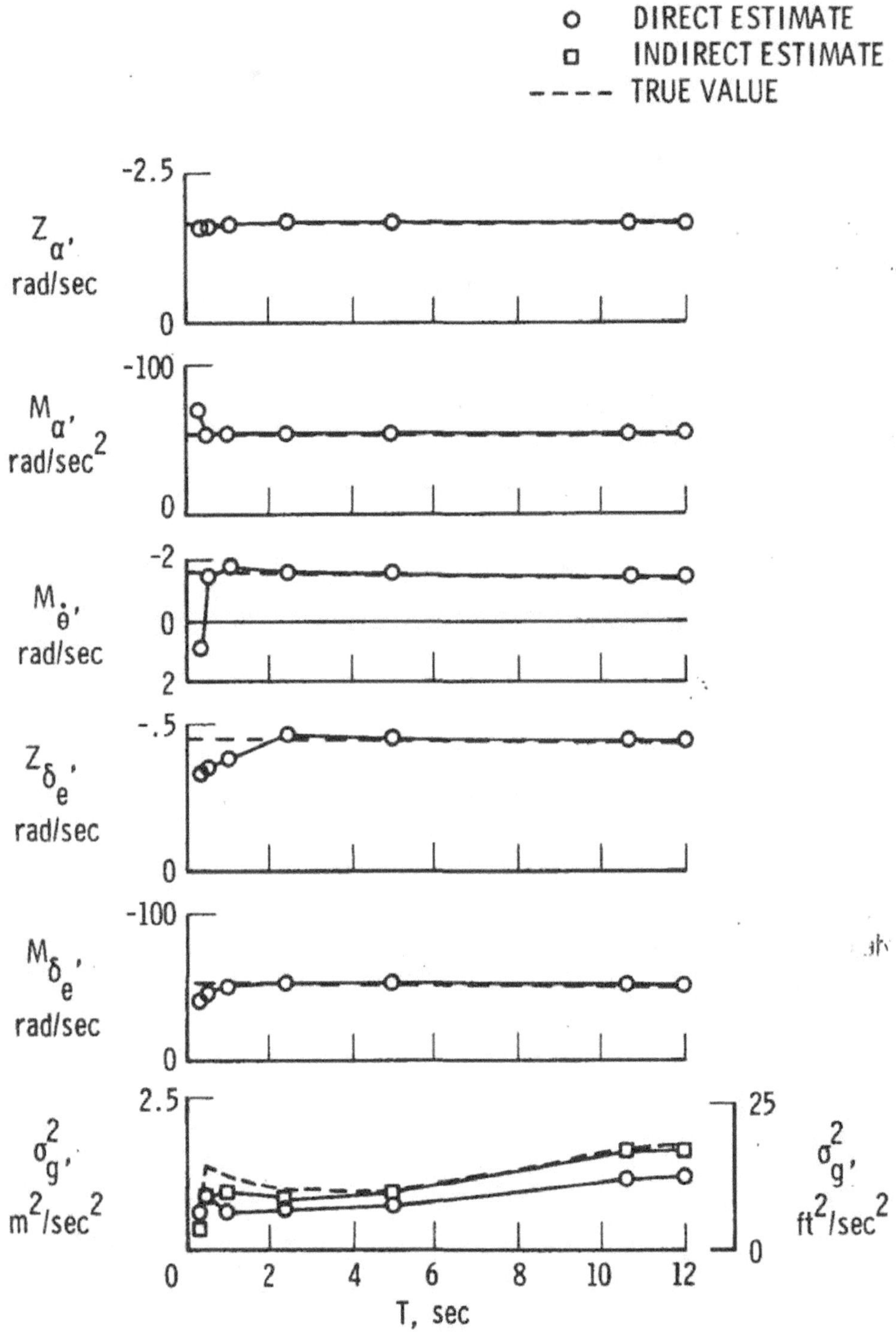

Figure 7. Effect of observation time interval on estimated coefficients using Estimator III.

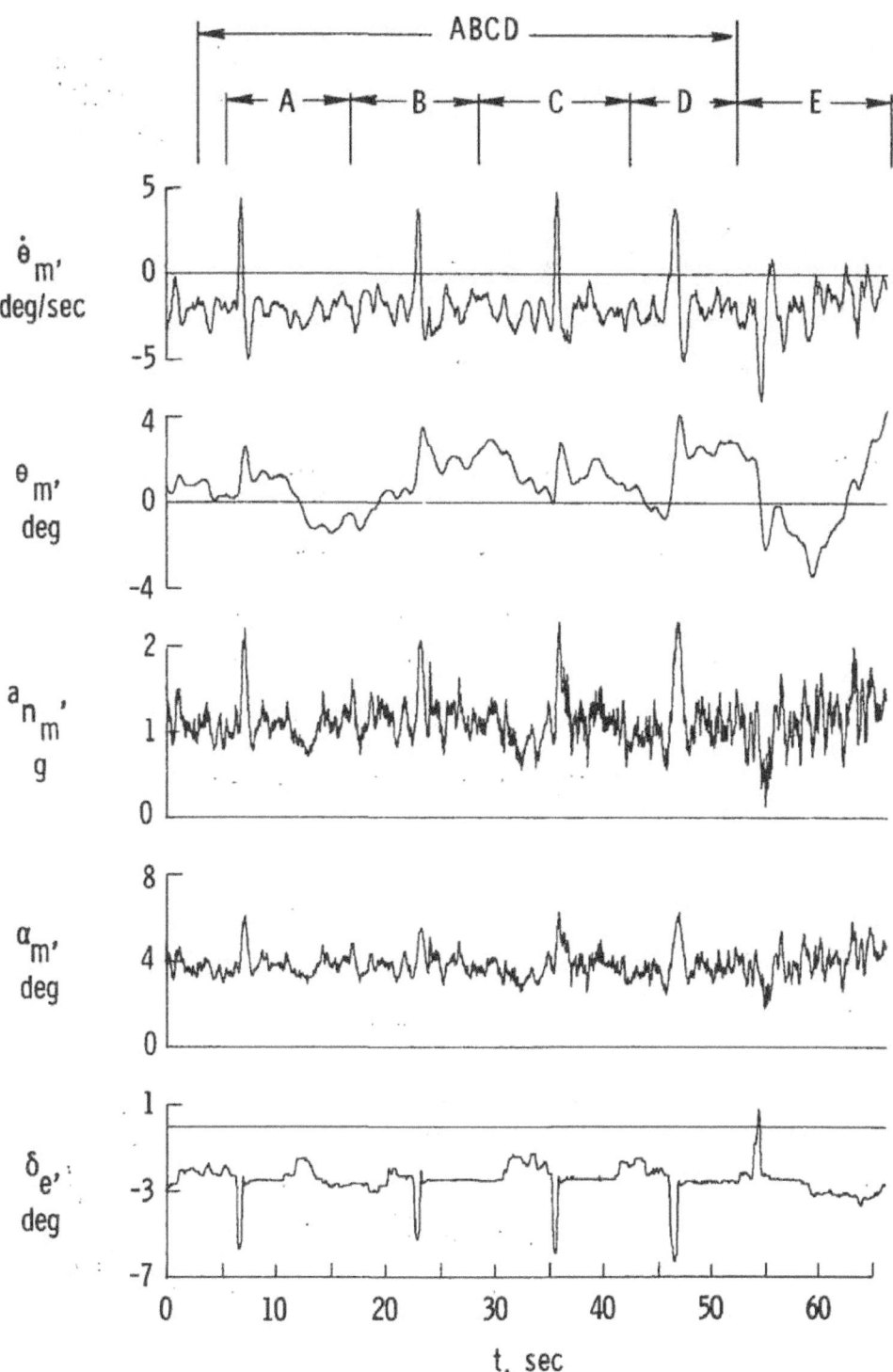

Figure 8. Total JetStar turbulence time history showing intervals of each maneuver.

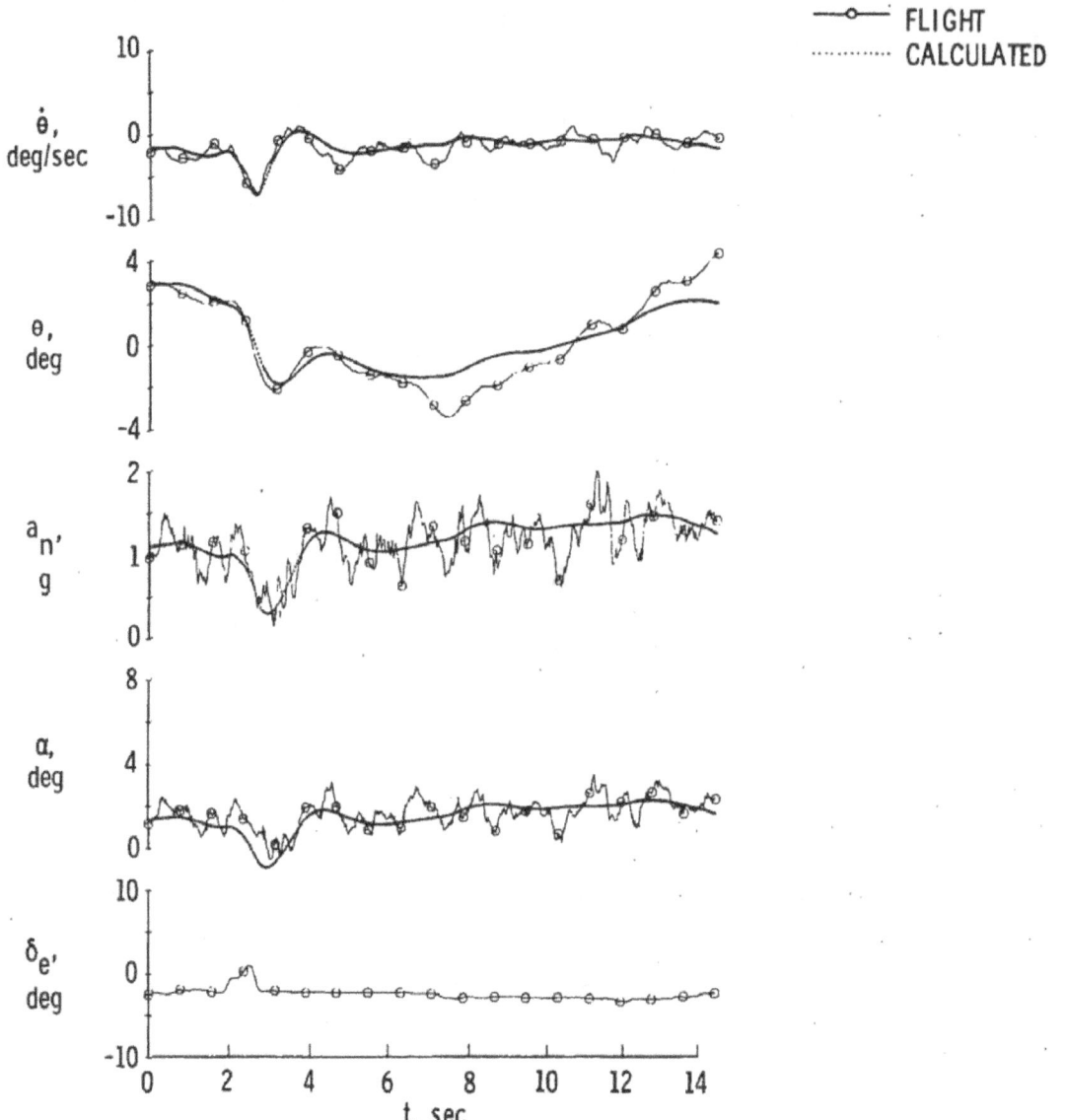

FLIGHT
CALCULATED

Figure 9. Comparison of flight data for maneuver E with estimated data obtained by Estimator I.

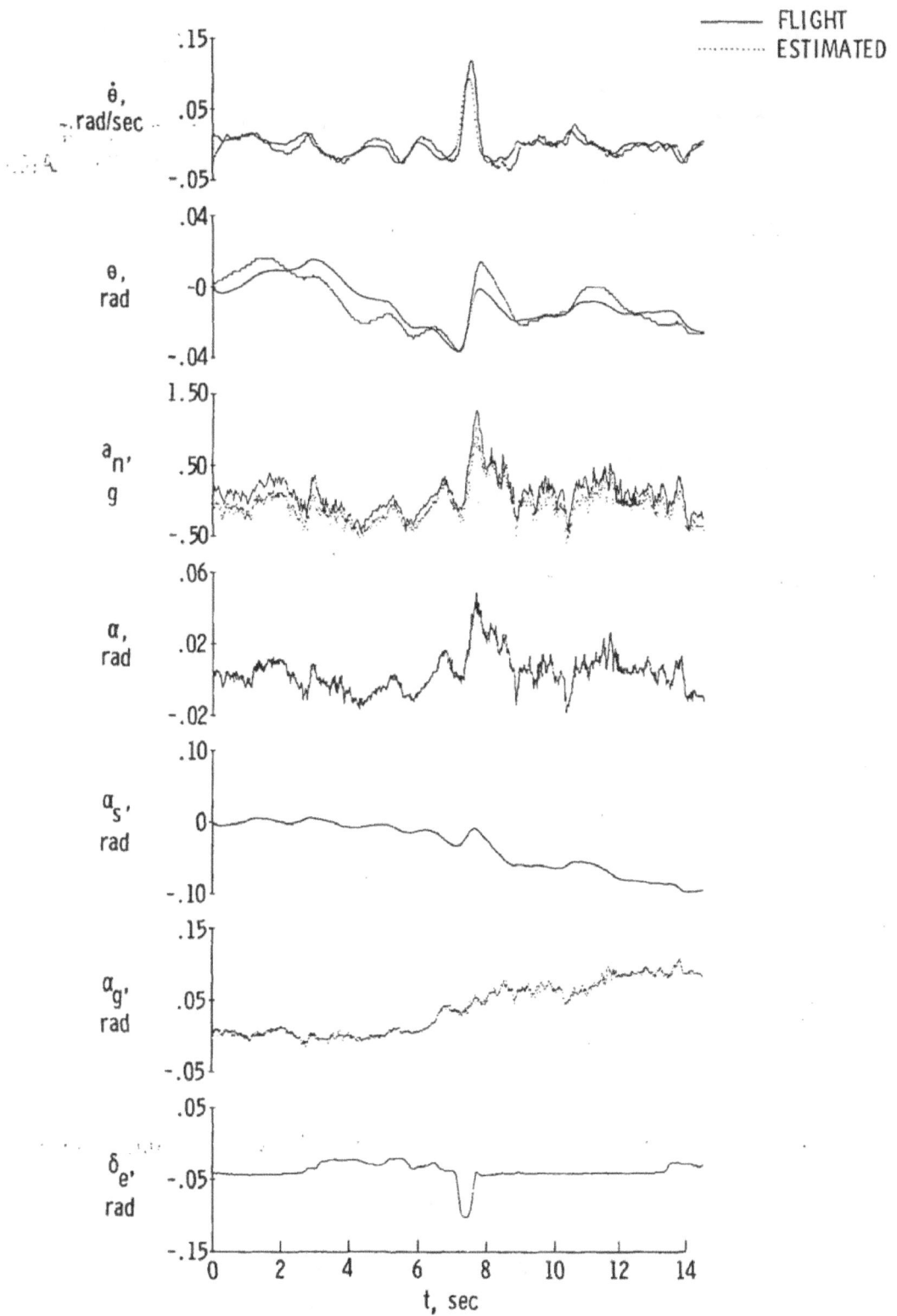

Figure 10. Comparison of flight data for maneuver C with estimated data obtained by Estimator II.

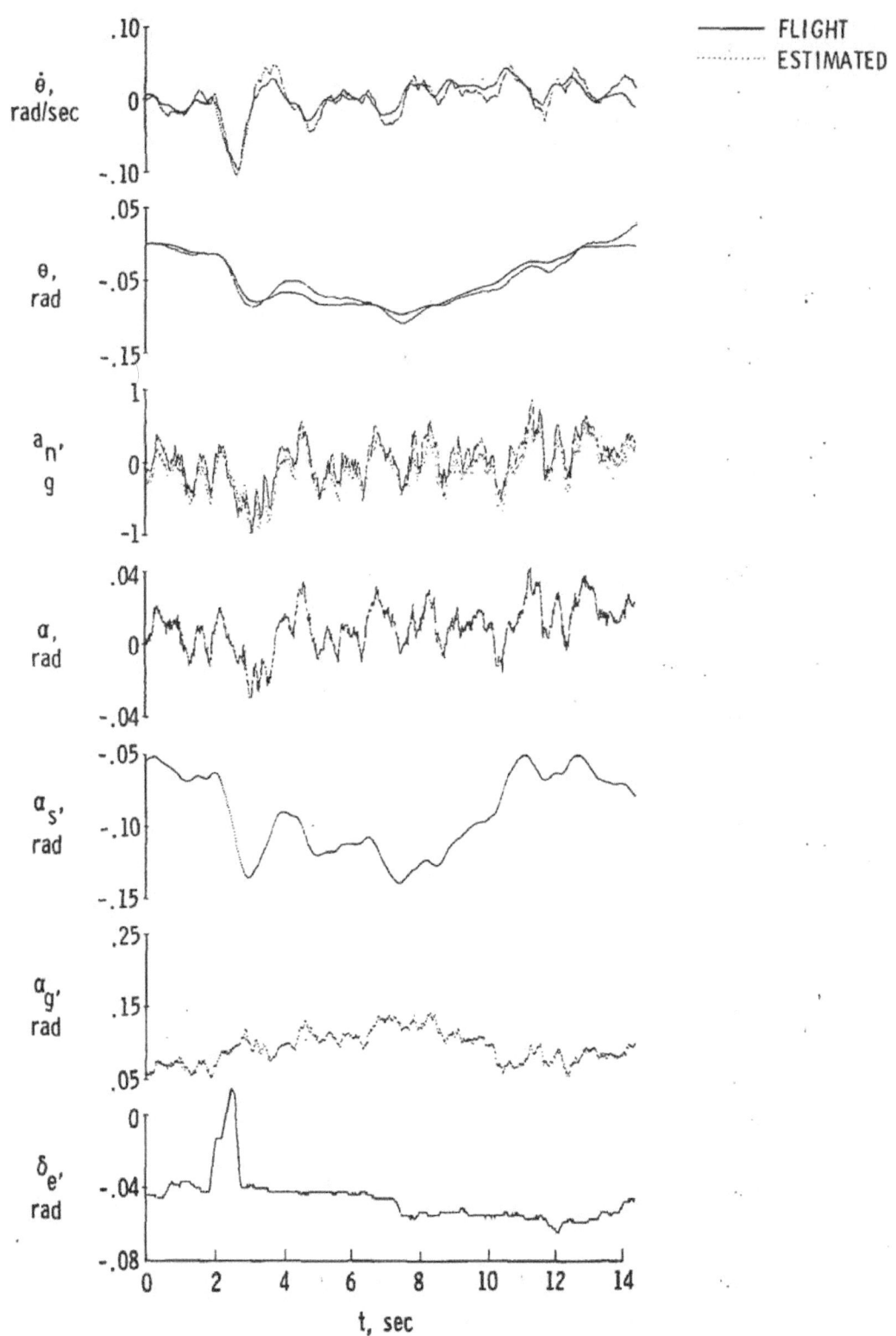

Figure 11. Comparison of flight data for maneuver E with estimated data obtained by Estimator II.

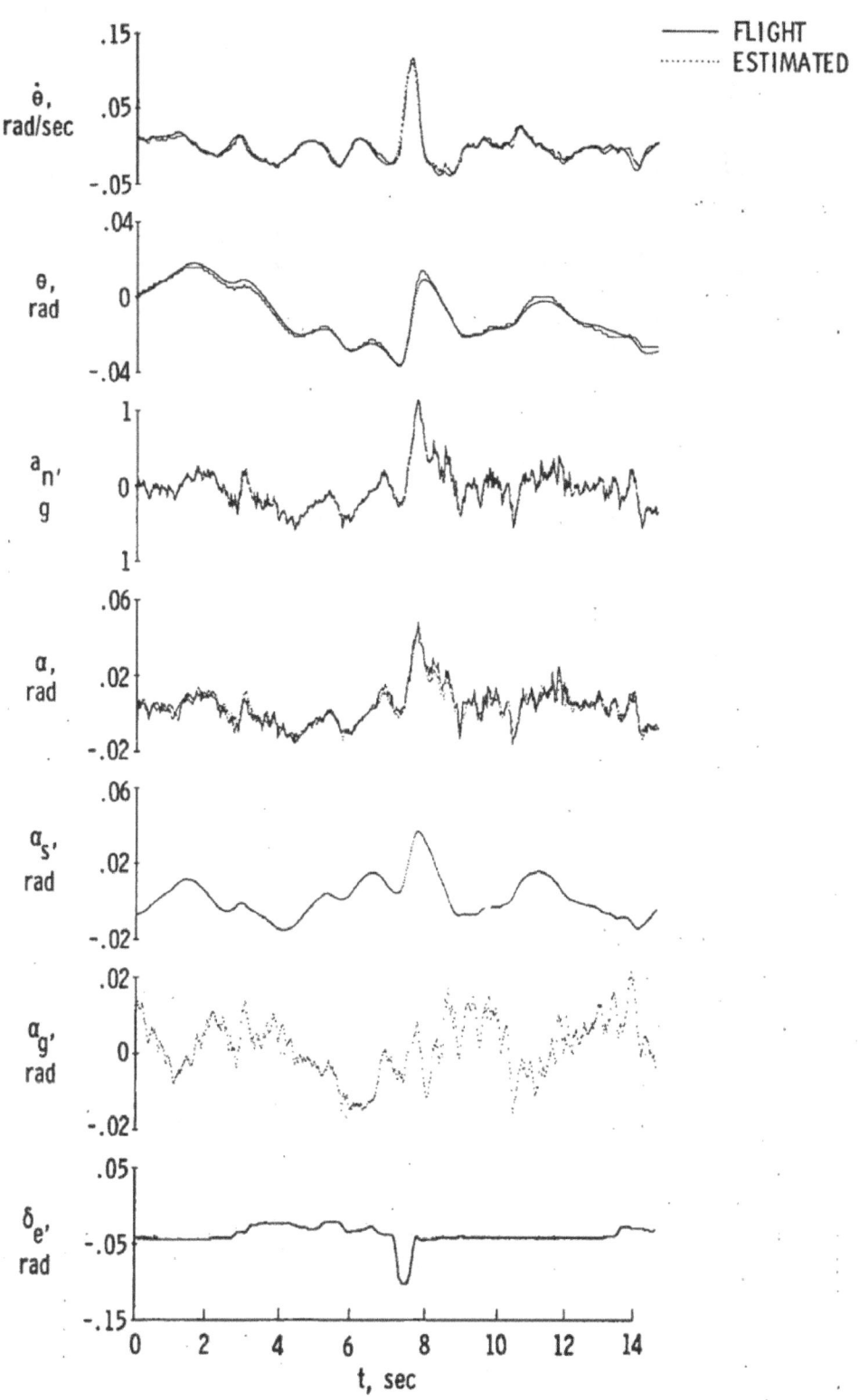

Figure 12. Comparison of flight data for maneuver C with estimated data obtained by Estimator III.

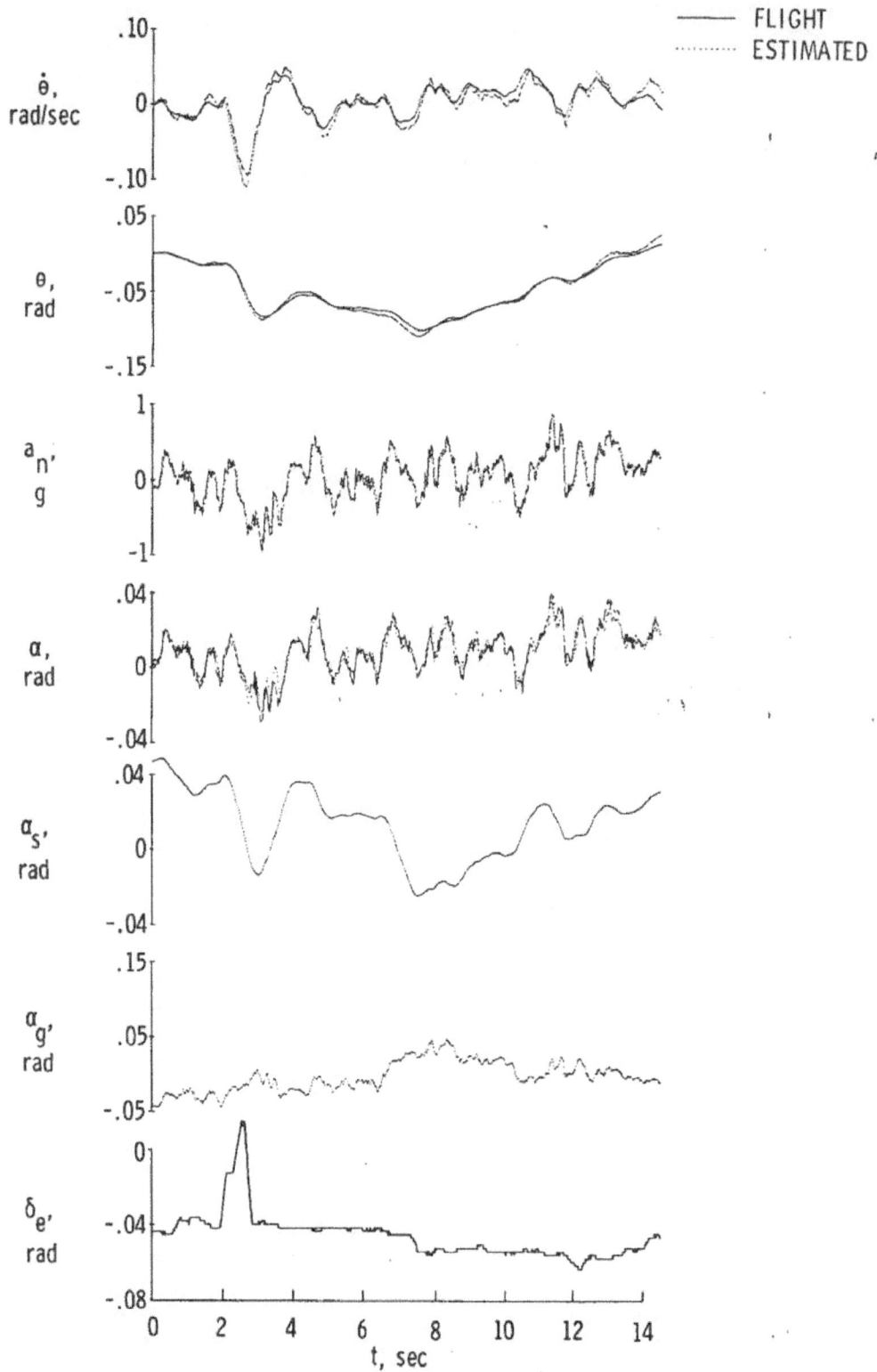

Figure 13. Comparison of flight data for maneuver E with estimated data obtained by Estimator III.

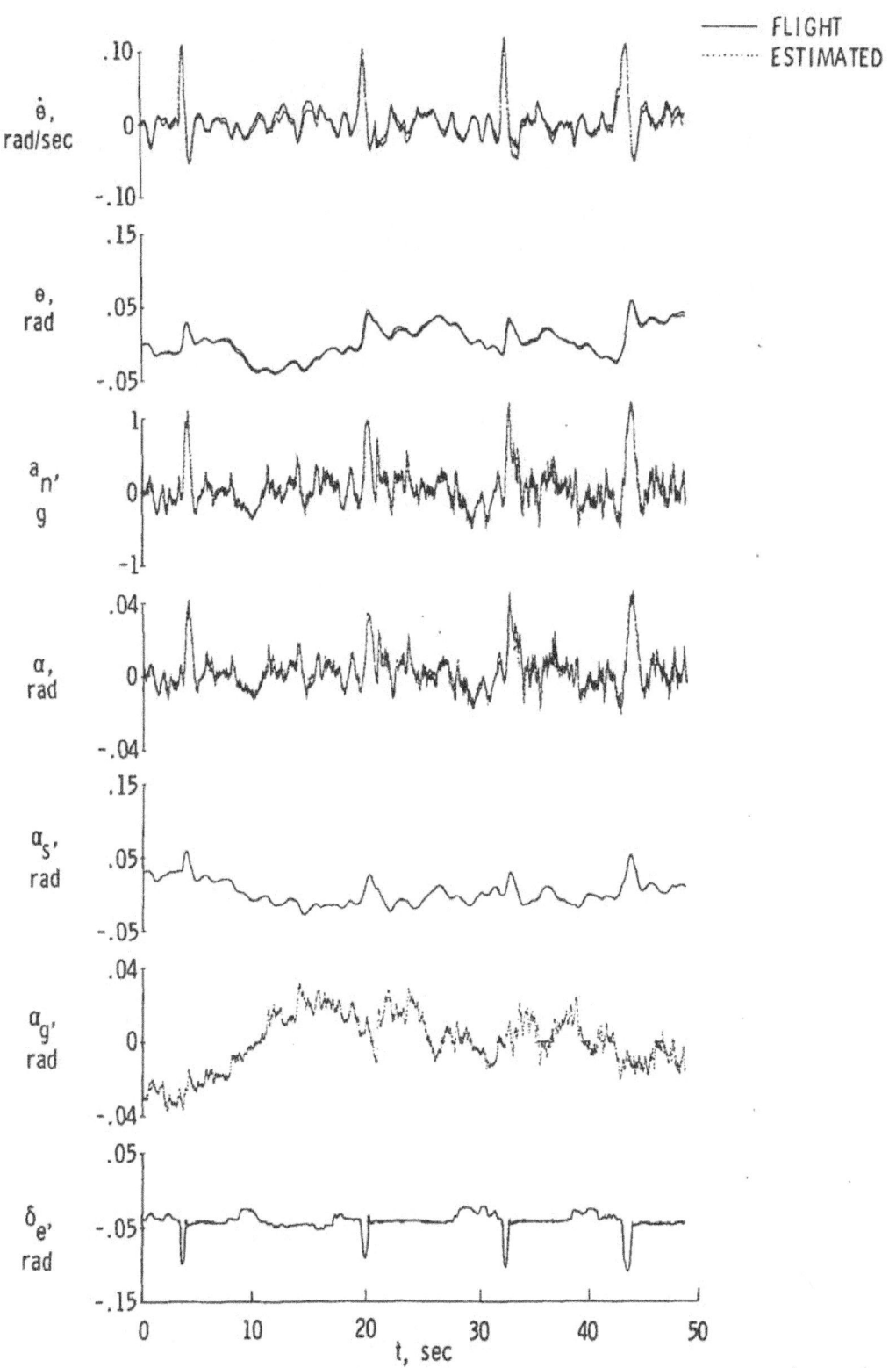

Figure 14. Comparison of flight data for maneuver ABCD with estimated data obtained by Estimator III.

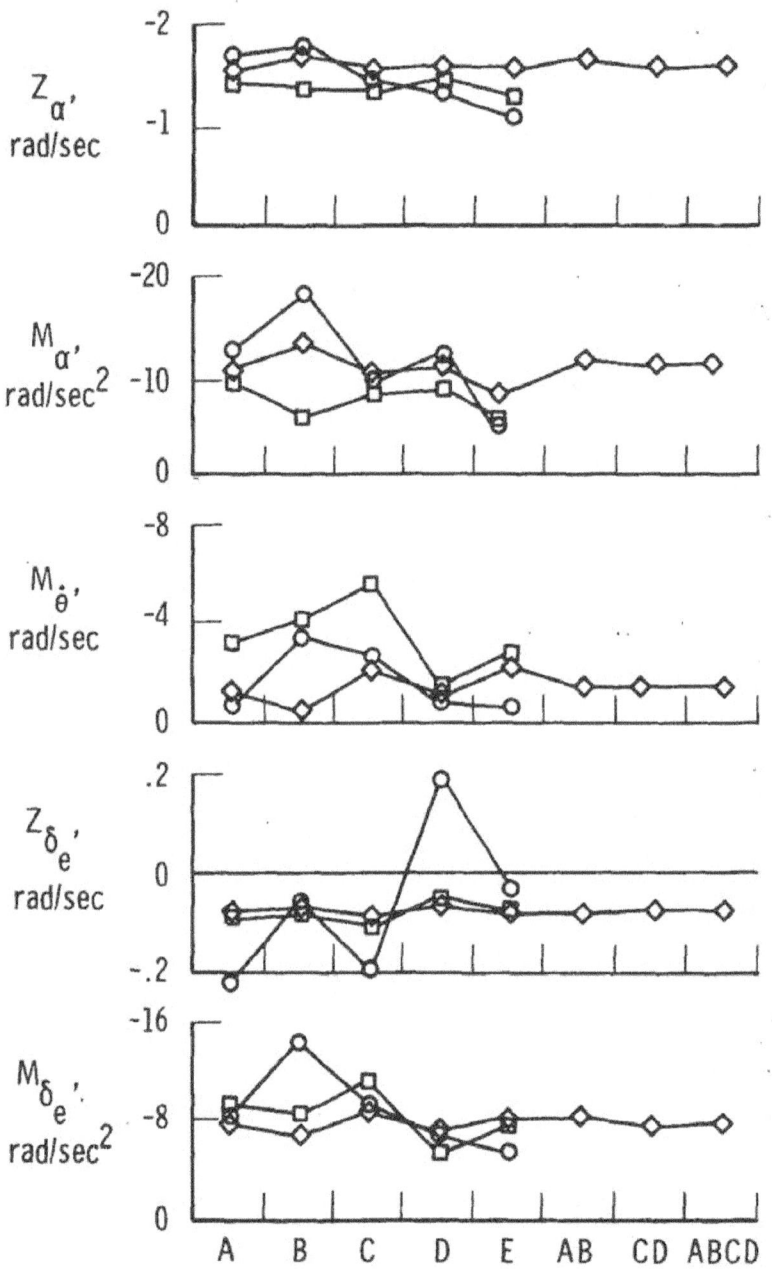

Figure 15. Summary of estimates of coefficients obtained by three maximum likelihood estimators for various maneuvers.

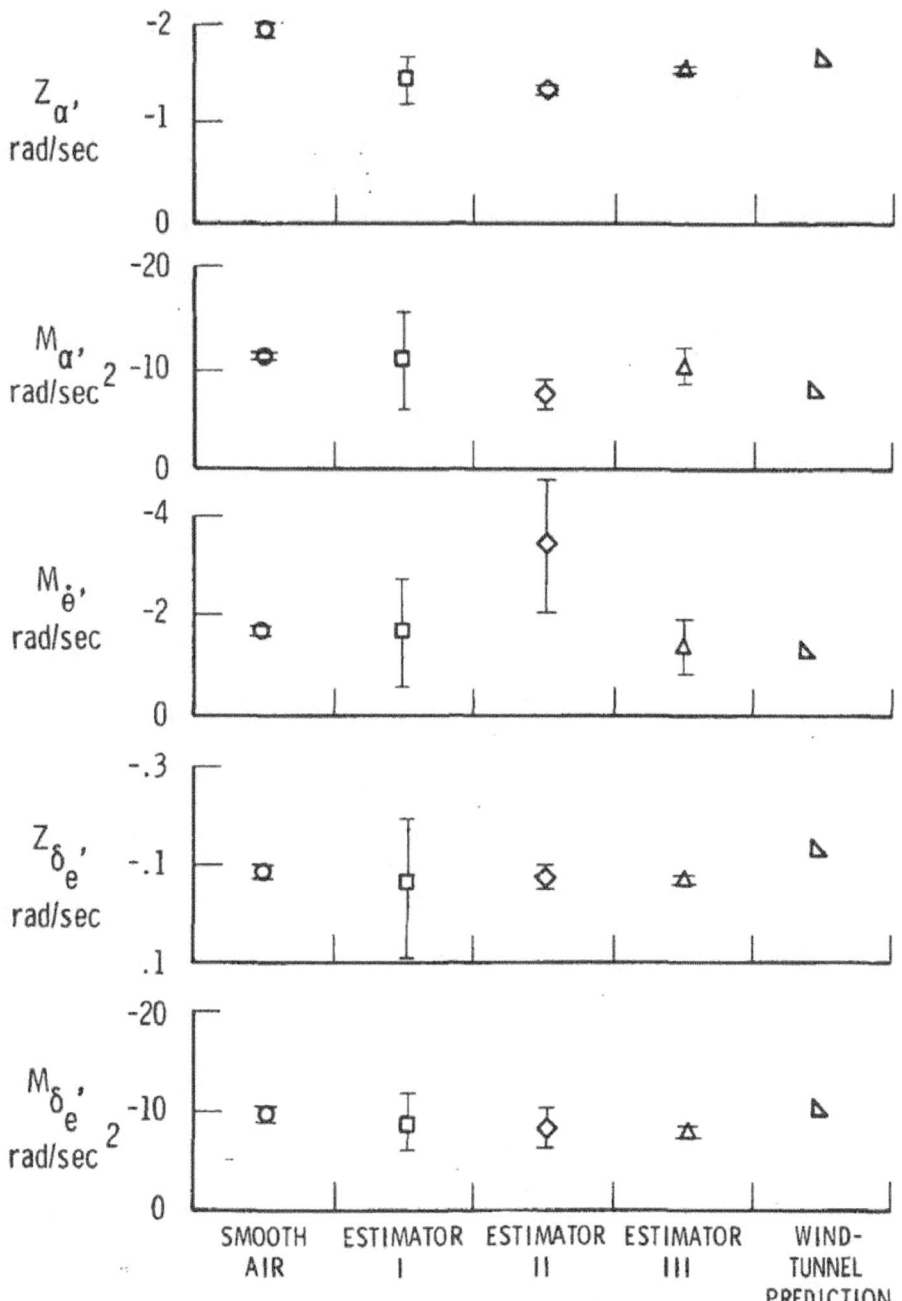

Figure 16. Means and standard deviations for five methods of estimating stability and control derivatives.

IDENTIFICATION OF M2/F3 STABILITY AND CONTROL

DERIVATIVES FROM FLIGHT DATA CONTAINING GUST EFFECTS*

By David E. Stepner and Raman K. Mehra

Systems Control, Inc.
260 Sheridan Ave.
Palo Alto, Ca 94306

SUMMARY

This paper discusses the application of the maximum likelihood identification technique to M2/F3 lifting body flight data containing wind gust effects. With the objective of this effort being the identification of the stability and control derivatives, it is shown first that the output error technique (or modified Newton-Raphson) fails to fit the recorded data accurately. The means of applying the maximum likelihood technique to this problem are then discussed and the results given which indicate an accurate fit to the data. The question of derivative signs opposite to the wind tunnel values is then addressed and the results of three techniques for dealing with this problem are presented. These techniques are a priori weighting, fixing parameter values and rank deficient inverses.

INTRODUCTION

The importance of extracting aircraft stability and control derivatives from flight test data has been recognized for a long time. Numerous methods for performing this parameter identification have been developed (see Taylor, et. al, Ref. 1) and usually they fall into one of two categories: (1) least squares or equation error methods, and (2) output error methods. However, all methods falling into either of these categories suffer from important deficiencies. Equation error methods lead to biased estimates in the presence of measurement noise and output error methods fail when process noise (wind gust distribution) is present. The development of the more general Maximum Likelihood (ML) Method [Ref. 2,3] for parameter identification has been motivated by several considerations. The ML method can handle both measurement and process noise. In cases where no process noise is present, it reduces to an output error method and where no measurement noise is present, to an equation error method. The ML method can also yield realistic values of the variances of the parameter estimates and it can be used to estimate the noise covariances.

M2/F3 FLIGHT TEST DATA

The data supplied on flight 21, case 6 of the M2/F3 lifting body contained very evident effects of wind gusts on the time histories of the sideslip angle and lateral acceleration. Nothing was known about the statistics (correlation

---

*This work was supported by NASA under Contract No. NAS1-10700

time, covariance) of these gusts, nor was the data processing instruments responsible for the quantization and clipping effects known. A total of 401 data points were supplied, representing 8.02 seconds of flight at an angle-of-attack of 1.57° and speed of .468 M. Previous attempts to fit this data had provided inaccurate results.

The model used to fit the observed data was the linearized lateral equations of motion with the gust effect, β_n, entering as sideslip angle. These equations were

$$
C
\begin{bmatrix} \dot{p} \\ \dot{r} \\ \dot{\beta} \\ \dot{\phi} \end{bmatrix}
=
\underbrace{\begin{bmatrix}
L_p & L_r & L_\beta & 0 \\
N_p & N_r & N_\beta & 0 \\
\sin\alpha & \cos\alpha & Y_\beta & \dfrac{g\cos\theta}{v} \\
1 & \tan\theta & 0 & 0
\end{bmatrix}}_{F}
\underbrace{\begin{bmatrix} p \\ r \\ \beta \\ \phi \end{bmatrix}}_{x}
+
\underbrace{\begin{bmatrix}
L_{\delta_a} & L_{\delta_r} & L_o \\
N_{\delta_a} & N_{\delta_r} & N_o \\
Y_{\delta_a} & Y_{\delta_r} & Y_o \\
0 & 0 & 0
\end{bmatrix}}_{G}
\underbrace{\begin{bmatrix} \delta_a \\ \delta_r \\ 1 \end{bmatrix}}_{u}
+
\underbrace{\begin{bmatrix} L_\beta \\ N_\beta \\ Y_\beta \\ 0 \end{bmatrix}}_{\Gamma}
\beta_n
$$

where (all quantities in body axes)

    p is roll rate (°/sec)

    r is yaw rate (°/sec)

    β is sideslip angle (°)

    ϕ is roll angle (°)

    δ_a is aileron deflection (°)

    δ_a is rudder deflection (°)

    C is a transformation matrix

$$
C =
\begin{bmatrix}
1 & -\dfrac{I_{xz}}{I_x} & 0 & 0 \\
-\dfrac{I_{xz}}{I_z} & 1 & 0 & 0 \\
0 & 0 & 1 & 0 \\
0 & 0 & 0 & 1
\end{bmatrix}
$$

The inertias, angle of attack, flight path angle and velocity were all assumed to be known and constant over the data record. All parameters in F and G were to be identified.

The measurement equations were given as

$$
\begin{bmatrix}
y_1 \\
y_2 \\
y_3 \\
y_4 \\
y_5
\end{bmatrix}
=
\begin{bmatrix}
p \\
r \\
\beta + \beta_n \\
\phi \\
a_y \left(= Y_\beta (\beta + \beta_n) + Y_{\delta_a} \delta_a + Y_{\delta_r} \delta_r + Y_o \right)
\end{bmatrix}
+
\begin{bmatrix}
n_1 \\
n_2 \\
n_3 \\
n_4 \\
n_5
\end{bmatrix}
$$

where n_i, $i = 1, \ldots, 5$ are independent white Gaussian measurement errors.

OUTPUT ERROR – NO WIND GUSTS INCLUDED

A first processing of the M2/F3 flight data was performed with the ML algorithm in the output error mode, i.e., no wind gust effects included ($\beta_n \equiv 0$). Even in this mode, the ML method is different from the more familiar Newton-Raphson technique in that the weighting matrix for the measurements is adaptive. The time histories of the measurements are given in Figure 1. As these traces indicate, the worst fits are to sideslip angle and lateral acceleration, although none of the fits to the observed data are very good. The parameter estimates obtained for the output error processing and their standard deviations are given in Table 1.

PERFECT SIDESLIP ANGLE MEASUREMENT

For this processing of the data, it was assumed that the measurement noise on the sideslip angle measurement is much smaller than the gust. There is then perfect correlation between the gust and the sideslip angle measurement disturbance, both being β_n. The Kalman filter for the complete 4 state model can be explicitly derived as

$$
C \dot{\hat{x}} = F \dot{\hat{x}} + Gu + \Gamma(y_3 - \beta)
$$

where Γ is the Kalman gain. y_3 is treated as a deterministic control and the model order can be reduced to 3 by canceling the β state for the above equation. The time histories of the fit are shown in Figure 2 and the parameter estimates in Table 1. There is excellent agreement to the observed data and the residuals of the lateral acceleration approached white noise, which is indicative of a perfect fit.

As the estimates in Table 1 indicate, some of the derivatives have changed sign from the wind tunnel values. These were attributed to the small excitation provided by the maximum of -.2° rudder deflection, the closed loop identifiability problems caused by the yaw damper and the possibly inadequate model structure.

Once a complete set of parameter estimates was obtained, it was possible to recover the time history of the gust, β_n, and its statistical properties.

117

The β_n time history, which is shown in Figure 3, was easily fit with a first order model. In addition, the sample covariance indicated that its intensity was at least two orders of magnitude stronger than the sideslip angle measurement noise.

A PRIORI WEIGHTING

With the ML technique providing excellent fits, efforts were then focused by eliminating the numerical problems causing the incorrect derivative signs. The first attempt used the a priori weighting technique in which a weighted quadratic term in the difference between the estimated and a priori values is added to the likelihood function to improve the conditioning of the information matrix. Unfortunately, such a method requires extensive tuning of the weights in order to obtain good fits to the observed data. This tuning was not done but rather weights supplied for an HL-10 lifting body were used. The resulting fits to the observed data were poor, especially for lateral acceleration, although, as shown in Table 1, many of the derivative estimates were in excellent agreement with the wind tunnel values.

FIXED PARAMETER VALUES

The basic causes of the incorrect signs for some of the parameters were that either the sensitivity of the output to changes in those parameters were small or that there was high correlation among the sensitivities of several of the parameters. Both these problems can be identified by noting the size of the diagonal elements of the information matrix and by noting its normalized off-diagonal coefficients. A technique often used to treat these problems has been to fix one or more of the parameters at a priori (e.g., wind tunnel) values and proceeding with the identification of the remainder.

Several different combinations of fixed parameters were investigated for obtaining an accurate fit to the observed data and maintaining the wind tunnel value signs. The best results were obtained with the L_p, L_r, L_β, N_p, N_r and δ_r derivatives at fixed values. Although the fits were improved over those from the a priori weighting technique (especially for lateral acceleration), they were still far below those obtained by the ML method alone. The parameter estimates, as given in Table 1, are all seen to have the same sign as the wind tunnel values.

It was finally determined that the parameter fixing technique did not improve the identification because: (1) the correlation may not simply be between pairs of parameters, but may involve an entire set of unknown parameters, and (2) it is not usually possible to choose correctly a set of parameters to fix or the values at which to fix them.

RANK DEFICIENT INVERSE

The solution to the parameter dependency problem is to find the directions in parameter space corresponding to combinations of parameters which cannot be identified. A perfect dependency among the parameters would, strictly speaking, result in a zero eigenvalue of the information matrix, causing it to be singular.

118

However, since round-off and other numerical errors prevent the information matrix from being exactly singular, all the eigenvalues will be non-zero with a spread between the smallest and largest eigenvalue being many orders of magnitude. In such a case, it is better to use a rank deficient solution for the inverse rather than a full rank solution. That is, the inverse to the information matrix should be computed leaving out one or more of the smallest eigenvalues.

The fits to the observed data are shown in Figure 4. Comparing these with Figure 2, it can be seen that the fits are only slightly degraded. The parameter values obtained for this third order rank deficient solution (3 eigenvalues neglected) are given in Table 1. Only one of the parameters still has an opposite sign from the wind tunnel values. Many of the parameters, such as N_{δ_r}, now have the correct sign from physical considerations, where before they did not. It is clear that further development work on this rank-deficient solution approach will improve the estimates even more.

CONCLUDING REMARKS

The generalized Maximum Likelihood (ML) Identification Method, which includes the output error and equation error methods as special cases, has been applied to flight test data with gust effects from an M2/F3 lifting body. With the ML method accurate fits to the observed measurements were obtained whereas the output error method failed to adequately match the data. Several of the derivatives of the identified linearized lateral equations of motion were of opposite sign from the wind tunnel values. Three techniques were investigated for correcting this problem: a priori weighting, parameter fixing, and rank deficient inverses. Of those, only the rank deficient technique maintained good fit to the data while correcting most of the sign reversals. This technique used with the maximum likelihood method seems to offer the best technique for automatically eliminating non-identifiable prameter combinations and identifying those that remain, even in the presence of gusts.

ACKNOWLEDGMENT

The authors would like to thank Dr. H. Rediess, Ken Iliff and Alex Sims of NASA FRC, Edwards for supplying data on the M2/F3 and for many informative discussions.

REFERENCES

[1] Taylor, L, et al, "A Comparison of Newton-Raphson and Other Methods for Determining Stability Derivatives from Flight Data", Third Technical Workshop on Dynamic Stability Problems, Ames Research Center, 1968.

[2] Mehra, R.K., Stepner, D.E. and Tyler, J.S., "A Generalized Method for the Identification of Aircraft Stability and Control Derivatives from Flight Data," 1972 JACC, Stanford, California.

[3] Stepner, D.E. and Mehra, R.K., "Maximum Likelihood Identification and Optimal Input Design for Identifying Aircraft Stability and Control Derivatives", Final Report to NASA LRC, October 1972.

Parameter	Wind Tunnel & Theoretical	Max. lik. estimate – output error mode (with St'd dev.)	Max. lik. estimate Assuming perf. β Meas. (with St'd dev.)	Max. lik. estimates With a priori weighting (with St'd dev.)	Max. lik. estimates with dependent params. fixed (with St'd dev.)	Max. lik. with Rank Deficient Solution
L_p	-0.4673	-1.548 (0.0935)	0.679 (0.035)	-0.461 (0.0182)	*	-.531 (.0189)
L_r	0.8878	2.008 (1.187)	10.49 (0.547)	4.154 (0.140)	*	4.268 (.144)
L_β	-75.140	-54.49 (2.45)	-97.79 (1.615)	-67.95 (1.02)	*	-103.35 (.105)
N_p	.0802	.102 (0.006)	-.0203 (0.00393)	.00475 (0.00349)	*	.0397 (.00682)
N_r	-.6876	-.0307 (0.078)	-1.675 (0.0590)	-.764 (0.0134)	*	-.989 (.0672)
N_β	7.5342	2.876 (0.136)	7.324 (0.152)	6.763 (.0876)	4.435 (.113)	7.568 (.306)
Y_β	-.2001	-.0476 (0.125)	-1.249 (.0597)	-.202 (0.00392)	-1.36 (.0594)	-1.19 (.0590)
$L_{\delta a}$	14.04	14.82 (0.301)	9.804 (0.109)	10.96 (0.161)	9.66 (.169)	10.25 (.0845)
$L_{\delta r}$	10.03	73.97 (8.59)	-109.28 (5.519)	-42.18 (3.13)	*	-5.539 (.0257)
L_o	0	11.14 (1.828)	-10.46 (0.328)	-.572 (0.115)	-9.004 (.141)	-10.89 (.280)
$N_{\delta a}$	.83	.596 (0.0223)	.719 (.0104)	.762 (0.111)	.756 (.0134)	.561 (.0254)
$N_{\delta r}$	-4.06	-12.874 (0.578)	6.844 (0.643)	-4.37 (0.106)	*	-.512 (.651)
N_o	0	-.345 (0.121)	.177 (0.0357)	-.233 (0.0433)	-.00239 (.0320)	.587 (.0833)
$Y_{\delta a}$	0	-.00033 (0.0151)	-.0363 (0.00669)	-.0847 (0.00867)	-.0275 (.00634)	-.0360 (.00660)
$Y_{\delta r}$	0	.0301 (0.363)	-.874 (0.222)	-1.932 (0.286)	*	-.737 (.219)
Y_o	0	.0179 (0.354)	.378 (0.0299)	-.0974 (0.378)	.456 (.0189)	.408 (.0296)
ϕ_{bias}			-.281 (0.0531)	-6.01 (0.0933)	-.667 (.108)	-.164 (.0428)

TABLE 1 M2/F3 PARAMETER ESTIMATES AND STANDARD DEVIATIONS

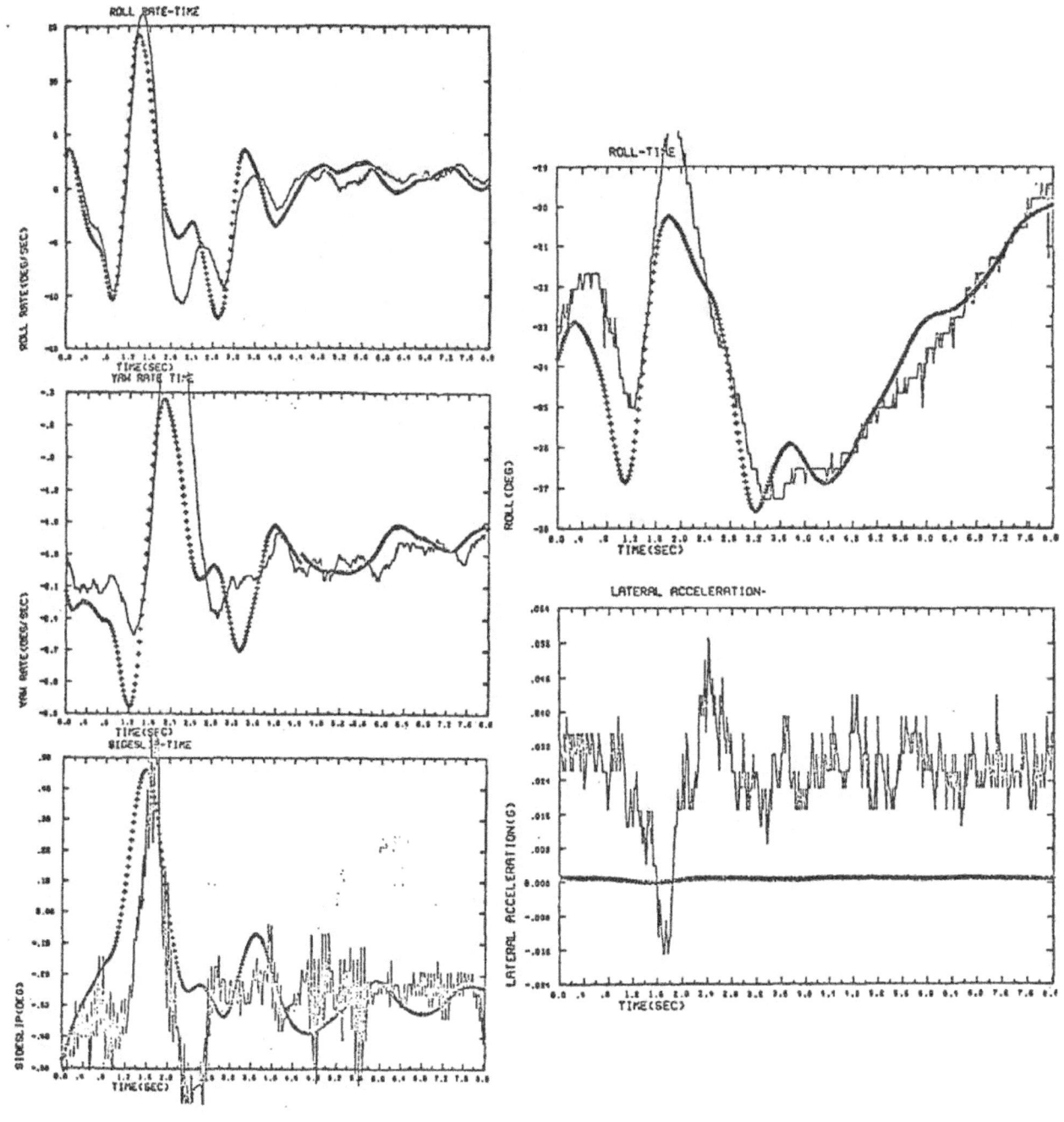

Figure 1   M2/F3: OBSERVATIONS AND ESTIMATES - OUTPUT ERROR

121

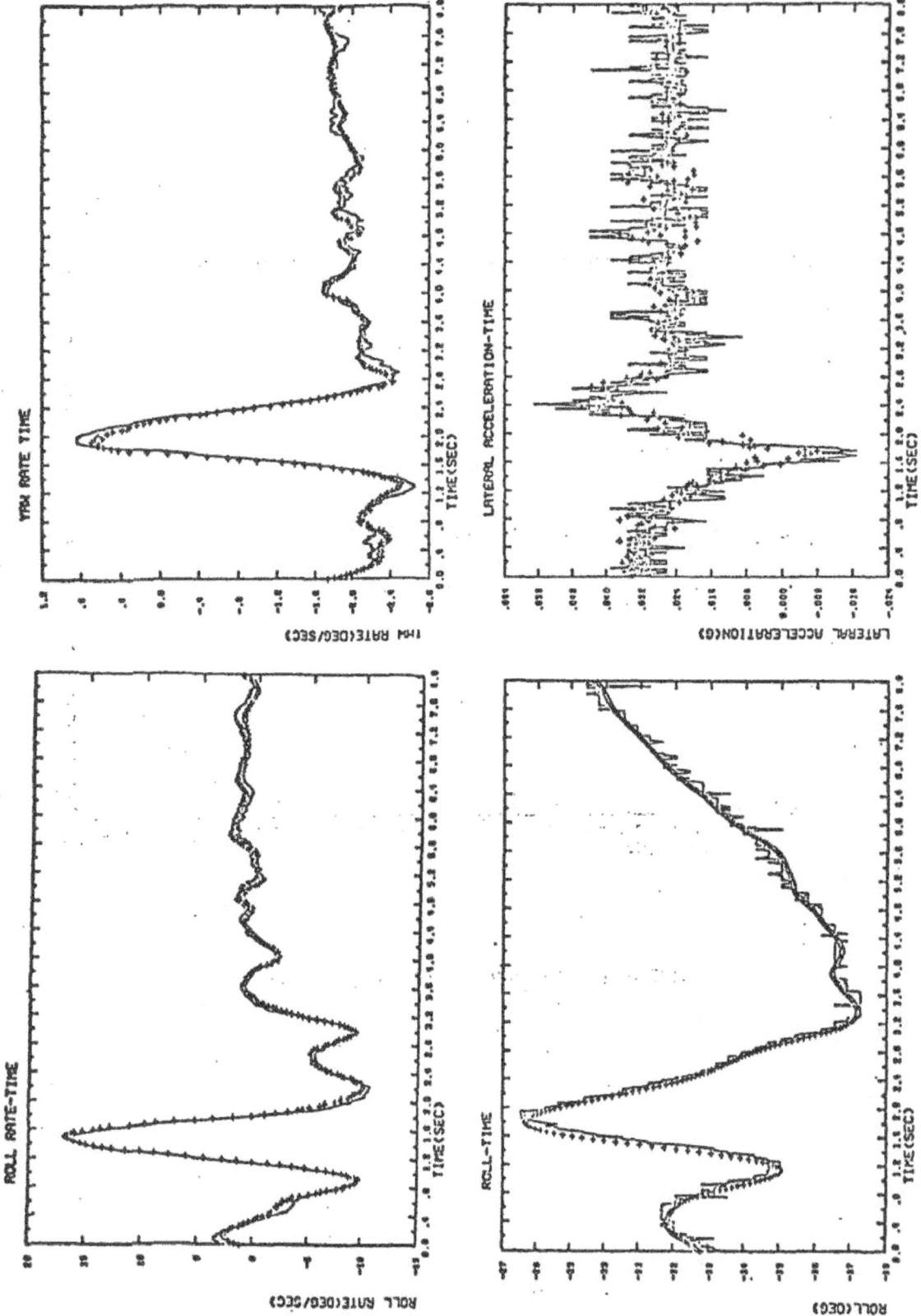

Figure 2 M2/F3: OBSERVATIONS AND ESTIMATES - KALMAN FILTER WITH $z_\beta = \beta + \beta_N$

122

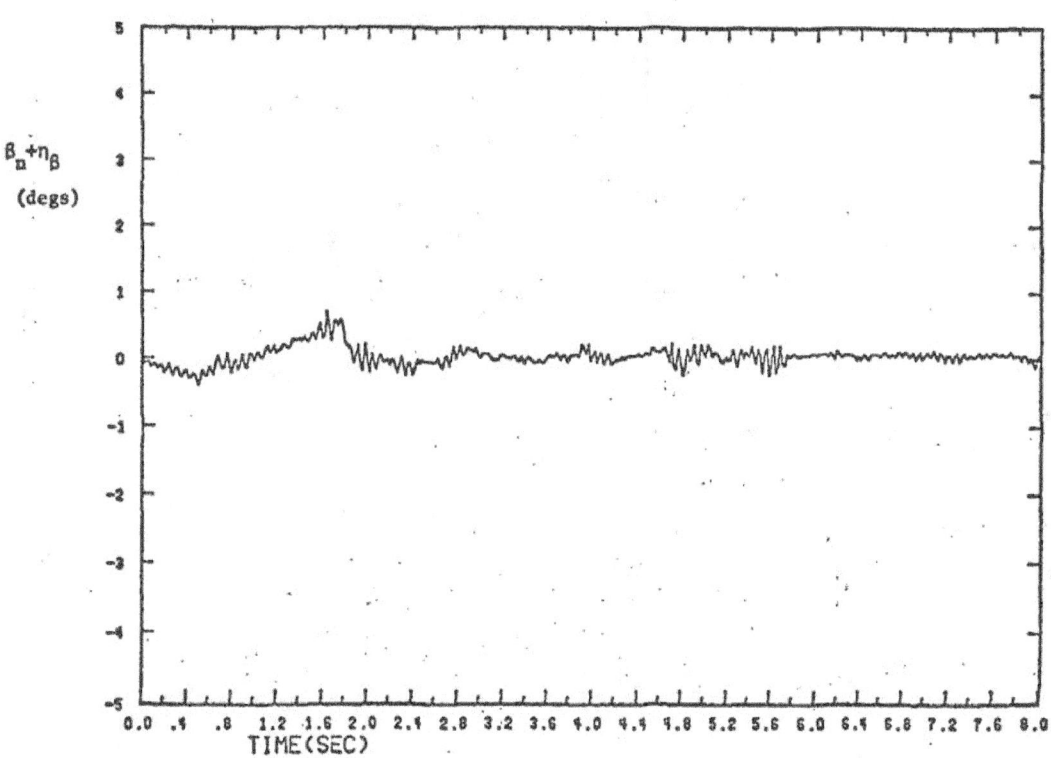

FIGURE 3   TIME HISTORY OF $\beta_n + n_\beta$

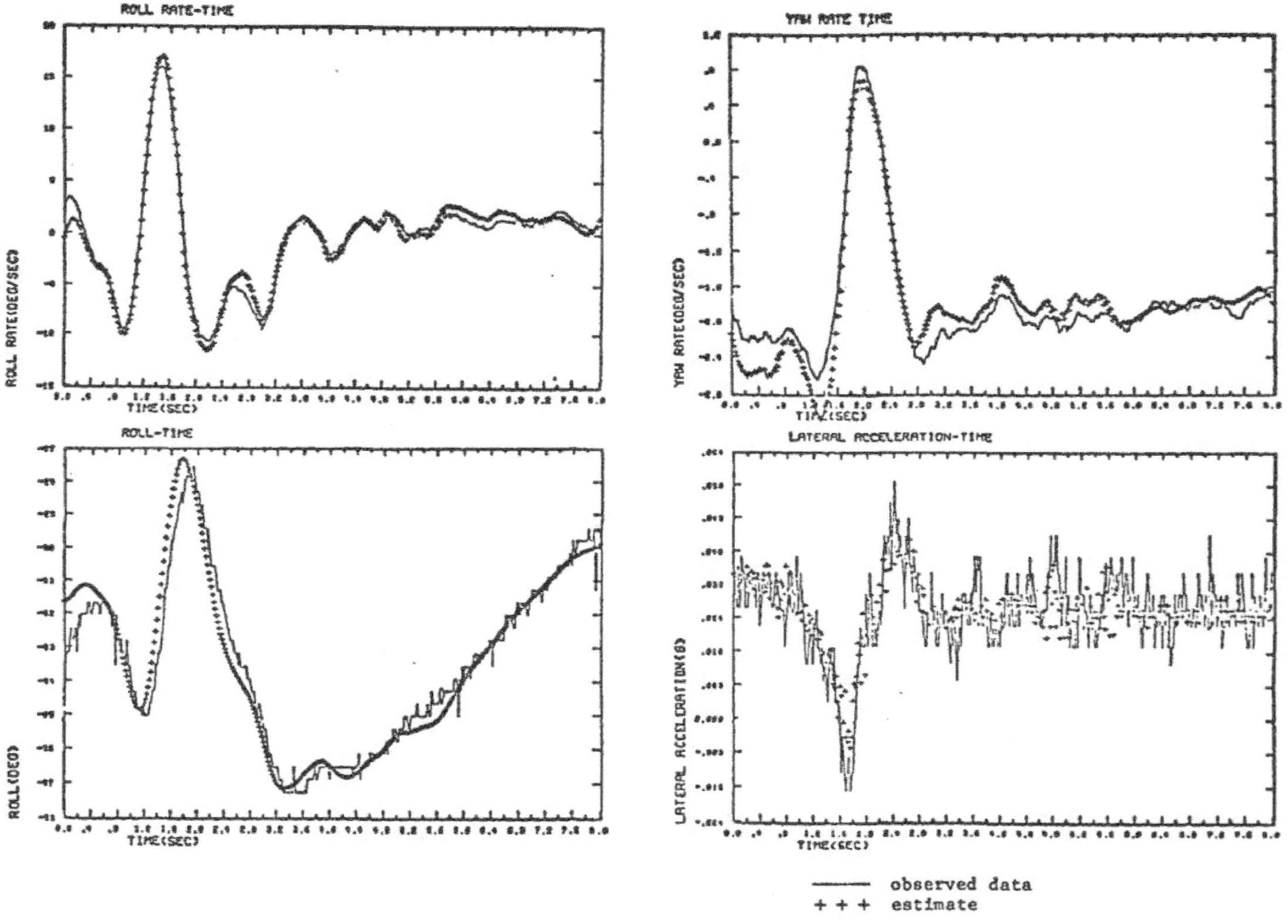

FIGURE 4   M2/F3 TIME HISTORIES WITH RANK DEFICIENT SOLUTION

ESTIMATION OF LONGITUDINAL AERODYNAMIC COEFFICIENTS

AND COMPARISON WITH WIND-TUNNEL VALUES

By Rodney C. Wingrove

NASA Ames Research Center

SUMMARY

This paper reviews some recent experience at Ames Research Center in the estimation of aerodynamic coefficients for the Lear-Jet and the Augmentor Wing Jet STOL Research Aircraft. The coefficients estimated from flight data are compared with values based on large-scale wind-tunnel tests. The results obtained by the regression and quasilinearization identification techniques are also compared. The regression method generally provides the lower standard deviation in the coefficient estimates and provides the better fit to the wind-tunnel values. The addition of nonlinear terms in the aerodynamic equations decreases the difference between the estimated and measured time histories but also increases the standard deviation in the estimated coefficient values.

INTRODUCTION

This paper considers the problems in estimating aerodynamic stability and control characteristics from recorded flight-test data. Although many identification methods have recently been developed (refs. 1-8), several problems remain. The problems include determining the form of aerodynamic equations required to mathematically model each type of aircraft and obtaining unbiased estimates, with small standard deviations, for the parameters in these models (refs. 3 and 7-11). This paper will review some recent flight experience at Ames in these problem areas.

Several recent studies (refs. 1-4) have compared different identification algorithms for estimating aircraft parameters and have found that the results may depend on the technique used. These identification techniques fall generally into two categories: equation error and output error. This paper will present results for both a regression technique (equation error) and a quasilinearization technique (output error).

With noise in the measured aircraft states, the regression technique can produce biased estimates of the coefficient values (refs. 1-4). The quasilinearization technique can reduce the bias error; however, it may produce the larger standard deviations in the estimated coefficient values (ref. 1). The effects of errors in modeling the aerodynamics (e.g., uncertainties in the proper type and number of terms) have not been studied in detail and it is not yet clear how these errors may affect the results obtained from each method.

This paper first describes the subject aircraft and presents the equations used to represent their aerodynamic characteristics. The identification techniques used to extract the coefficient values from the flight data are next outlined and the results obtained by each method are compared. Comparisons are then made between these coefficients derived from flight data and those obtained from wind-tunnel data. The final section reviews one of the more important problem areas, that of determining the number and type of nonlinear terms to be used in modeling the aerodynamics.

SYMBOLS

a_m	pitching acceleration, rad/sec^2
a_x	acceleration measured along X axis, g units
a_z	acceleration measured along Z axis, g units
B	constant bias term
$\bar{c}$	mean aerodynamic chord, m (ft)
C	aerodynamic coefficient
g	acceleration of gravity, m/sec^2 (ft/sec^2)
I_{yy}	inertia about the Y axis
J	cost function
M	aircraft mass
N	number of time series data points
q	pitching rate, rad/sec (or noted otherwise)
Q	dynamic pressure
S	aircraft wing area, m^2 (ft^2)
T	thrust component
u	velocity along X axis, m/sec (ft/sec)
V	total velocity, m/sec (ft/sec)
V_c	calibrated airspeed, m/sec (knots/hr)
w	velocity along Z axis, m/sec (ft/sec)
W	weighting matrix

Y   vector of aircraft states and accelerations

α angle of attack, rad (or noted otherwise)

δ elevator deflection, rad (or noted otherwise)

θ pitch angle, rad (or noted otherwise)

ρ atmospheric density

σ standard deviation (rms)

$\hat{\ }$ estimated value

AERODYNAMIC CONFIGURATIONS AND EQUATIONS

Results are presented from flight-test data obtained for the Lear-Jet and Augmenter Wing Jet STOL Research Aircraft (AWJSRA) (fig. 1). The Lear-Jet is a conventional light twin-jet transport; the AWJSRA, however, is unconventional in that it is a powered-lift configuration.

Aerodynamic coefficients for these two aircraft have been determined previously from wind-tunnel tests conducted in the Ames 40- by 80-Foot Wind Tunnel. For the Lear-Jet, a full-size airframe (unpowered) was tested in the tunnel (refs. 12 and 13); for the AWJSRA, a half-scale model (powered) was tested in the tunnel (refs. 14 and 15). Adjustments in the AWJSRA data to account for differences between the tunnel and flight configurations are discussed in reference 16, and the coefficient values are summarized in references 17 and 18.

Standard flight-test instrumentation was used for the flight investigation. The instrumentation included nose booms with pitot-static systems and vanes, body-mounted accelerometers and rate gyros, vertical gyros, and position transducers on the control surfaces. (The AWJSRA also included instrumentation to measure propulsive characteristics.)

The mathematical model that represents the forces acting on the airframes is described in a body-axis coordinate system. The linear accelerations (a_x, a_z) and the pitching acceleration (a_m) acting on the Lear-Jet and on the AWJSRA are taken as follows:

Lear-Jet:

$$\hat{a}_x = (QS/M)[\hat{C}_{x_0} + \hat{C}_{x_\alpha}\alpha + \hat{C}_{x_\delta}\delta + \hat{C}_{x_q}(q\bar{c}/2V) + \hat{C}_{x_{\alpha^2}}\alpha^2] + \hat{T}_x \tag{1}$$

$$\hat{a}_z = (QS/M)[\hat{C}_{z_0} + \hat{C}_{z_\alpha}\alpha + \hat{C}_{z_\delta}\delta + \hat{C}_{z_q}(q\bar{c}/2V)] + \hat{T}_z \tag{2}$$

$$\hat{a}_m = (QS\bar{c}/I_{yy})[\hat{C}_{m_0} + \hat{C}_{m_\alpha}\alpha + \hat{C}_{m_\delta}\delta + \hat{C}_{m_q}(q\bar{c}/2V)] + \hat{T}_m \tag{3}$$

127

AWJSRA:

$$\hat{a}_x = (QS/M)[\hat{C}_{x_o} + \hat{C}_{x_\alpha}\alpha + \hat{C}_{x_\delta}\delta + \hat{C}_{x_q}(q\bar{c}/2V) + \hat{C}_{x_{\alpha^2}}\alpha^2] + T_x \qquad (4)$$

$$\hat{a}_z = (QS/M)[\hat{C}_{z_o} + \hat{C}_{z_\alpha}\alpha + \hat{C}_{z_\delta}\delta + \hat{C}_{z_q}(q\bar{c}/2V) + \hat{C}_{z_{C_j}}C_j] + T_z \qquad (5)$$

$$\hat{a}_m = (QS\bar{c}/I_{yy})[\hat{C}_{m_o} + \hat{C}_{m_\alpha}\alpha + \hat{C}_{m_\delta}\delta + \hat{C}_{m_q}(q\bar{c}/2V)] \qquad (6)$$

The coefficient terms for the Lear-Jet are those used in most studies of aircraft longitudinal motion. The only nonlinear coefficient is a C_x term due to a nonlinear (α^2) variation with angle of attack. The model for the AWJSRA includes an additional C_z term due to the powered lift function C_j (C_j = thrust of cold air/QS).

The thrust terms T_X and T_Z for the AWJSRA were measured by the aircraft instrumentation. However, for the Lear-Jet, the thrust was not measured and the thrust terms $\hat{T}_X$, $\hat{T}_Z$, and $\hat{T}_m$ were taken as unknown parameters.

Equations (1) through (6) represent a standard form of the coefficient terms used for the data analysis in the first part of this paper. Some of the effects of using alternative coefficient terms and the general problem of choosing the number and type of coefficients will be discussed in the last section.

IDENTIFICATION METHODS

The aerodynamic coefficients were estimated using the regression and quasilinearization identification algorithms. This application of regression (also called equations of motion or least squares) is similar to the applications described in references 7 and 8. The application of quasilinearization (also called modified Newton-Raphson) is somewhat different from that used in previous studies (refs. 1-4). Most applications of quasilinearization have incorporated equations linearized about a nominal path; however, this study incorporates the full nonlinear aerodynamic and kinematic equations.

Regression Method

This technique determines the set of constant coefficient values that minimizes the least-squares functions:

$$J_X = \sum^N (a_X - \hat{a}_X)^2$$

$$J_Z = \sum^N (a_Z - \hat{a}_Z)^2$$

$$J_m = \sum^N (a_m - \hat{a}_m)^2$$

where a_x, a_z, and a_m represent the measurements and $\hat{a}_x$, $\hat{a}_z$, and $\hat{a}_m$ represent the estimated model outputs (eqs. (1)-(6)). From these formulations the unknown coefficients are determined by the well-known matrix inversion procedure (refs. 7 and 8).

<h2 style="text-align:center">Quasilinearization Method</h2>

This technique, in contrast to the regression method, integrates the kinematic equations to obtain estimated time histories of the aircraft states along with estimates for the aerodynamic coefficients. This technique minimizes a weighted least-squares function of the form

$$J = \sum_{}^{N} [\hat{Y} - Y]^T W [\hat{Y} - Y]$$

where $\hat{Y}$ represents the vector of estimated variables

$$\hat{Y}^T = [\hat{u}, \hat{w}, \hat{q}, \hat{\theta}, \hat{a}_x, \hat{a}_z, \hat{a}_m]$$

and Y represents the vector of measured variables

$$Y^T = [u, w, q, \theta, a_x, a_z, a_m]$$

The positive diagonal weighting matrix W is used to account for the relative confidence placed on each of the measured variables. (For this study, the weighting values were taken as $W_{ii} \approx 1/\sigma_i^2$; see table I.) The equations of motion used with the quasilinearization method are of the form

$$\dot{\hat{u}} = g\hat{a}_x - \hat{q}\hat{w} - g \sin \hat{\theta} + \hat{B}_u \qquad , \hat{u}(o) = u_o \tag{7}$$

$$\dot{\hat{w}} = g\hat{a}_z + \hat{q}\hat{u} + g \cos \hat{\theta} + \hat{B}_w \qquad , \hat{w}(o) = w_o \tag{8}$$

$$\dot{\hat{q}} = \hat{a}_m + \hat{B}_q \qquad , \hat{q}(o) = q_o \tag{9}$$

$$\dot{\hat{\theta}} = \hat{q} + \hat{B}_\theta \qquad , \hat{\theta}(o) = \theta_o \tag{10}$$

$$\hat{a}_x = (\rho \hat{V}^2 S/2M)[\hat{C}_{x_0} + \hat{C}_{x_\alpha}\hat{\alpha} + \hat{C}_{x_\delta}\delta + \hat{C}_{x_q}(\hat{q}\bar{c}/2\hat{V}) + \hat{C}_{x_{\alpha^2}}\hat{\alpha}^2] + \hat{T}_x \tag{11}$$

$$\hat{a}_z = (\rho \hat{V}^2 S/2M)[\hat{C}_{z_0} + \hat{C}_{z_\alpha}\hat{\alpha} + \hat{C}_{z_\delta}\delta + \hat{C}_{z_q}(\hat{q}\bar{c}/2\hat{V})] + \hat{T}_z \tag{12}$$

$$\hat{a}_m = (\rho \hat{V}^2 S\bar{c}/2I_{yy})[\hat{C}_{m_0} + \hat{C}_{m_\alpha}\hat{\alpha} + \hat{C}_{m_\delta}\delta + \hat{C}_{m_q}(\hat{q}\bar{c}/2\hat{V})] + \hat{T}_m \tag{13}$$

where $\hat{V} = \sqrt{\hat{w}^2 + \hat{u}^2}$, $\hat{\alpha} = \sin(\hat{w}/\hat{V})$, and B represents unknown bias terms. (Equations (11) through (13) represent the Lear-Jet; slight modifications as noted in equations (3) to (6) are incorporated for the AWJSRA.) With this formulation, initial estimates for the unknown parameter values are made (e.g., from the regression results) and then the estimates are successively improved, in an iterative manner, using the quasilinearization algorithm (refs. 1-6).

FLIGHT-TEST RESULTS

The aerodynamic coefficients were estimated using data recorded during maneuvers in which the aircraft motions were excited by elevator inputs. Representative maneuvers are illustrated in figure 2 for the Lear-Jet and in figure 3 for the AWJSRA. These maneuvers start from steady-state trimmed conditions and include elevator step inputs to excite the long-period phugoid motions and elevator doublet inputs to excite the short-period motions.

The aerodynamic coefficients were estimated both for the total record and for individual short records (see figs. 2 and 3). The short records (selected to be nearly identical) were analyzed in an attempt to gain some indication of the relative standard deviations in the estimated coefficient values. For the Lear-Jet, 13 short records (8 sec long) were available and, for the AWJSRA, 3 short records[1] (40 sec long) were available.

Comparison of Estimated and Measured Time Histories

A comparison of measured flight time histories with those computed using the estimated coefficient models for the Lear-Jet is presented in figures 4 and 5. Figure 4 presents the results obtained by the regression method and figure 5, those obtained by the quasilinearization method. The time histories illustrate that the aerodynamic coefficient model for the Lear-Jet provides an excellent fit to the measured flight data. Values for rms difference (residual) between the estimated and measured time histories are listed in table I. The regression and quasilinearization techniques produce somewhat different results. The regression method provides a significantly better fit to the measured linear accelerations a_z and a_x. The rms of the residuals (σ_{a_z} and σ_{a_x} in table I) with the regression methods are about 40 percent of the corresponding residual values with the quasilinearization method. The rms of the residuals for the pitching acceleration (σ_{a_m}) are about the same for each method.

Time history comparisons for the AWJSRA are presented in figure 6 (regression results) and figure 7 (quasilinearization results). These data illustrate that the aerodynamic coefficient models for the AWJSRA provide a generally good fit to the measured data. Examination of the recorded flight data for the AWJSRA indicates significant high-frequency noise in the measured values of a_x and a_z, which is believed to be associated with the higher levels of vibration on the AWJSRA as compared with the Lear-Jet. The rms values of the residuals for the other flight data, a_m, u, w, q, and θ, appear to be quite similar for both aircraft (table I).

---

[1] Although three samples do not represent a good basis for statistical analysis, a comparison of these three samples does give some indication of the scatter to be expected in the estimated coefficient values.

Comparison of Estimated Coefficients

The individual coefficient values estimated from the flight data are presented for the Lear-Jet (table II) and for the AWJSRA (table III). Figures 8 and 9 present for each aircraft (in bar chart form), the mean values, along with the standard deviations, normalized with respect to the average coefficient values. These bar charts are intended to compare (in graphical form) the results obtained by the regression and quasilinearization techniques.

In general, the agreement between the coefficients, as measured by each identification technique, is better for the Lear-Jet (fig. 8) than with the AWJSRA (fig. 9). Also, the agreement (and standard deviation) for the moment (C_m) terms are generally better than for the translational (C_x) and (C_z) terms.

For both aircraft, the more important aerodynamic coefficients such as C_{z_α}, C_{x_α} (or $C_{x_{\alpha^2}}$), C_{m_α}, C_{m_δ}, and C_{m_q} are in good agreement. The standard deviations of these estimated parameters are also relatively small. All the other parameters have higher standard deviations in their estimated values, and generally, the mean values as measured by each method do not show such good agreement. These comparisons also show that for those parameters with significant standard deviations (e.g., greater than about 10 percent of their mean values), the regression estimates have less standard deviation than the quasilinearization estimates.

Comparison With Wind–Tunnel Values

The estimated coefficients values (tables II and III) are presented graphically in figure 10, normalized with respect to their corresponding values as determined from the wind-tunnel tests. This figure illustrates that the agreement between the flight and wind-tunnel values depends somewhat on the identification technique. For almost all the coefficients, the regression values better agree with the predicted values. A majority of the coefficients estimated by the regression technique are within about ±10 percent of the wind-tunnel values. The exceptions are C_{z_δ}, C_{m_α}, and C_{m_δ} for the Lear-Jet along with C_{z_δ} and $C_{x_{\alpha^2}}$ for the AWJSRA. Possible reasons for some of these differences between the flight and wind-tunnel values have been discussed in previous studies.

The terms C_{z_δ} and C_{x_δ} account for the effects of the elevator on the linear forces. These contributions of the elevator are quite small relative to the other forces acting on the airframe and as such are difficult to measure accurately in flight. (Also, these terms are strongly coupled with C_{z_q} and C_{x_q}.) Previous studies (e.g., refs. 7, 19-21) have also noted the large standard deviations associated with estimating these terms from flight-test data.

The C_{m_α} term, representing the aircraft static stability, depends strongly on the location of the center of gravity during the maneuver. For instance, an uncertainty of the cg location in the Lear-Jet of 11 cm (5 in.) could account for the 30-percent difference in the predicted C_{m_α} (wind tunnel C_{m_α} corrected for cg location) as compared to the flight measured value. However, there is probably more than just an uncertainty in the aircraft cg to account for this difference. The C_m terms measured for the Lear-Jet are consistently lower than predicted. Several other studies (e.g., refs. 12, 19-22) have also found consistently lower values for the C_m terms measured in flight as compared with wind-tunnel values.

131

Discussion of Flight-Test Results

The results presented above show that both the regression and quasilinearization methods provided similar results for the rotational mode. However, for the translational modes, the regression method generally provides better results than the quasilinearization method. For instance, the regression method provides a better fit to the measured accelerations, less standard deviation in the estimated coefficient values, and better agreement with the wind-tunnel values.

The differences to be expected between the regression and quasilinearization methods depend, to a large extent, on the amount of measurement noise. Any noise in the measurement of the states α (or w), V (or u), and q could cause bias errors with the regression method. Although the amount of noise cannot be determined with certainty, the recorded data in figures 5(b) and 7(b) show very little of what may be termed white or near white measurement noise (e.g., there is a low noise-to-signal ratio). Apparently, for the flight-test situations considered in this study, there are no large amounts of measurement noise that could cause significant errors with the regression method.

Any inaccuracy in estimating the states u, w, q, and θ can cause errors with the quasilinearization technique. To minimize this type of error, an accurate mathematical model of the forces and moments acting on each type of aircraft is required. For the flight-test situations in this paper, some modeling errors are undoubtedly present which cause estimation errors with the quasilinearization technique. As discussed next, the best form of the mathematical model to use for each aircraft is not always obvious.

EFFECTS OF NONLINEAR TERMS IN THE MATHEMATICAL MODEL

The complexity of the model describing the aerodynamic characteristics depends on the aircraft configuration, the flight conditions, the angle-of-attack region, the power-lift levels, etc. A best choice of the number and type of coefficients to be used in the model is not always easy to determine. If the purpose of the investigation is to estimate only the linear model of the aircraft, then there is finite number of possible terms that need be considered. However, if the purpose of the investigation is to provide a best fit to measured flight data and provide an estimate of the nonlinear aerodynamic characteristics, an unlimited number of nonlinear terms (e.g., polynomials, cross products, etc.) could conceivably be used. Although some studies (e.g., refs. 3, 8, and 11) have considered the problems of choosing the "best" form of the mathematical model for different types of aircraft, no effective procedure has been established.

For the purposes of this paper, the choice of the number and types of nonlinear terms included in the mathematical models was based on three factors:

(1) An examination of the nonlinear terms required to provide a good fit to the available wind-tunnel or predicted aerodynamic data

(2) An examination of the nonlinear terms required to provide a good fit to the flight-test data

(3) An examination of the standard deviation in the nonlinear terms as estimated from the flight data.

132

These three factors are illustrated by the following example.

Figure 11 illustrates the effect of using successively higher order polynomials to model the variation of C_X with α (for the Lear-Jet). Evaluation of the three factors yields:

(1) An examination of the wind-tunnel data (right side of fig. 11) shows that the use of only a first-order polynomial does not provide a good fit to the data. Use of either the second-order or third-order polynomial provides an excellent fit.

(2) An examination of the flight-test data (left side of fig. 11) shows results similar to the wind tunnel, that is, a first-order polynomial does not provide a good fit to the data; the second-order and third-order polynomials provide an excellent fit.

(3) An examination of the standard deviations in the estimated coefficient values (center of fig. 11) illustrates that there is a slight increase in the standard deviations when the second-order rather than the first-order polynomial is used. However, there is a large increase in the standard deviations when a third-order polynomial is used.

For this example, the second-order polynomial form was chosen because it provides a satisfactory fit to both the wind-tunnel and flight-test data and it provides estimated coefficient values with a relatively small standard deviation. One may be tempted to include higher order polynomial terms in identification but, as this example illustrates, the use of higher order nonlinear terms (e.g., third order or above for this example) can produce unsatisfactory large standard deviations in the estimated coefficient values.

For this example (fig. 11), the selection of a good form of the mathematical model is fairly obvious. However, for some of the other aerodynamic terms, the best form to use for the model is more difficult to determine. Figure 12 illustrates the effects of using quadratic (α^2) terms in each of the aerodynamic equations.[2]

The effects of using quadratic terms on the standard deviations of the estimated coefficient values are illustrated on the left side of figure 12. For each case, use of the quadratic terms results in higher standard deviations in the estimated coefficient values. Some terms are affected much more than others. With the Lear-Jet, for instance, the $C_{X_{\alpha^2}}$ term can be estimated with a standard deviation of about 10 percent of its mean value whereas the $C_{m_{\alpha^2}}$ has a 930 percent standard deviation about its mean value. (The mean value for $C_{m_{\alpha^2}}$ is quite small.)

The effects of using quadratic terms on the rms difference between the estimated and measured time histories are illustrated on the right side of figure 12. For each case, use of the quadratic terms provides a better fit to flight data. For the Lear-Jet, only one of the residuals, σ_{a_X}, is significantly reduced by using a quadratic term. For the AWJSRA, none of the residuals show significant reductions with use of the quadratic terms. (However, the high noise level on the measured a_X tends to mask out the improvement due to the addition of nonlinear terms.)

---

[2] The results presented in figures 11 and 12 were obtained by the regression method. Quasilinearization will produce higher standard deviations, particularly with the less important nonlinear terms.

In choosing the terms to be used in the aerodynamic equations, there is a tradeoff in lowering the rms of the residuals or lowering the standard deviation of the estimated coefficient values. For the flight-test data considered in this paper, the aerodynamic equations (eqs. (1)-(6)) are believed to represent a reasonable compromise between these two factors. For other flight conditions with these aircraft (i.e., higher angles of attack, different power-lift levels, etc.) and for other aircraft configurations, the use of additional nonlinear terms in the aerodynamic equations must be considered.

CONCLUDING REMARKS

This paper has reviewed some recent flight experience at Ames for the identification of longitudinal aerodynamic coefficients. Results were presented for the Lear-Jet and the Augmentor Wing Jet STOL Research Aircraft. Comparisons were made between results obtained by regression and quasilinearization identification techniques. Also, the coefficients estimated by these techniques were compared with values obtained from wind-tunnel experiments.

The results show that both identification methods provide nearly identical results for the rotational mode (C_m), but the regression method provides better results for the translational modes (C_X and C_Z). The regression method provides less standard deviations in the estimated coefficient values and provides better agreement with the wind-tunnel values.

A majority of the coefficients estimated by the regression method are within about ± 10 percent of the predicted values based on wind-tunnel tests. The exceptions are C_{z_δ}, C_{m_α}, and C_{m_δ} for the Lear-Jet with C_{z_δ} and $C_{x_{\alpha^2}}$ for the AWJSRA.

This paper illustrates that including nonlinear terms in the aerodynamic equations effect both the standard deviation of the estimated coefficient values and the rms difference between the estimated and measured time histories. The choice of the number and type of terms to be used in the aerodynamic equations represents a compromise between these two factors.

ACKNOWLEDGMENT

The author thanks P. D. Talbot of the US ARMY AMRDL/Ames and C. T. Jackson, G. W. Stinnett, and R. F. Vomaske of Ames for their aid in obtaining the flight-test data used in this paper.

REFERENCES

1. Iliff, K. W.; and Taylor, L. W.: Determination of Stability Derivatives From Flight Data Using a Newton-Raphson Minimization Technique. NASA TN D-6579, 1972.

2. Denery, D. G.: Identification of System Parameters From Input-Output Data With Applications to Air Vehicles. NASA TN D-6468, 1971.

3. Chen, R. T. N.; Eulrich, B. J.; and Lebacgz, V. J.: Development of Advanced Techniques for the Identification of V/STOL Aircraft Stability and Control Parameters. CAL Rept. BM-2820-F-1, Aug. 1971.

4. Mehra, R. K.; Stepner, D. E.; and Tyler, J. S.: A Generalized Method for the Identification of Aircraft Stability and Control Derivatives From Flight Test Data. Proceedings of 1972 Joint Automatic Control Conference, Aug. 1972, pp. 525-534.

5. Wingrove, Rodney C.: Applications of a Technique for Estimating Aircraft States From Recorded Flight Test Data. AIAA Paper 72-965, 1972.

6. Grove, Randall D.; Bowles, Roland L.; and Mayhew, Stanley C.: A Procedure for Estimating Stability and Control Parameters From Flight Test Data by Using Maximum Likelihood Methods Employing a Real-Time Digital System. NASA TN D-6735, 1972.

7. Gerlach, O. H.: The Determination of Stability Derivative and Performance Characteristics From Dynamic Maneuvers. AGARD-CP-85, May 1971.

8. Gerlach, O. H.: The Determination of Performance and Stability Parameters from Non-Steady Flight Test Maneuvers. SAE Paper 700236, March 1970.

9. Jonkers, H. L.; Mulder, J. A.; and van Woerkom, K.: Measurements in Nonsteady Flight: Instrumentation and Analysis. Proceedings of 7th International Aerospace Instrumentation Symposium, March 1972, pp. 21.1-21.10.

10. Sorensen, J. A.; and Tyler, J. S., Jr.: Evaluation of Flight Instrumentation for the Identification of Stability and Control Derivative. AIAA Paper 72-963, Sept. 1972.

11. Aubrun, Jean-Noël: Nonlinear Systems Identification in the Presence of Nonuniqueness. NASA TN D-6467, 1971.

12. Neal, Ronald D.: Correlation of Small-Scale and Full-Scale Wind Tunnel Data With Flight Test Data on the Lear Jet Model 23. SAE Paper 700237, March 1970.

13. Soderman, Paul T.; and Aiken, Thomas N.: Full-Scale Wind Tunnel Tests of a Small Unpowered Jet Aircraft With a T-Tail. NASA TN D-6573, 1971.

14. Koenig, D. G.; Corsiglia, V. R.; and Morelli, J. P.: Aerodynamic Characteristics of a Large-Scale Model With an Unswept Wing and Augmented Jet Flap. NASA TN D-4610, 1968.

15. Cook, A. M.; and Aiken, T. N.: Low-Speed Aerodynamic Characteristics of a Large-Scale STOL Transport Model With an Augmented Jet Flap. NASA TM X-62,017, 1971.

16. Quigley, Hervey C.; and Nark, Theodore C., Jr.: A Progress Report on the Development of an Augmentor Wing Jet STOL Research Aircraft. SAE Paper 710757, Sept. 1971.

17. Spitzer, R. E.: Predicted Flight Characteristics of the Augmentor Wing Jet STOL Research Aircraft. NASA CR-114463, 1972.

18. Cleveland, William B.; Vomaske, Richard F.; and Sinclair, S. R. M.: Augmentor Wing Jet STOL Research Aircraft Digital Simulation Model. NASA TM X-62,149, 1972.

19. Suit, W. T.: Aerodynamic Parameters of the Navion Airplane Extracted From Flight Data. NASA TN D-6643, 1972.

20. Steinmetz, George G.; Parrish, Russell V.; and Bowles, Roland L.: Longitudinal Stability and Control Derivatives of a Jet Fighter Airplane Extracted From Flight Test Data by Utilizing Maximum Likelihood Estimation. NASA TN D-6532, 1972.

21. Williams, James L.: Extraction of Longitudinal Aerodynamic Coefficients From Forward-Flight Conditions of a Tilt Wing V/STOL Airplane. NASA TN D-7114, 1972.

22. Black, Ernest L.; and Booth, George C.: Correlation of Aerodynamic Stability and Control Derivatives Obtained From Flight Tests and Wind Tunnel Tests on the XC-142A Airplane. AFFDL-TR-68-167.

TABLE I.– RMS DIFFERENCE BETWEEN ESTIMATED AND MEASURED TIME HISTORIES; LONG RECORDS

		Lear-Jet		AWJSRA	
		Regression	Quasilinearization	Regression	Quasilinearization
σ_{a_z}	, g units	0.00885	0.0244	0.0127	0.0176
σ_{a_x}	, g units	.00105	.00260	.00605	.00788
σ_{a_m}	, deg/sec^2	1.34	1.12	.708	.726
σ_u,	, m/sec	- - -	.829	- - -	.552
σ_w	, m/sec	- - -	.263	- - -	.230
σ_q	, deg/sec	- - -	.276	- - -	.295
σ_θ	, deg	- - -	.411	- - -	.630

TABLE II.– COMPARISON OF ESTIMATED COEFFICIENTS AND WIND–TUNNEL VALUES FOR THE LEAR–JET; FLAPS AND LANDING GEAR UP, LOW–SUBSONIC MACH NUMBER, $1° \lesssim \alpha \lesssim 7°$.

| | Regression | | | Quasilinearization | | | Wind Tunnel (Ref. 13) |
| | Long Record | Short Records | | Long Record | Short Records | | |
		Mean	Standard Deviation		Mean	Standard Deviation	
C_{z_0}	-0.111	-0.089	0.081	-0.164	-0.077	0.134	-0.117
C_{z_α}	-5.118	-5.131	.038	-5.216	-5.271	.060	-5.140
C_{z_δ}	-.342	-.373	.117	-.701	-.481	.240	-.470
C_{z_q}	-21.781	-22.700	3.231	-24.363	-20.988	8.723	(-7)*
C_{x_0}	-.027	-.028	.005	-.030	-.030	.016	-.027
C_{x_α}	.153	.142	.041	.173	.180	.046	.161
$C_{x_{\alpha^2}}$	2.879	2.973	.294	2.858	2.776	.325	2.558
C_{x_δ}	.057	.062	.016	.085	.097	.033	.054
C_{x_q}	1.214	1.363	.634	1.117	1.861	1.021	(.7)*
C_{m_0}	.066	.074	.028	.069	.071	.035	
C_{m_α}	-.810	-.817	.030	-.816	-.817	.030	-1.12**
C_{m_δ}	-1.036	-1.084	.062	-1.052	-1.086	.048	-1.34**
C_{m_q}	-16.460	-17.816	1.835	-17.638	-18.444	1.320	(-18)*

*Predicted tail contribution only, $q + \dot{\alpha}$.

**Adjusted for cg location and $C_{m_{\dot{\alpha}}}$.

TABLE III.– COMPARISON OF ESTIMATED COEFFICIENTS AND WIND–TUNNEL VALUES FOR THE AWJSRA; FLAPS = 67°; LANDING GEAR DOWN, LOW–SUBSONIC MACH NUMBER, $0.3 \lesssim C_j \lesssim 0.4$, $-2° \lesssim \alpha \lesssim 10°$.

| | Regression | | | Quasilinearization | | | Wind Tunnel (Ref. 18) |
	Long Record	Short Records Mean	Short Records Standard Deviation	Long Record	Short Records Mean	Short Records Standard Deviation	
C_{z_0}	-1.197	-1.193	0.029	-1.465	-1.403	0.094	-1.162
C_{z_α}	-5.046	-5.071	.122	-6.195	-5.767	.137	-4.794
C_{z_δ}	-.692	-.683	.006	-.165	-.338	.202	-.550
C_{z_q}	-17.282	-17.016	.792	10.006	-3.107	2.941	(-10)*
C_{x_0}	-.327	-.327	.001	-.329	-.327	.013	-.294
C_{x_α}	1.318	1.323	.085	1.380	1.377	.273	1.329
$C_{x_{\alpha^2}}$	2.376	2.225	.603	2.273	2.395	1.144	3.560
C_{x_δ}	.163	.184	.040	.163	.111	.061	
C_{x_q}	1.232	1.573	.571	3.481	1.348	1.982	(1)*
C_{m_0}	.057	.056	.003	.041	.037	.005	
C_{m_α}	-.560	-.553	.038	-.501	-.450	.067	-.52**
C_{m_δ}	-2.036	-2.050	.006	-1.902	-1.969	.046	-2.05**
C_{m_q}	-29.261	-29.718	1.231	-31.593	-32.168	.722	(-35)*
$C_{z_{C_j}}$	-3.001	-3.003	.106	-1.958	-2.156	.328	-3.45

*Predicted tail contribution only, $q + \dot\alpha$.

**Adjusted for cg location and $C_{m_{\dot\alpha}}$.

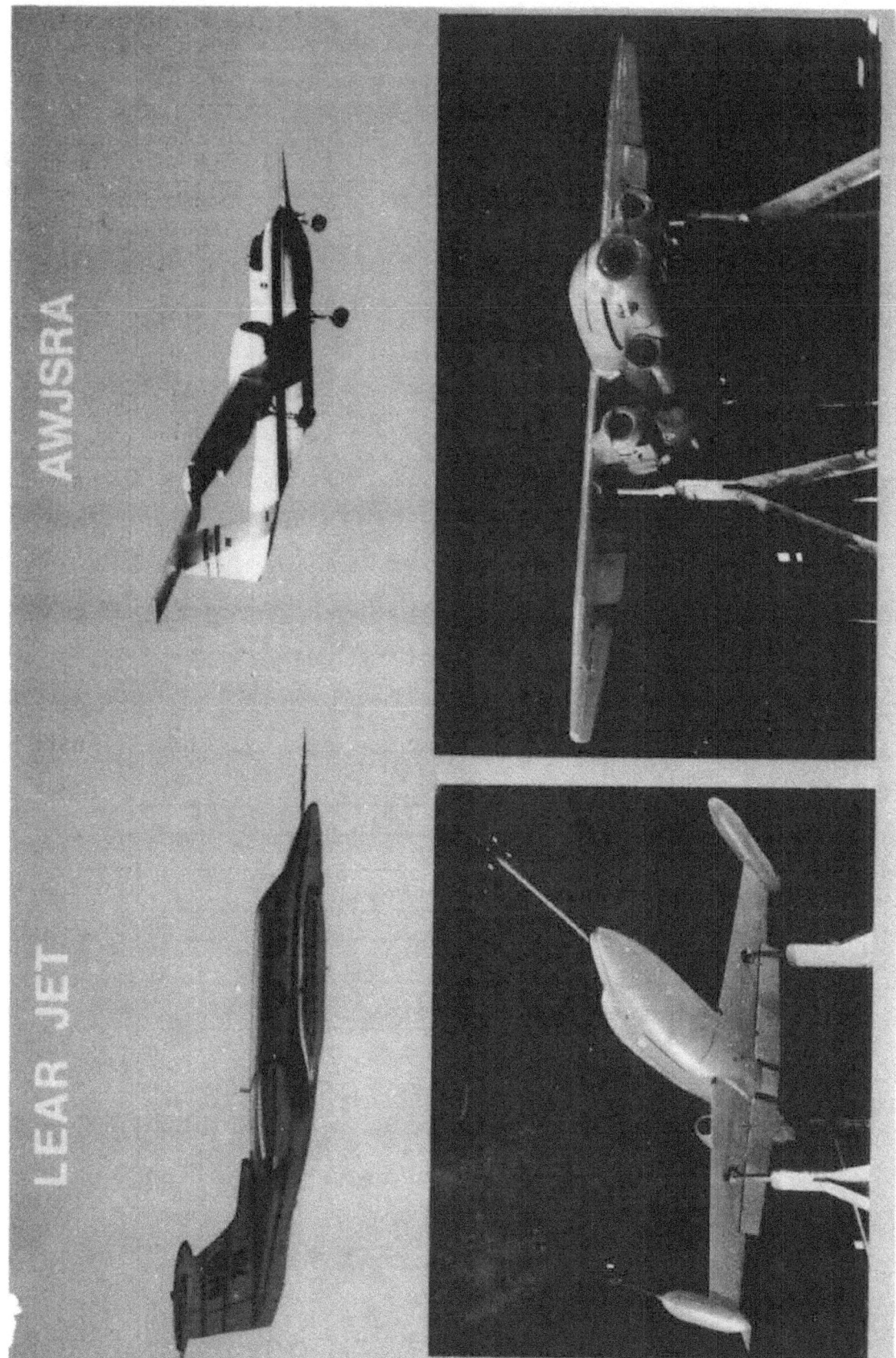

Figure 1.— Flight and wind-tunnel configurations.

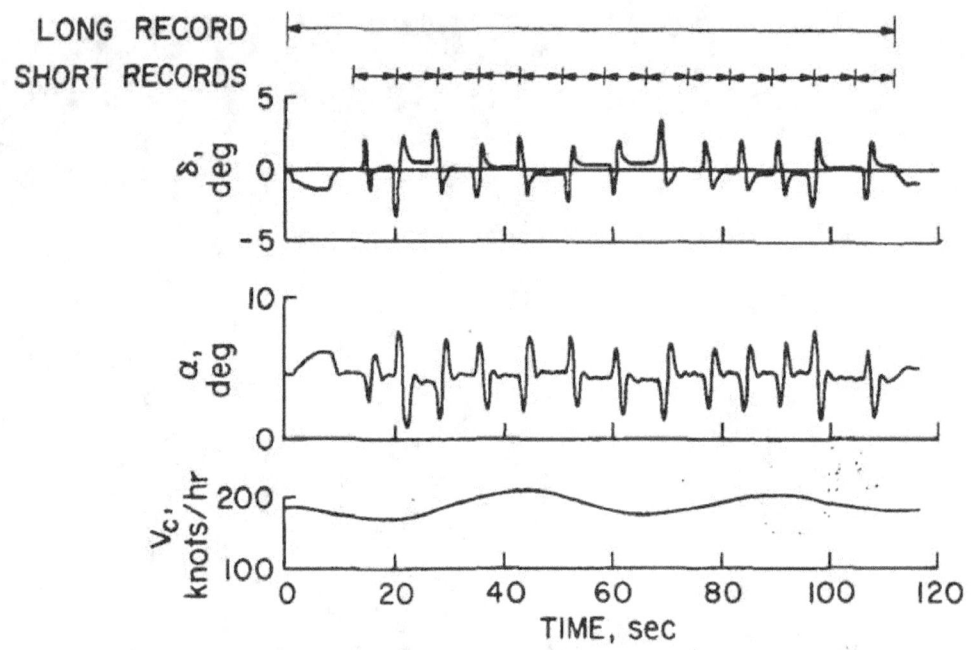

Figure 2.— Flight maneuver for the Lear-Jet.

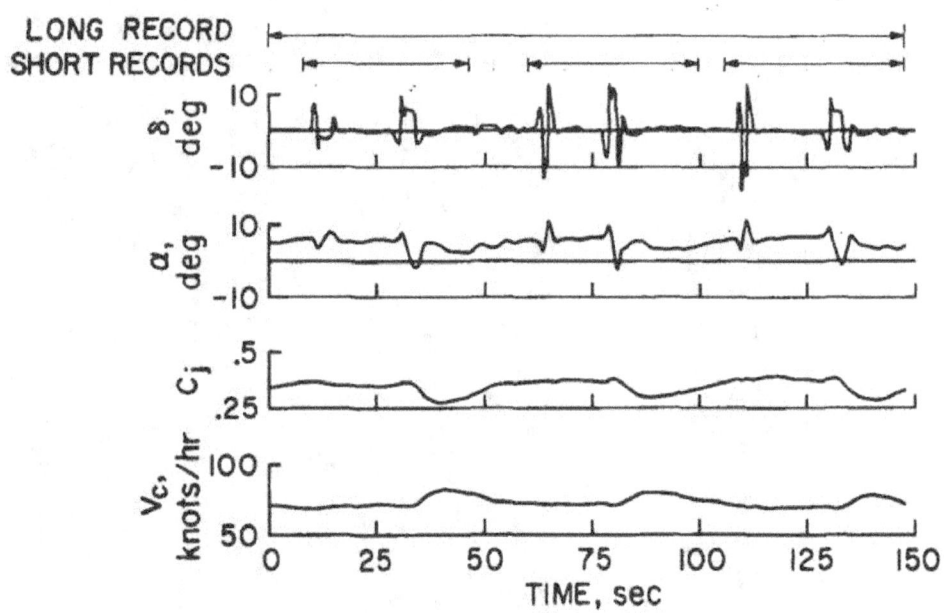

Figure 3.— Flight maneuver for the AWJSRA.

140

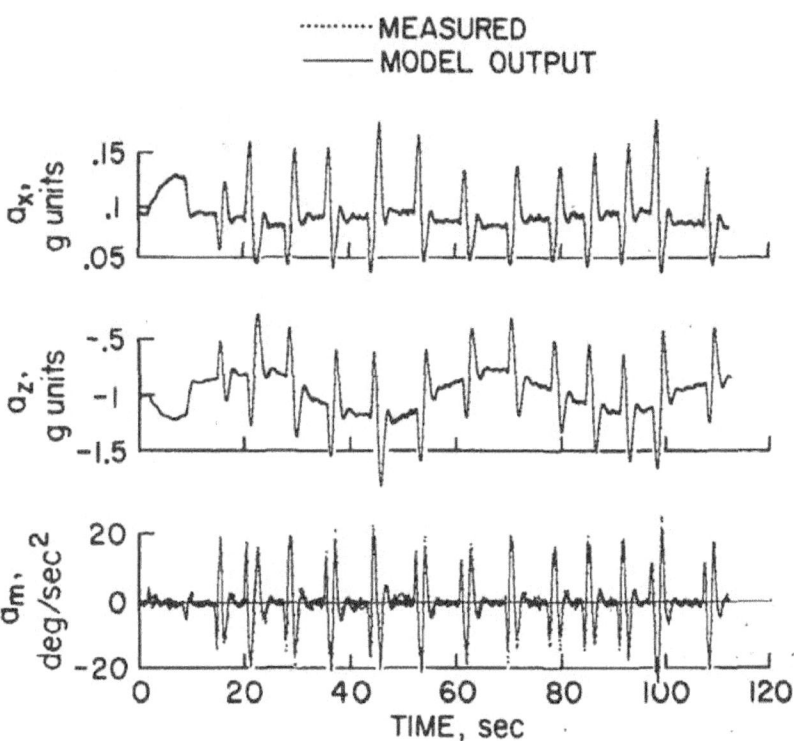

Figure 4.– Estimated model outputs compared with direct measurements for the Lear-Jet; regression method.

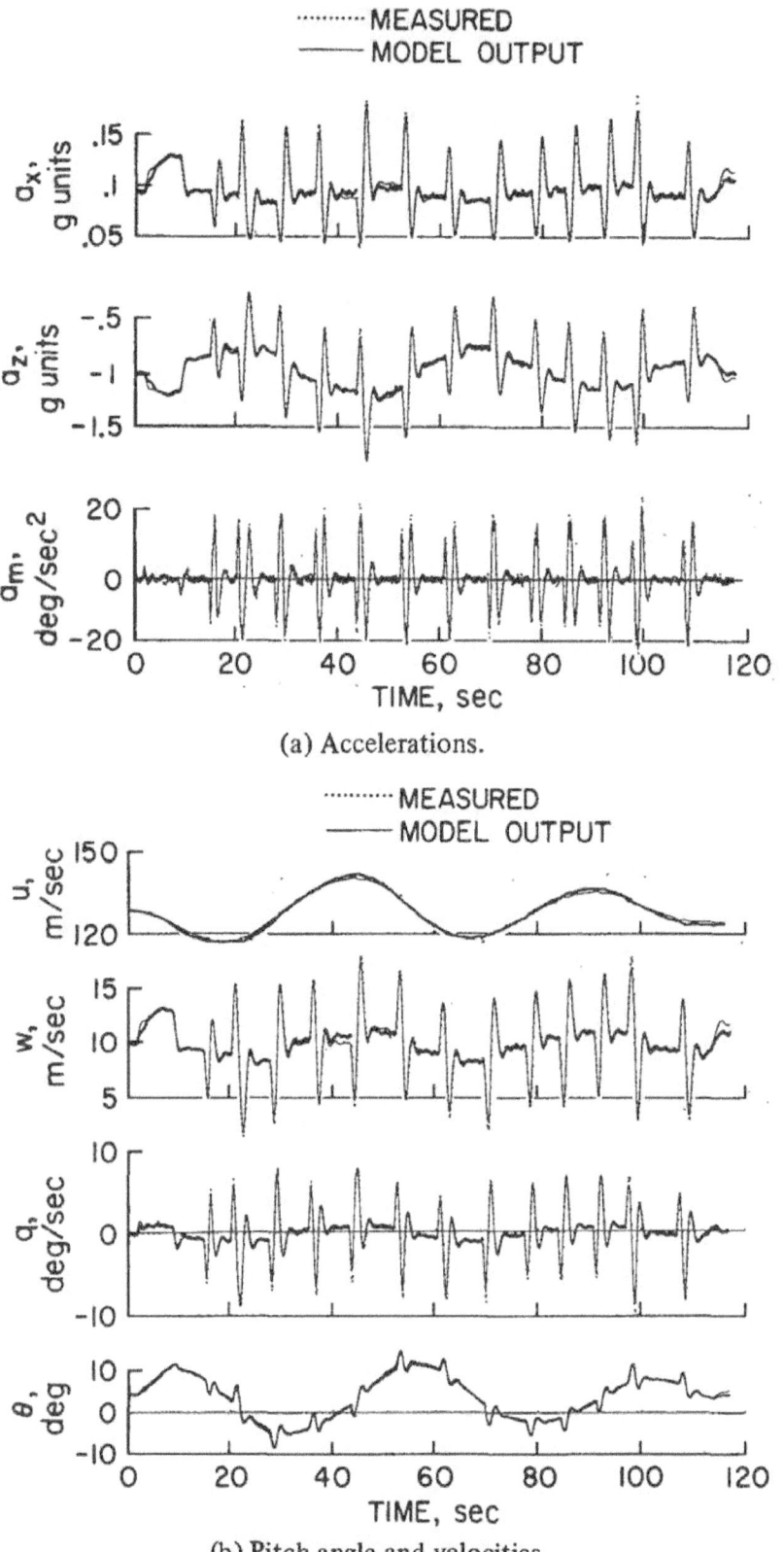

(a) Accelerations.

(b) Pitch angle and velocities.

Figure 5.– Estimated model outputs compared with direct measurements for the Lear-Jet; quasilinearization method.

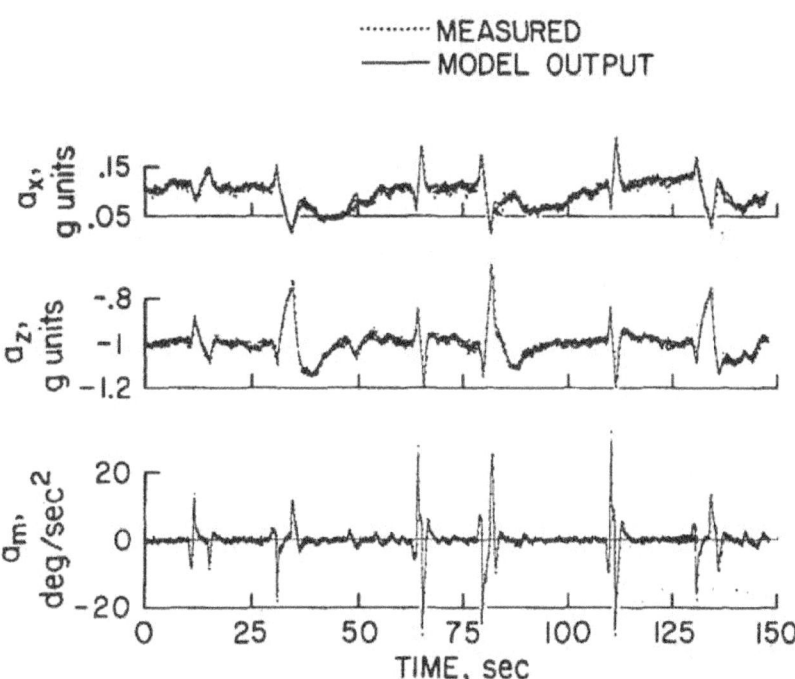

Figure 6.— Estimated model outputs compared with direct measurements for the AWJSRA; regression method.

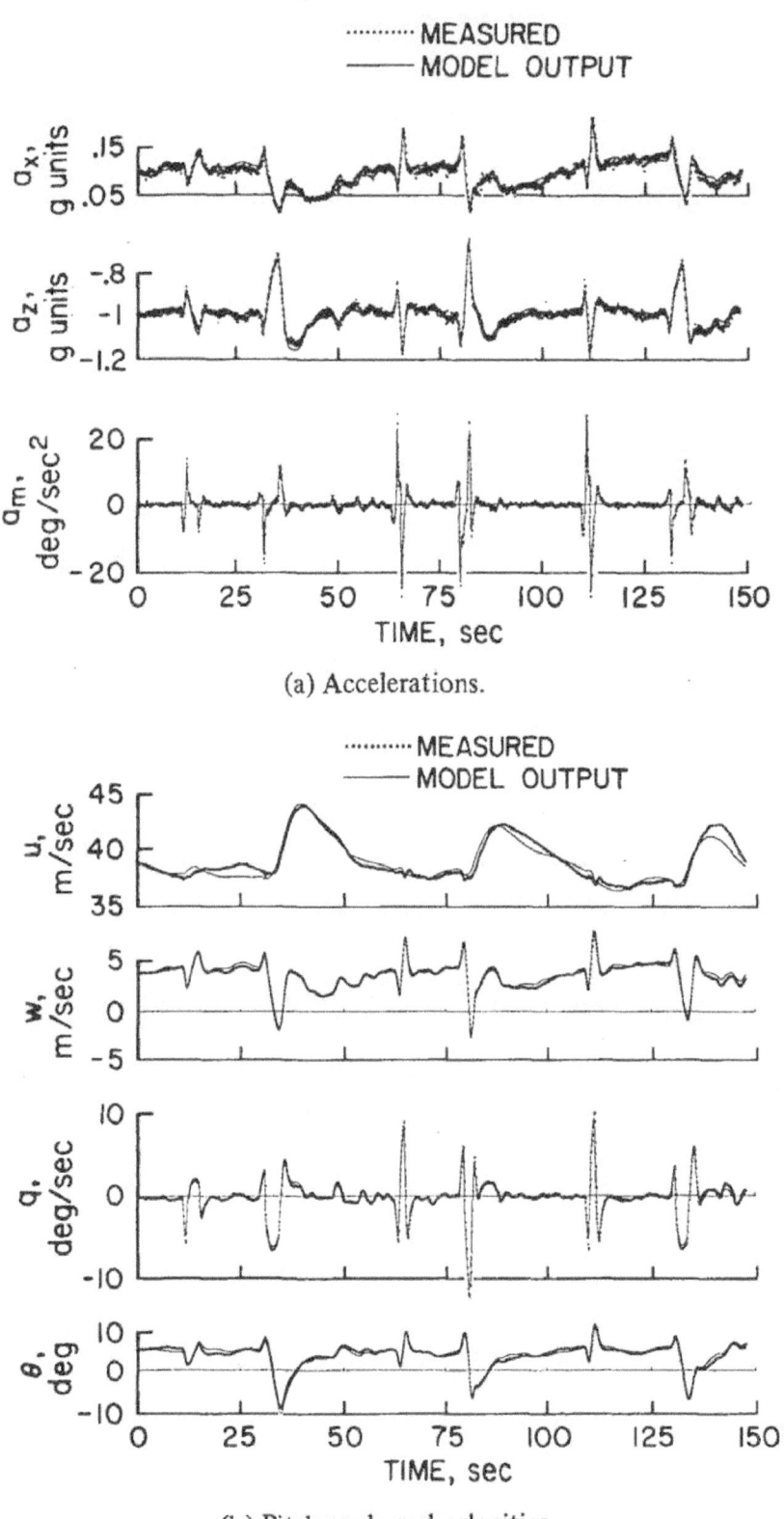

(a) Accelerations.

(b) Pitch angle and velocities.

Figure 7.— Estimated model outputs compared with direct measurements for the AWJSRA; quasilinearization method.

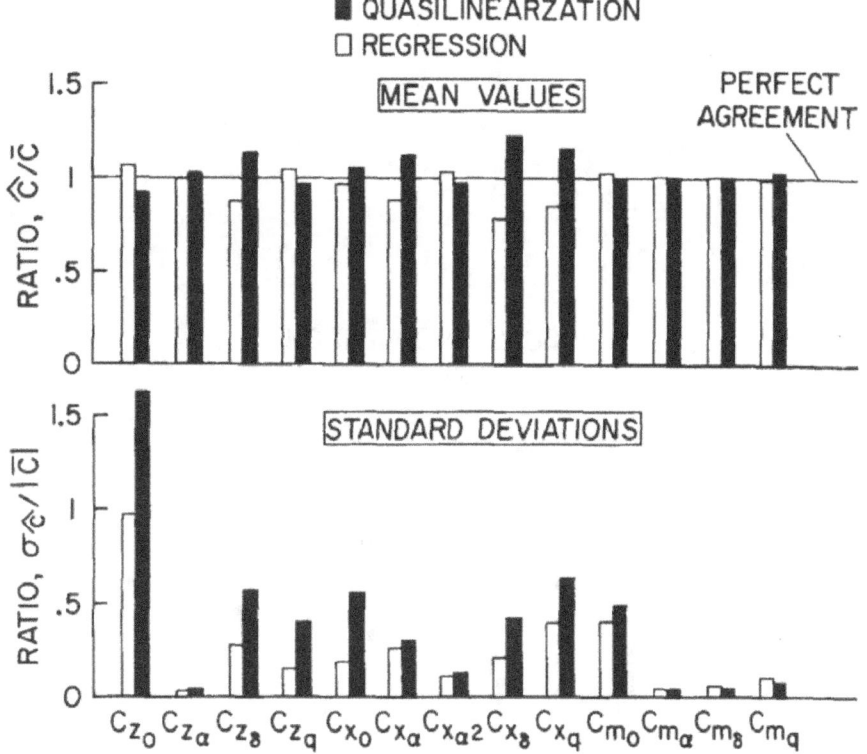

Figure 8.– Comparison of coefficient mean values and standard deviations estimated for the Lear-Jet; 13 short records, 8 sec each.

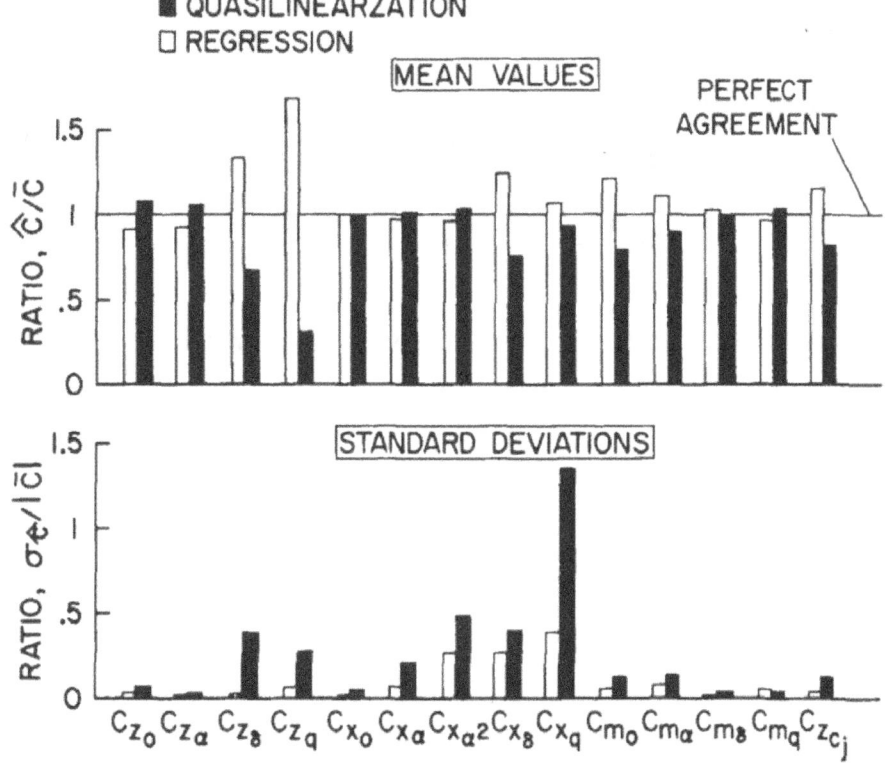

Figure 9.– Comparison of coefficient mean values and standard deviations estimated for the AWJSRA; 3 short records, 40 sec each.

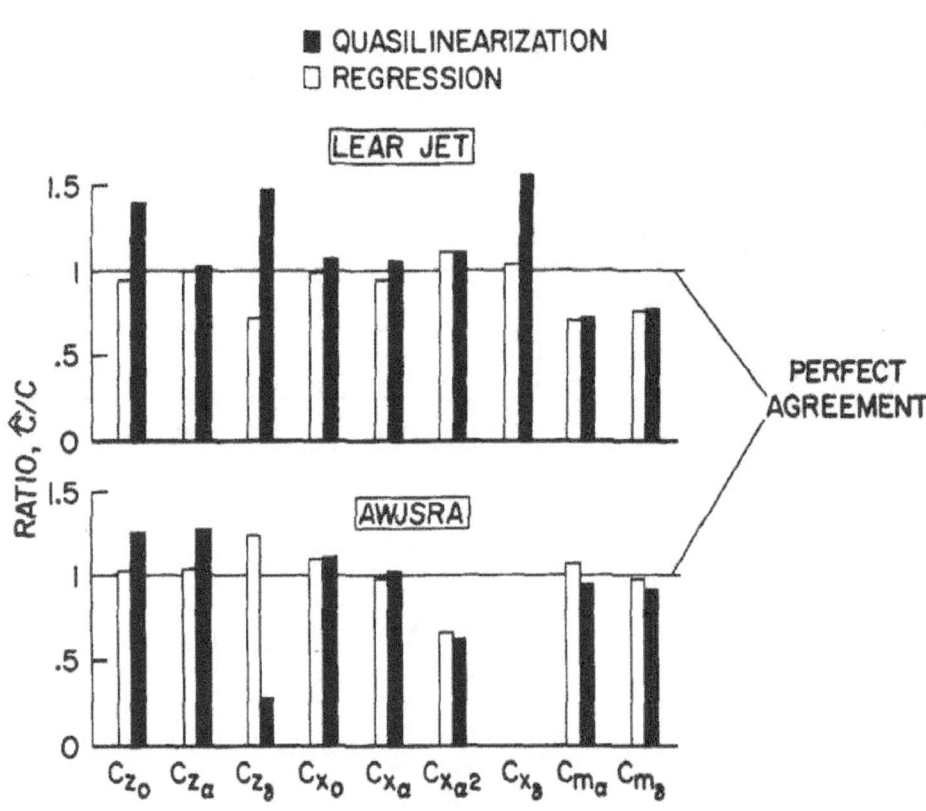

Figure 10.— Estimated coefficients compared with wind-tunnel values; long records.

146

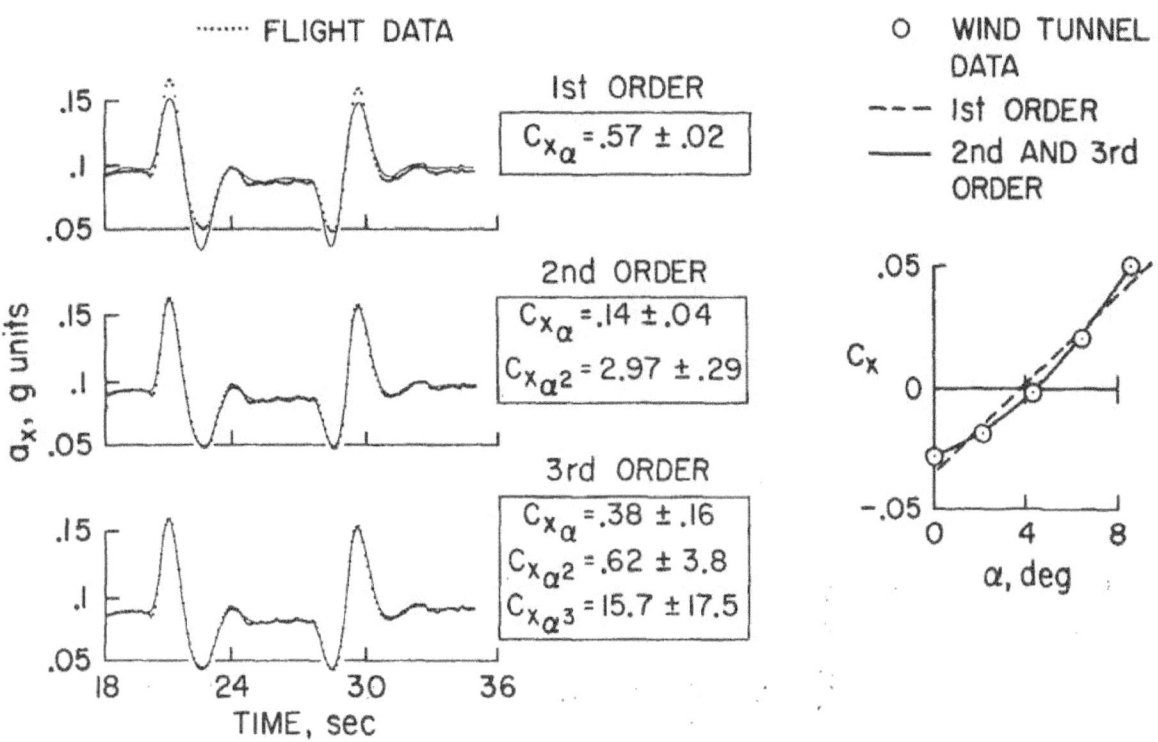

Figure 11.— Effect of nonlinear terms in the modeling of C_X with α.

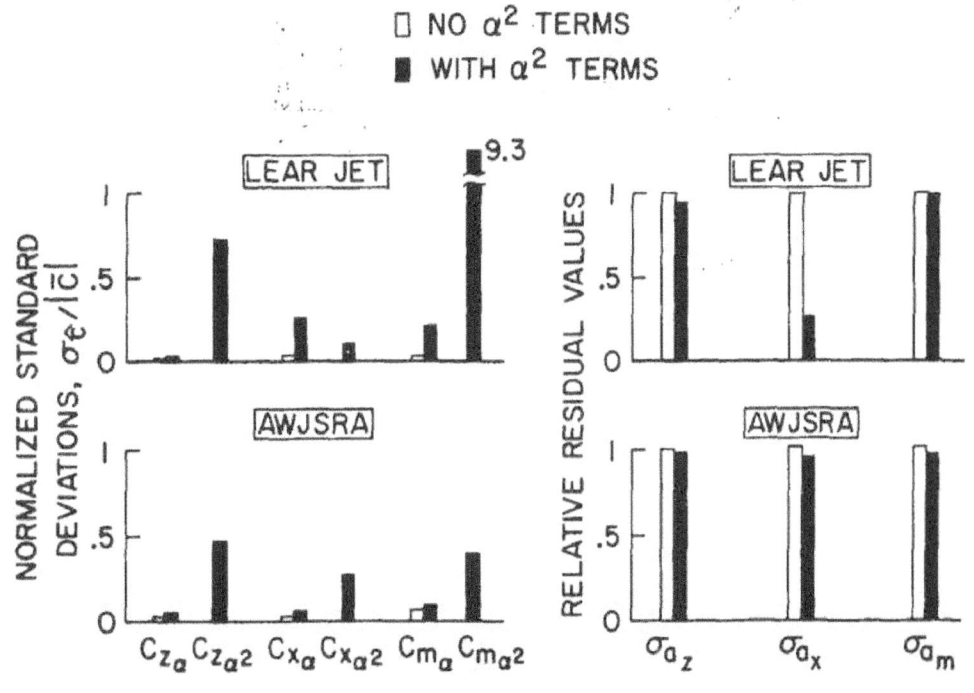

Figure 12.— Effect of quadratic terms in α.

APPLICATION OF A KALMAN FILTER IDENTIFICATION TECHNIQUE
TO FLIGHT DATA FROM THE X-22A VARIABLE STABILITY V/STOL AIRCRAFT[*]

J. Victor Lebacqz[+]

Calspan Corporation
Buffalo, New York 14221

ABSTRACT

A digital identification technique based on Kalman filter theory has been developed for the estimation of V/STOL aircraft stability and control parameters from flight data. The emphasis of this paper is on the application of this technique to flight data from flying qualities experiments using the variable stability X-22A V/STOL aircraft. The estimation algorithm is briefly reviewed, experimental and data acquisition procedures used in the X-22A flight programs are outlined, and specific problem areas such as the determination of noise statistics and selection of pilot inputs to enhance identifiability are discussed. Results are presented for a wide range of simulated dynamic configurations.

SYMBOLS

I_x moment of inertia about body X-axis, ft-lb sec^2

I_y moment of inertia about body Y-axis, ft-lb sec^2

I_z moment of inertia about body Z-axis, ft-lb sec^2

I_{xz} product of inertia in body axes, ft-lb sec^2

K_L $= (I_y - I_z)/I_x$ nondimensional inertia coupling in roll

K_N $= (I_x - I_y)/I_z$ nondimensional inertia coupling in yaw

$L_{()}$ $= 1/I_x \ \partial L/\partial()$ dimensional roll moment derivative, (rad/sec^2)/( )

$L'_{()}$ $= (1 - I_{xz}^2/I_x I_z)^{-1}(L_{()} + \frac{I_{xz}}{I_x} N_{()})$, (rad/sec^2)/( )

---

[*] The development of the identification technique discussed herein was supported by the Naval Air Systems Command under Contract N00019-69-C-0534. The two X-22A flying qualities research programs that are discussed were sponsored by the Naval Air Systems Command, the National Aeronautics and Space Administration (LRC), the Federal Aviation Agency, and the Air Force Flight Dynamics Laboratory under Contracts N00019-71-C-0044 and N00019-72-C-0417.

[+] Associate Aeronautical Engineer

$M_{()}$ $= 1/I_y \ \partial M/\partial( )$ dimensional pitch moment derivative, $(\text{rad/sec}^2)/( )$

$N_{()}$ $= 1/I_z \ \partial N/\partial( )$ dimensional yaw moment derivative, $(\text{rad/sec}^2)/( )$

$N'_{()}$ $= (1 - I'^2_{xz}/I_x I_z)^{-1}(N_{()} + \frac{I_{xz}}{I_z} L_{()}), \ (\text{rad/sec}^2)/( )$

n_y body Y-axis acceleration, 57.3 ft/sec^2

p body axis roll rate, deg/sec

q body axis pitch rate, deg/sec

r body axis yaw rate, deg/sec

t time, seconds

u body X-axis velocity, ft/sec

V true velocity, ft/sec

w body Z-axis velocity, ft/sec

$X_{()}$ $= 1/M \ \partial X/\partial( )$ dimensional X-force derivative, $(\text{ft/sec}^2)/( )$

$Y_{()}$ $= 1/M \ \partial Y/\partial( )$ dimensional Y-force derivative, $(\text{ft/sec}^2)/( )$

$Z_{()}$ $= 1/M \ \partial Y/\partial( )$ dimensional Z-force derivative, $(\text{ft/sec}^2)/( )$

α angle of attack, degrees

β angle of sideslip, degrees

γ glide slope angle, degrees

δ_{ES} longitudinal stick position, positive aft, inches

δ_{AS} lateral stick position, positive right, inches

δ_{RP} rudder pedal position, positive right, inches

ζ_d damping ratio of Dutch roll characteristic response

ζ_{ST} damping ratio of short term longitudinal response

θ pitch attitude, degrees

$\sigma_{()}$ variance of ( ) in units of ( )

τ_R roll mode time constant, seconds

ϕ roll attitude, degrees

$|\phi/\beta|_d$ magnitude of roll-to-sideslip ratio in Dutch roll component

ω_d Dutch roll undamped natural frequency, rad/sec

ω_{ST} longitudinal short term undamped natural frequency, rad/sec

$()_0$ reference or initial condition of ( )

$(\dot{\ })$ time rate of change of ( ), ( )/sec

INTRODUCTION

A prerequisite in the use of response-feedback variable stability aircraft to obtain flying qualities data is an accurate method for estimating stability and control parameters from flight data. In general, experimental flying qualities investigations seek to correlate dynamic characteristics of an aircraft in the performance of a prescribed task to pilot opinion of the suitability of the characteristics for that task. Variable stability aircraft incorporate electronic implementation of control laws that vary the response characteristics of the aircraft in a prescribed manner. The most prevalent mechanization of this capability, and the one used in the X-22A V/STOL variable stability aircraft, is the response-feedback system. With this technique response variables of the aircraft are sensed directly and used to command control deflections proportionally, thereby changing the closed-loop aircraft characteristics; by varying the matrix of feedback gains, a wide variety of aircraft characteristics can be simulated for piloted evaluations. Unlike a ground simulator or model-following variable stability aircraft, however, the resulting dynamic characteristics are not accurately known a priori; it is therefore mandatory to have an accurate and efficient means of identifying the characteristics obtained from flight records.

Since the inception of the variable stability aircraft in the early 1950's by the NACA and Cornell Aeronautical Laboratory (now Calspan Corporation), this problem of identification of the simulated dynamics, both for calibration purposes and for the correlation of pilot opinion ratings with the achieved dynamic configurations, has been of extreme theoretical and practical concern. Early methods included various analytic treatments based on hand measurement of recorded responses to prescribed inputs and matching the responses with the outputs of an analog computer (References 1 and 2). With the advent of the digital computer, it became feasible to handle large amounts of data that might require numerical analyses. This capability led first to equation-error techniques (Reference 3) and then to response-error methods (References 4, 5, and 6), which were applied with various degrees of success to the aircraft identification problem. As is by now well known, however, accurate identification of aircraft parameters requires advanced methods that can treat both equation errors (process noise) and response errors (measurement noise). Methods which have this capability include various techniques predicated upon maximizing a likelihood function (References 7 and 8) and techniques which extend Kalman filter theory to nonlinear situations (Reference 9).

This paper discusses the application of the identification technique developed in Reference 9 to flight data from the X-22A variable stability V/STOL aircraft (Figure 1). The X-22A aircraft is a unique research tool which is capable of reproducing a wide range of vehicle dynamic characteristics at many fixed-operating STOL flight conditions as well as through a complete V/STOL transition (120 kts ⇒ 0 kts). To date, the aircraft has been used in two STOL flying qualities programs, one investigating longitudinal short-term dynamic characteristics in the landing approach (Reference 10), and the other studying lateral-directional dynamic characteristics and roll control power requirements for landing approach (Reference 11). The large variety of dynamic situations that have been simulated, coupled with the relative inaccuracy of aerodynamic data for the basic aircraft and the requirement for accurate yet economically efficient identification, provides an extensive data base for the evaluation of the practical usefulness of an advanced identification technique.

The paper is organized as follows. The next two sections give a brief review of the estimation technique and a summary of the data acquisition and handling procedures used in the X-22A flight programs. The succeeding two sections discuss the selection of the input information required by the identification algorithms and the importance of the aircraft control inputs to enhance identifiability. Representative results from both the longitudinal and lateral-directional experiments are then presented, followed by some concluding remarks.

IDENTIFICATION TECHNIQUE

The identification technique used on the X-22A flight data has been described in detail elsewhere (References 9, 12), and hence will be discussed only briefly here. The central idea is to obtain a suboptimal minimum variance estimate of the parameters (and states) from the measured data for generally nonlinear systems by extension of Kalman filter theory. To this end, we consider an augmented state consisting of the aircraft states and the parameters to be identified; the resulting state equation is, of course, nonlinear even if the unaugmented state equation is linear, and therefore some form of approximation to the optimal nonlinear filter is required. A common approximation is to use an extended Kalman filter to estimate the states; this technique, however, has been shown to yield biased estimates, the cause for which may be viewed as inaccuracies in the reference trajectory about which the linearization takes place. To improve the reference trajectory, therefore, a locally iterated filter-smoother is used (References 9, 13, 14), which is possible because of the recursive nature of the technique. This procedure updates the reference trajectory between every two time points through alternate one-stage extended Kalman filtering and one-stage smoothing, the iterations continuing until there is negligible change in the reference trajectory between successive iterations. It can be shown formally that this procedure reduces the bias caused by state and measurement nonlinearities (Reference 9).

To further improve the parameter estimates, a suboptimal fixed-point smoothing algorithm is available as an option in the technique. The fixed-point smoother is desirable as a means of improving the initial estimates of the states (i.e., at t_0); since the parameters are assumed to be random variables with constant mean, the improved initial estimate (using all of the data) from the smoother may result in improved estimates of both the parameter values and their variances. The fixed-point smoother uses the final reference trajectory from the filter-smoother for its linearization, and in fact is mechanized to work in conjunction with it in an "on-line" fashion.

152

The salient features of this identification technique may be summarized for reference as follows:

1. The technique seeks minimum variance estimates (i.e., the conditional mean) of general systems described by nonlinear state and measurement equations including both process and measurement noise. The formulation of the algorithms is predicated on this generality; hence for example, the implicit nonlinearity introduced by augmenting the state with the parameters to be identified does not compromise the formulation. Strictly speaking, the technique may be theoretically imperfect for systems which can be described by linear state and measurement equations, since we still approximate the nonlinear augmented system with a linearized procedure. From a practical point of view, however, the local iteration procedure appears to achieve a sufficiently good reference trajectory to obviate this imperfection in applications.

2. The technique is recursive in nature -- both the iterated filter-smoother and the fixed-point smoother operate on the data points in a sequential fashion. It is theoretically possible to apply the algorithms on line in real time, although this capability is not included at present in the X-22A data acquisition equipment. This recursive nature of the technique tends to eliminate the need for digital or optimal processing of flight data prior to identification to obtain accurate initial parameter estimates, a need which is characteristic of methods which use the entire data record initially and require approximate gradient techniques to perform the required optimization. This insensitivity to the initial estimates will be demonstrated by example in a later section of this paper.

3. The technique as currently employed does not estimate the measurement and process noise covariances. Methods which directly maximize the likelihood function do perform this estimation well for linear systems (Reference 15), and the lack of this capability is somewhat of a drawback of the technique. For the application of the technique to X-22A flying qualities flight data, however, the deficiency is minor, as (1) the model structure is generally well defined and calibration flight records are obtained in relatively smooth air, both of which decrease the process noise in the system, and (2) the quality of the data acquisition procedures and measuring sensors is high enough to obtain valid a priori measurement noise statistics from the flight records. The determination of the process and measurement noise covariances for the X-22A is discussed in more detail in a later section.

DATA ACQUISITION EQUIPMENT AND PROCEDURES

The data acquisition systems and procedures used for X-22A flight programs are described in References 10 and 16, and only those aspects which bear on identification of the flight records are repeated here.

A schematic of the digital data acquisition system is shown in Figure 2. Sensors in the aircraft measure all pertinent quantities, such as rigid body responses, control deflections, and variable stability system command signals. This information is sampled 200 times per second and telemetered via an L-band pulse-code-modulated telemetry link to a mobile ground station, where it is decoded and recorded on line on the "bit-stream" recorder. For post flight data analyses, the bit-stream information is processed through the digital mini-computer to produce an IBM 370/65 compatible digital tape.

The data on this digital tape are then processed and edited to be compatible with the identification computer programs. In the first X-22A flight program, the data were initially digitally filtered by a third-order Butterworth filter in order to reduce the sampling rate to the 1/0.08 samples/second of the identification technique without introducing aliasing errors. It has been ascertained experimentally, however, that this filtering is not necessary; hence, on the second program, no digital or other filtering of the telemetered data was performed.

The digital data are also transformed from measured variables to equations-of-motion variables at the center of gravity. These transformations are not strictly required for the identification algorithms, but result in more efficient (i.e., less costly) identification. The primary transformations required are on the aerodynamic motions, since angle of attack and sideslip angle are measured on a boom in front of the aircraft; in addition, the longitudinal equations of motion are written in terms of body-axis vertical velocity, to which angle of attack is converted.

CHOICE OF REQUIRED INPUT INFORMATION

In common with any technique based on Kalman filter theory, the following input information is required for the algorithm:

1. Initial estimates of the parameters.

2. Variances of the initial estimates.

3. Reference conditions of the states.

4. Measurement noise variances.

5. Process noise variances.

The initial parameter estimates are obtained from a conventional least-squares equation error method, which also produces estimates of the parameter variances. It has been observed experimentally that the variance estimates obtained by this method do not correctly represent either the absolute or the relative accuracy of the initial parameter estimates. Two alternatives may be

154

taken. The first is to multiply the initially estimated variances equally by an arbitrary factor. The second, which is more appealing theoretically, is to use an independent technique to calculate them. Since it is clear that the initial variances should reflect the identifiability of the parameters to some extent, which is in turn dependent on the control input used, one means of obtaining the variances is to obtain the Cramer-Rao lower bound on the covariance of the parameter estimates for the given data. The variances obtained by this method would be expected to be more correct in their ratios to each other, and again can be multiplied by an arbitrary factor and used as the initial variances. For the results presented in this paper, the method of multiplying the equation-error computed variances by a constant factor was used. This choice was dictated by operational considerations: the need to process large quantities of data in a rapid fashion during calibration flights results in eliminating if possible intermediate computing steps, such as the separate calculation of the Cramer-Rao lower bound. In general, experimental experience has demonstrated that, for the type and quality of the identification records for the X-22A, the more direct method of uniformly increasing the equation-error method variances appears to be adequate.

The measurement noise statistics are obtained by visual examination of the flight records. Generally, the "hash" on the records is assumed to equal the variance of the measurement noise, which provides a conservative value. This estimate is then checked qualitatively by comparing plots of the residual sequences of the filter operation with the assumed noise statistics, and readjusting the statistics if required. The X-22A data acquisition system provides data with excellent signal-to-noise ratios in general, and therefore this method of estimating the measurement noise variances is sufficiently precise. Again, in the interests of rapid and efficient identification procedures, the measurement noise statistics are kept the same for all data records if possible. For the bulk of the results presented in this paper, these statistics are:

Longitudinal*

σ_u = 1.0 (ft/sec)
σ_w = 0.25 (ft/sec)
σ_θ = 0.15 (deg)
σ_q = 0.1 (deg/sec)

Lateral-Directional

σ_β = 0.2 (deg)
σ_p = 0.1 (deg/sec)
σ_r = 0.1 (deg/sec)
σ_ϕ = 0.1 (deg)
σ_{n_y} = 10.0 (57.3 x ft/sec^2)
$\sigma_{\dot{p}}$ = 3.0 (deg/sec^2)
$\sigma_{\dot{r}}$ = 2.0 (deg/sec^2)

In addition to selecting the measurement noise statistics from visual examination of the data, the reference (or initial) conditions of the states are chosen to be the first datum points ($t \triangleq 0$) on each record tape. Since calibration identification records of the evaluation configurations are usually obtained about trimmed flight, the first point on the data tape is

---

* Acceleration measurements were not used during the longitudinal flying qualities program, as will be discussed in the section on results.

generally a valid reference condition. The fixed point smoother may be used to obtain an estimate of the initial conditions if necessary, but this computation is not generally required for the X-22A data.

The most difficult choice of required input information is that of the process noise statistics. To some degree, the process noise covariance matrix Q is a "fiddle parameter" in the algorithm which may be used to improve its performance for a given data record. On the other hand, the requirement for rapid post-flight identification as nearly automatic as possible leads to a desire to hold these statistics at a fixed value for all flight records. To make this tradeoff, then, it is important to define precisely what the sources of process noise might be. For the X-22A data, there are essentially three sources of process noise:

1. Gust or turbulence inputs.

2. The variable stability system.

3. Modeling errors.

Of these, the gust inputs are of the least significance for the records that are analyzed, because the majority of calibration identification records are obtained in turbulence-free air to facilitate rapid checks on the frequency and damping of prevalent rigid-body modes of motion. The variable stability system is the source of "noise" both as a result of its dynamics not being included in the model and through its operation on noisy measurement signals. The primary source of modeling errors, however, is the fundamental restriction that we seek the best linear model for the aircraft dynamics that will fit the data, as most flying qualities parameters are defined in terms of linear systems.

With regard to the choice of process noise statistics, therefore, the following considerations are relevant. For simulated aircraft that are highly augmented with regard to the X-22A (e.g., higher rigid body frequencies and dampings), the assumption of a linear model becomes increasingly valid, but the process noise added by the variable stability system increases. For simulated aircraft whose rigid body motions are similar to the X-22A (very little augmentation), the effects of the variable stability system are reduced but nonlinearities may start to become important. The magnitude of the process noise in these two cases may be considered approximately the same. The worst case is one in which the X-22A must be highly de-augmented, as linear aerodynamic terms may approach zero, thereby accentuating nonlinearities, and the variable stability system effects again become larger. For this case, it may be necessary to assume more process noise.

For identification of the X-22A data, it is assumed that one set of process noise statistics is acceptable for all configurations save those which involve the de-augmentation of several stability derivatives, and this set is used for the rapid processing of the data. The values of the statistics are selected primarily by iteration on early data sets to achieve adequate performance, and then held constant. For example, most of the lateral-directional results presented in this paper used the following process noise variances:

$\dot{\beta}$ equation : 0.6 (deg/sec)

$\dot{p}$ equation : 0.2 (deg/sec^2)

$\dot{r}$ equation : 0.1 (deg/sec^2)

IDENTIFIABILITY OF DATA

It is well known that the control inputs can significantly affect the identifiability of a data record (References 9 and 17). For a given input, the best identification performance possible, in the sense of minimum mean square estimation error, is given by the Cramer-Rao lower bound: that is, the elements of the Cramer-Rao matrix are the lowest variances on the estimates that can be achieved. It is therefore possible, for example, to design inputs based on a minimization of this lower bound (Reference 17). The Cramer-Rao lower bound is the inverse of the Fisher information matrix, the elements of which are the sensitivity functions, and so maximization of some norm of this matrix may also be used to design inputs (Reference 9), although the two methods are not exactly equivalent. Implementation of such inputs in a flight program, however, is difficult, and hence approximations that provide at least some benefit to identifiability are sought.

In flying qualities experiments, inputs for identification records have historically been simple analytically and chosen to accentuate some particular features of the response. Examples include rudder doublets for the Dutch roll characteristics, and aileron steps for roll mode time constants and ϕ/δ_{AS} transfer function characteristics. It is easy to demonstrate that these inputs do indeed provide large sensitivities for the stability derivatives which have the primary influence on the characteristics of interest, but that other derivatives may not be identifiable with any accuracy at all. The usual procedure that is followed is to obtain several records with different inputs tailored heuristically to certain characteristics and thereby obtain in a composite fashion the total identification; this procedure was used for the longitudinal dynamics presented in this paper with the primary input being a longitudinal stick doublet. This method was justifiable for these data since the angle of attack stability and pitch-rate damping were essentially the only stability derivatives varied in the experiment, and a doublet maximizes the sensitivity of angle of attack and pitch attitude to these derivatives.

For the lateral-directional program, however, a majority of the stability derivatives was varied to achieve the desired dynamic configurations, and therefore a single simple input could not provide sufficient identifiability. In general, for amplitude-constrained inputs (which are necessary to aid the assumption of linearity), it can be shown that "switching" type inputs increase the sensitivity for most parameters, the frequency of switching being dependent on the dynamic characteristics of the system (References 17, 18). It was therefore decided to attempt to have the pilots provide this type of input in both yaw and roll. The advantages of using the pilot, rather than a programmed automatic input, include his capability to maintain the aircraft

157

responses within linear limits and his ability to sense to some degree the characteristic frequencies of the aircraft to provide switching cues. The disadvantage of using pilot inputs is primarily his tendency to act as a feedback controller; in that case, the inputs become linearly related to at least one aircraft output, which is inimical to good identification (Reference 9).

To check on the "goodness" of the inputs, two alternatives are available. First, as we have discussed, the Cramer-Rao lower bound may be calculated for the record using the estimated stability derivatives. In a relative sense between several records, however, it is not necessary to perform this additional calculation. If we assume that the identification technique approaches an efficient estimator (unbiased, minimum variance), then the final variances of the parameters computed by the technique should approach the Cramer-Rao lower bound (Reference 9); therefore, a comparison of the magnitudes of the diagonal terms in the final covariance matrix provides some indication of the identifiability. It is also instructive to normalize this matrix and examine the normalized covariances between the parameters, as high value (e.g., > 0.9) indicates a strong degree of linear dependence. An example of two inputs for the same configuration is given in the next section; in general, the pilot inputs used in the lateral-directional program provide good identifiability.

APPLICATION TO FLIGHT DATA

Longitudinal Dynamics

The first X-22A flying qualities program investigated the effect of short-term longitudinal dynamic characteristics for STOL landing approach (Reference 10). Sixteen combinations of short-term frequency and damping were evaluated during visual and simulated instrument approaches at a representative glide slope angle of $\gamma = -9^{\circ}$. The primary derivative variations made with the variable stability system in this program were angle-of-attack stability (M_w) and pitch rate damping (M_q), as these derivatives have a major influence on the short-term characteristics.

The assumed equations of motion for the identification process are written in body axes, and include nonlinear kinematic and gravitational terms but only linear aerodynamic terms; they are:

$$\dot{u} + wq + g \sin\theta = X_0 + X_u(u - u_0) + X_w(w - w_0) + X_{\delta_{ES}}(\delta_{ES} - \delta_{ES_0})$$

$$\dot{w} - uq - g \cos\theta = Z_0 + Z_u(u - u_0) + Z_w(w - w_0) + Z_{\delta_{ES}}(\delta_{ES} - \delta_{ES_0})$$

$$\dot{q} = M_0 + M_u(u - u_0) + M_q(q - q_0) + M_w(w - w_0) + M_{\delta_{ES}}(\delta_{ES} - \delta_{ES_0})$$

Here, u_0, w_0, δ_{ES_0}, and q_0 are the reference conditions about which linearization is assumed valid, and the terms X_0, Z_0, M_0 are included to account for a possible off-trim condition.

Three representative sets of results of the identification of these configurations are shown in Figures 3 - 5. These results are obtained using the measurement noise statistics given in an earlier section. Neither acceleration measurements nor process noise were used on these records. In general, as is discussed in Reference 9, the use of the acceleration measurements tends to provide better estimates of the parameters; in particular, it is obvious that the control derivatives should be more accurately identified. In this X-22A experiment, the n_x accelerometer malfunctioned and the n_z accelerometer signal was compromised by a bias introduced by accelerometer stiction; therefore, the identifications were performed using only the state measurements. The assumption of no process noise was prompted primarily by the relative simplicity of the identification problem in this case (e.g., the interest in identifying primarily only two derivatives). With the absence of process noise, the option of the fixed-point smoother was not used; in addition, only one iteration of the locally iterated filter-smoother algorithm was required.

These three examples represent quite different levels of augmentation of the X-22A and resulting short-term dynamics. Configuration 6 in Figure 3 has a larger angle of attack stability than the basic airplane; Configuration 13 in Figure 4 has characteristics similar to the unaugmented X-22A; Configuration 15 in Figure 5 is de-augmented in both angle of attack and pitch rate stabilities. In all cases, the calculated time history from the identified parameters agrees well with the data. It is also worth noting that other means of identification, such as analog matching, were used on this program, and agreed well with the digital results.

Lateral-Directional Dynamics

The second X-22A flying qualities program, just completed, investigated lateral-directional flying qualities and roll control power requirements for STOL landing approach. This program used the variable stability system to a far greater extent than the first, as all of the derivatives in the roll and yaw moment equations were varied. Primary variables in the experiment were the roll mode time constant (3 values), the Dutch roll frequency (2 values), and the roll-to-sideslip ratio (2 values), out of which seven basic dynamic configurations were chosen. In addition, for each of these the yaw due to aileron (and thereby the zeros of the ϕ/δ_{AS} transfer function) was varied, and, for selected cases, the available roll control power was electrically limited. The demands on the identification technique were therefore considerably larger on this program, both in terms of required accuracy and, from an operational point of view, in terms of the vast quantity of calibration records to be analyzed during the set-up phase.

The identification of the lateral-directional data was performed using the following set of body-axis equations of motion (again, only linear aerodynamic terms are included):

$$\dot{\beta} = \frac{Y_\beta}{V}\,\beta + \left(\frac{Y_p}{V} + \frac{\alpha}{57.3}\right)p + \left(\frac{Y_r}{V} - 1\right)r + \frac{g}{V}\,\sin\phi + \frac{Y_0}{V}$$

$$\dot{p} = L'_\beta\,\beta + L'_p\,p + \left(L'_r + K_L\,\frac{q}{57.3}\right)r + L'_{\delta_{AS}}\,\delta_{AS} + L'_{\delta_{RP}}\,\delta_{RP} + L'_0$$

$$\dot{r} = N'_\beta\,\beta + \left(N'_p + K_N\,\frac{q}{57.3}\right)p + N'_r\,r + N'_{\delta_{AS}}\,\delta_{AS} + N'_{\delta_{RP}}\,\delta_{RP} + N'_0$$

$$\dot{\phi} = p + \frac{\theta q}{57.3}\,\sin\phi + \frac{\theta r}{57.3}\,\cos\phi$$

These equations assume "small" θ, α, and β, and account for the fact that the body axes used for the measurements are essentially the principal axes of the aircraft. In addition, for most of the calibration records the longitudinal aircraft motions are negligible, and hence the inertia coupling terms K_L and K_N do not appear in the identified results.

Three representative examples of the lateral-directional identification results which span the dynamics investigated in the program are shown in Figures 6, 7, and 8. Configuration 1 in Figure 6 has highly augmented roll damping and de-augmented directional stiffness; Configuration 4 in Figure 7 has augmented directional stiffness, approximately the same roll damping as the X-22A, and de-augmented dihedral effect; Configuration 6 in Figure 8 is similar to Configuration 4 except that the roll damping is highly de-augmented.

The results shown in these figures were obtained by the "production line" techniques dictated by the exigencies of a flight program as discussed earlier; that is, the set of measurement and process noise statistics presented in a previous section of this paper were used uniformly. The initial covariance matrix of the estimates was also obtained by simply multiplying the least-squares variances by a constant factor; the factor used is a very large 10^3 to ensure sufficient filter gain. The examples do not include the use of the fixed-point smoother algorithm; experimental experience has shown that the first data point provides suitable initial conditions, and that the locally iterated filter-smoother performance is sufficiently good to obviate the advantages of smoothing the estimates for these cases with linear aerodynamics.

Effect of Initial Estimate

As was discussed in an earlier section, the recursive nature of the locally iterated filter smoother algorithm appears to be advantageous in that the results of an identification run are fairly independent of the accuracy of the initial estimate. Figure 9 gives the results of identifying the same data as in Figure 7, but with the initial estimates set at zero. The initial variances, measurement noise statistics, and process noise statistics are identical. As can be seen, the derivative estimates and modal characteristics are identical. This demonstrated insensitivity to the initial estimates and resulting lack of non-uniqueness problems is extremely valuable to the "production line" identification required on flying qualities programs, as it eliminates the need for any optimal processing of the data prior to identification.

Effects of Control Inputs

As we have discussed, the time history of the control inputs is very important to obtain valid identification results. In Figure 10, the identification of the same configuration as in Figure 7 is shown, but from a different flight and hence with different control motions. The quality of the time history matches is roughly equal, and in fact the final variances of the estimates are generally within a factor of two of each other. It is useful, then, to compare the normalized covariances, a few of which are listed in the table below.

CONFIGURATION 4, COMPARISON OF NORMALIZED COVARIANCES

Parameters	Figure 7	Figure 10
$L_p' - L_r'$	-.151	.721
$N_p' - N_r'$	-.130	.608
$N_\beta' - L_\beta'$	-.376	-.437
$L_\beta' - L_p'$	.470	-.230
$L_{\delta_{AB}}' - L_p'$	-.769	-.397
$L_{\delta_{AG}}' - N_{\delta_{RP}}'$	.157	-.079

Clearly, each input increases the identifiability of some parameters at the expense of others. Neither can be considered an "optimum" for this configuration, but both yield covariances that indicate adequate identifiability.

CONCLUDING REMARKS

This paper has addressed the application of a digital identification technique based on Kalman filter theory to flight data from the X-22A variable stability aircraft. Over 300 flight data records were analyzed with this technique on the second X-22A flight program alone, and the emphasis of the paper has therefore been on the practical aspects of identifying many data covering a wide range of dynamic characteristics as simulated by the X-22A aircraft. A general conclusion that may be stated as a result of this unique experimental experience is that the technique provides a useful and efficient tool for identification of stability and control parameters from flight data, and, in fact, that it can be applied in the "production line" fashion required to

161

process large quantities of data rapidly during a flight program with little loss of accuracy.

Specific conclusions that may be drawn from the results discussed in this paper concerning the application of this technique to the X-22A flight data are:

- The locally iterated filter-smoother algorithm developed for nonlinear systems provides very good identification results for the quasilinear (linear aerodynamics) systems discussed in this paper.

- The recursive nature of the technique appears to offer the advantage of insensitivity to the initial parameter estimates for X-22A data, thereby eliminating any necessity for data processing prior to identification.

- The required input information to the algorithm (i.e., the covariance matrices) may be held essentially constant for "production line" identification during a flight program after an initial iteration period.

- The control input time histories are very critical to good identification results. Pilot inputs which attempt "switching" near characteristic frequencies provide good identifiability.

REFERENCES

1. Neal, T.P. : Frequency and Damping from Time Histories Maximum-Slope Method. Journal of Aircraft, Volume 4, No. 1, January-February 1967.

2. Hall, G.W. : A Method for Matching Flight Test Records with the Output of an Analog Computer. Paper presented at the National Electronics Conference, December 1969.

3. Di Franco, D.A. : In-Flight Parameter Identification by the Equations of Motion Technique -- Application to the Variable Stability T-33 Airplane. Cornell Aeronautical Laboratory Report No. TC-1921-F-3, December 1965.

4. Larson, D.B. : Identification of Parameters by the Method of Quasi-linearization. Cornell Aeronautical Laboratory Report No. 164, 1968.

5. Hall, G.W., Larson, D.B., and Martino, P.A.: A Quasilinearization Method for Determining Lateral-Directional Modal Parameters from Digitally Recorded Flight Test Data. Cornell Aeronautical Laboratory Report No. TM-2832-F-1, April 1970.

6. Grove, R.D., Bowles, R.L., and Mayhew, S.C.: A Procedure for Estimating Stability and Control Parameters from Flight Test Data by Using Maximum Likelihood Methods Employing a Real-Time Digital System. NASA TND-6735, May 1972.

7. Mehra, R.K.: Maximum Likelihood Identification of Aircraft Parameters. Paper 18-C presented at 1970 JACC, June 1970.

8. Molusis, J.A.: Helicopter Stability Derivative Extraction and Data Processing Using Kalman Filtering Techniques. Preprint No. 641, presented at the 28th Annual National Forum of the American Helicopter Society, May 1972.

9. Chen, R.T.N., Eulrich, B.J., and Lebacqz, J.V.: Development of Advanced Techniques for the Identification of V/STOL Aircraft Stability and Control Parameters. Cornell Aeronautical Laboratory, Inc. Report No. BM-2820-B-1, August 1971.

10. Smith, R.E., Lebacqz, J.V., and Schuler, J.M. : Flight Investigation of Various Longitudinal Short-Term Dynamics for STOL Landing Approach Using the X-22A Variable Stability Aircraft. CAL Report No. TB-3011-F-2 (Calspan Corporation), January 1973.

11. Lebacqz, J.V., Smith, R.E., and Radford, R.C.: An Experimental Investigation of STOL Lateral-Directional Flying Qualities and Control Power Requirements Using the Variable Stability X-22A Aircraft. Paper submitted for presentation at the Fifth Aircraft Design, Flight Test and Operations Meeting, August 1973.

12. Chen, R.T.N., and Eulrich, B.J.: Parameter and Model Identification of Nonlinear Dynamical Systems Using a Suboptimal Fixed-Point Smoothing Algorithm. Paper presented at 1971 JACC, August 1971.

13. Wishner, R.P. et.al.: A Comparison of Three Nonlinear Filters. Augomatica, Vol. 5, pp. 487-497, 1969.

14. Jazwinski, A.H.: Stochastic Processes and Filtering Theory. Academic Press, 1970.

15. Mehra, R.K., Stepner, D.E., and Tyler, J.S.: A Generalized Method for the Identification of Aircraft Stability and Control Derivatives from Flight Test Data. Paper 16-4 presented at 1972 JACC, August 1972.

16. Beilman, J.L.: An Integrated System of Airborne and Ground-Based Instrumentation for Flying Qualities Research with the X-22A Airplane. Paper presented at the 7th International Aerospace Instrumentation Symposium, March 1972.

17. Reid, D.B.: Optimal Inputs for System Identification. NASA CR-128173, May 1972.

18. Smith, R.E., et. al.: Final Report on X-22A Task II Program under Contract N00019-72-C-0417, in preparation. Calspan Corporation.

Figure 1   X-22A Variable Stability Aircraft

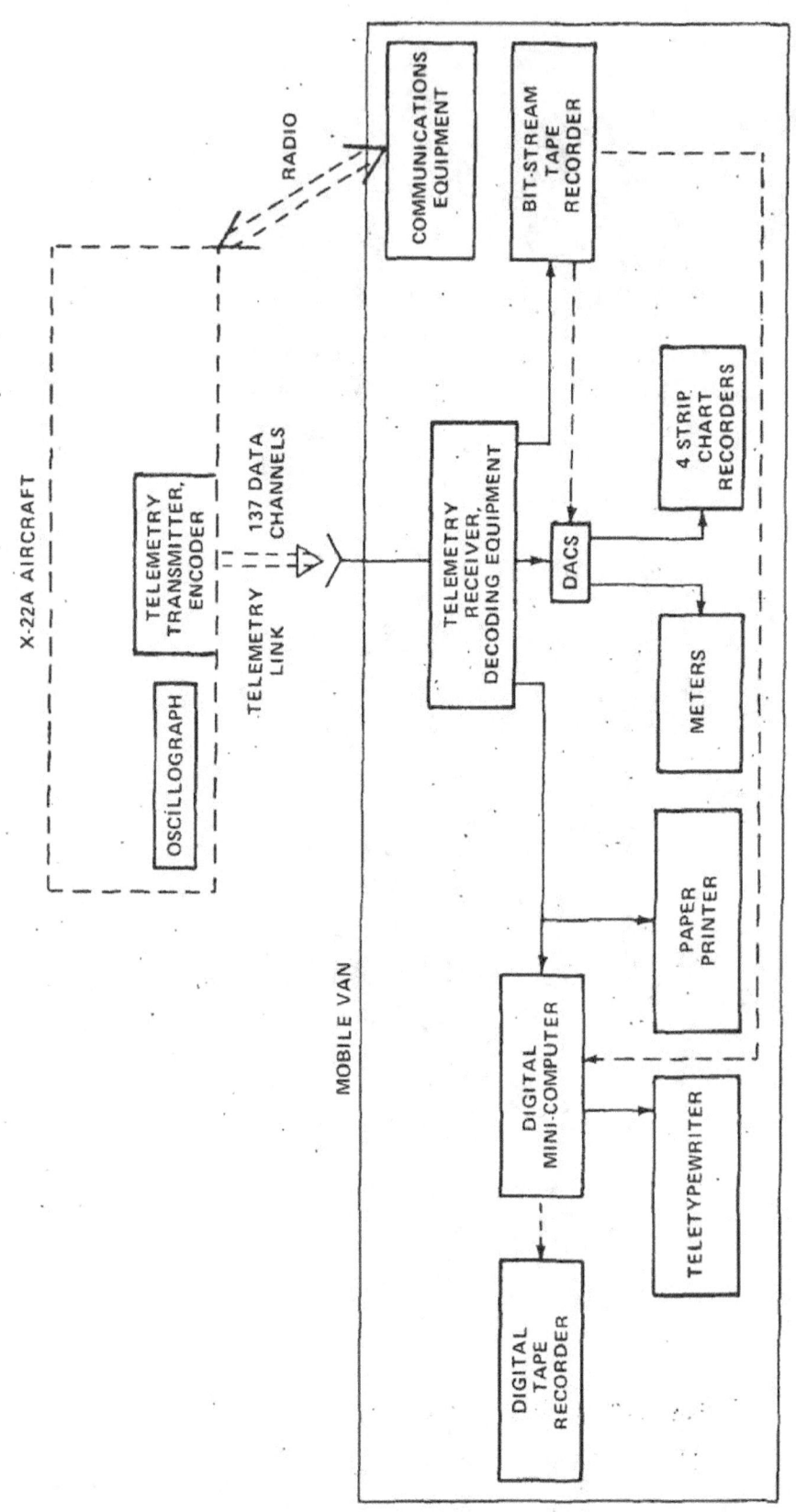

Figure 2 Schematic Diagram of Digital Data Acquisition System

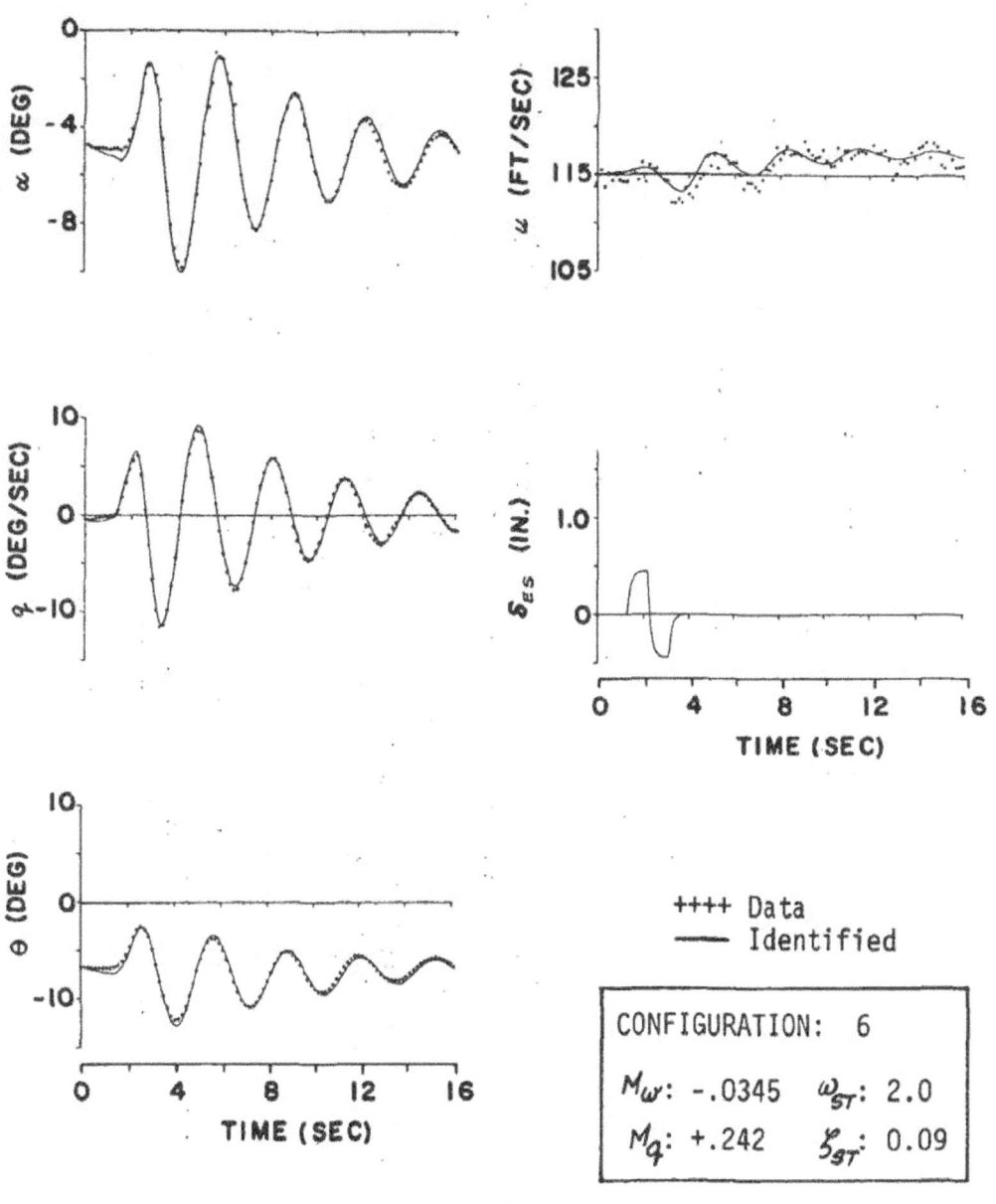

Figure 3   Identification of Configuration 6, Longitudinal

167

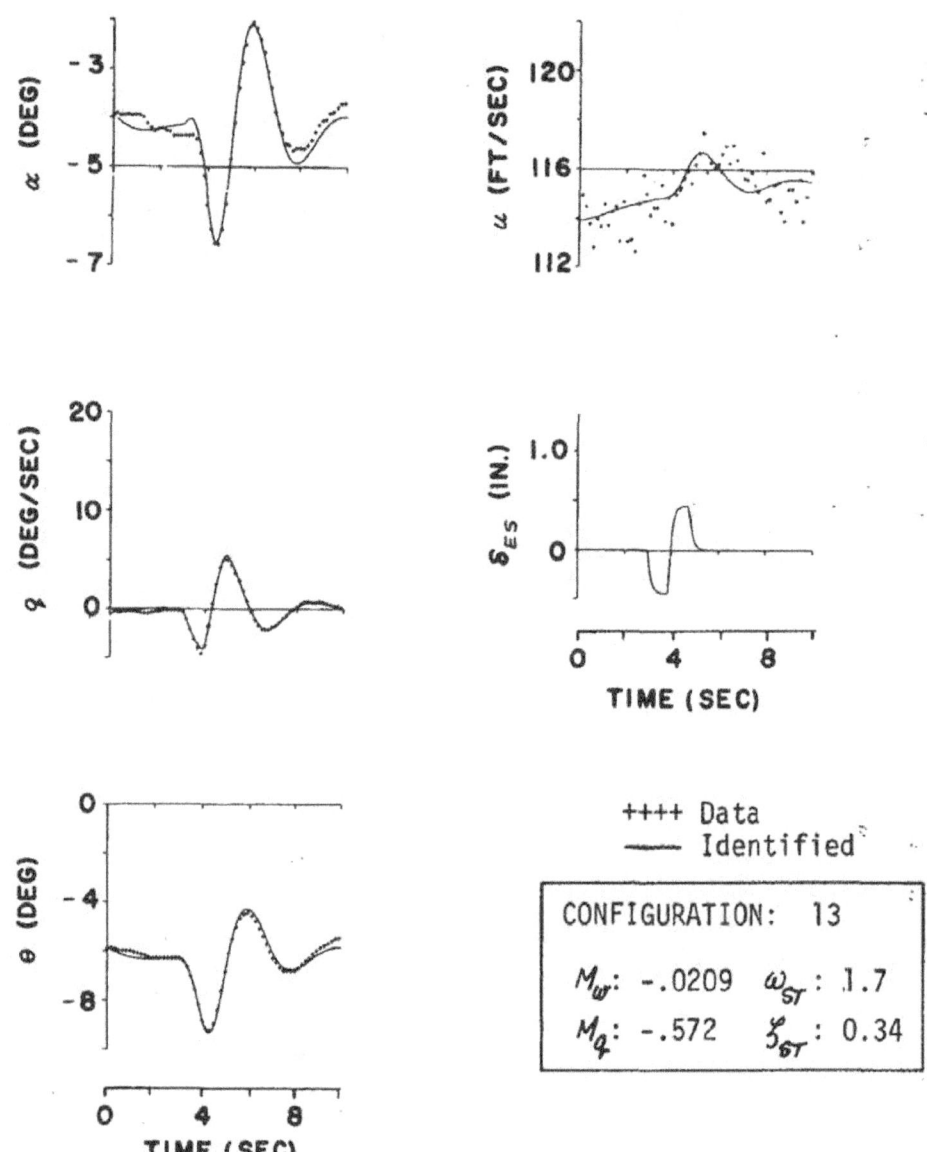

Figure 4   Identification of Configuration 13, Longitudinal

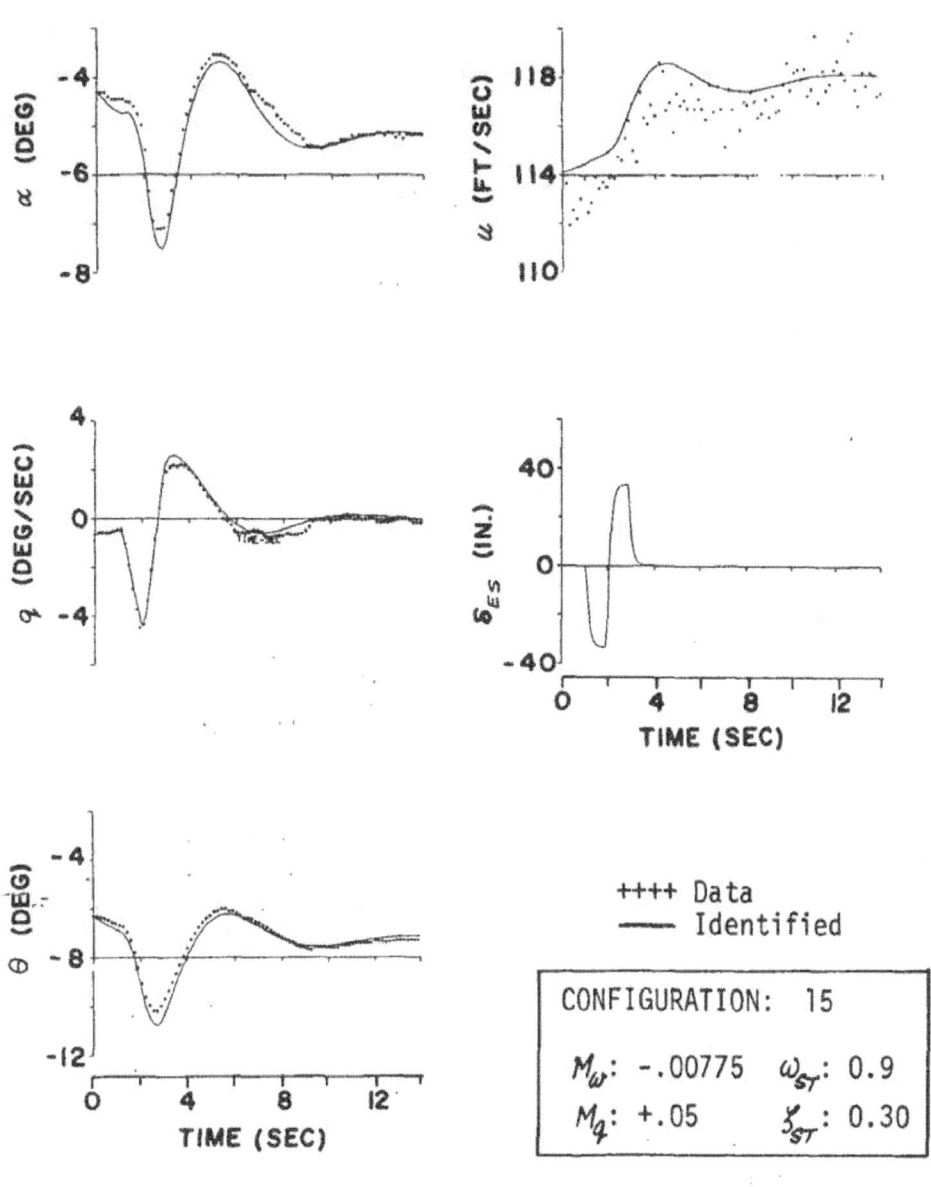

Figure 5   Identification of Configuration 15, Longitudinal

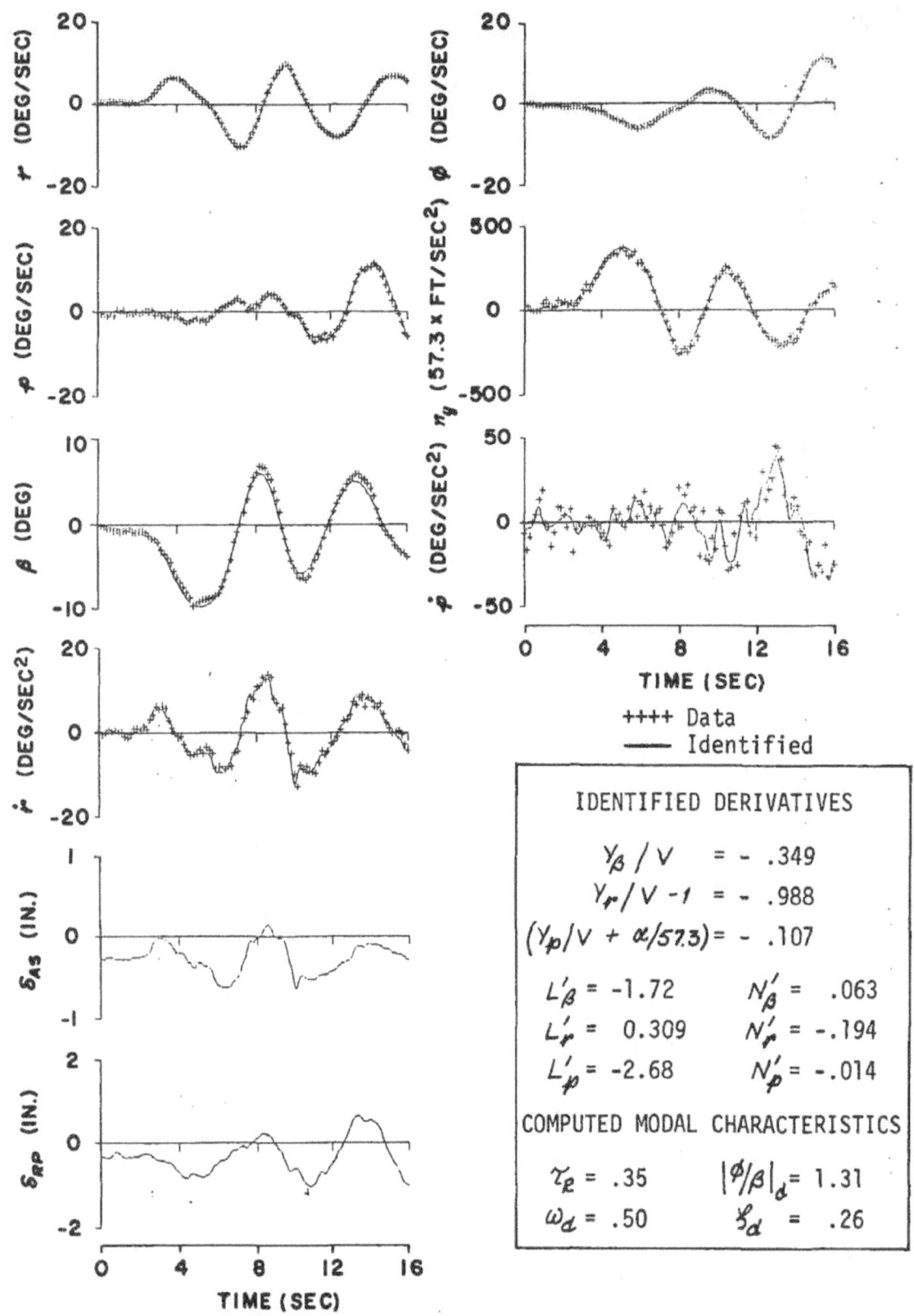

Figure 6   Identification of Configuration 1, Lateral-Directional

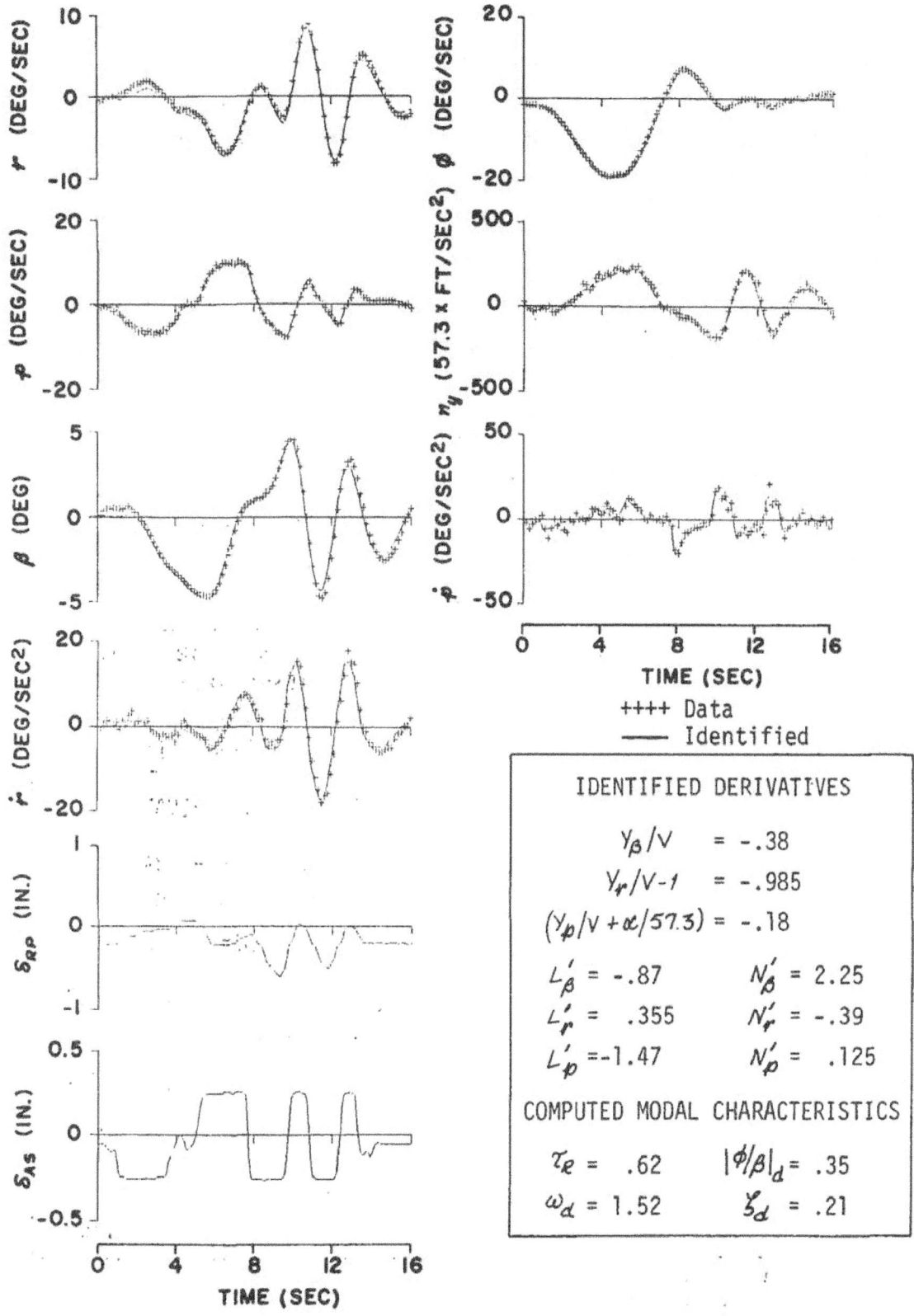

Figure 7   Identification of Configuration 4, Lateral-Directional

171

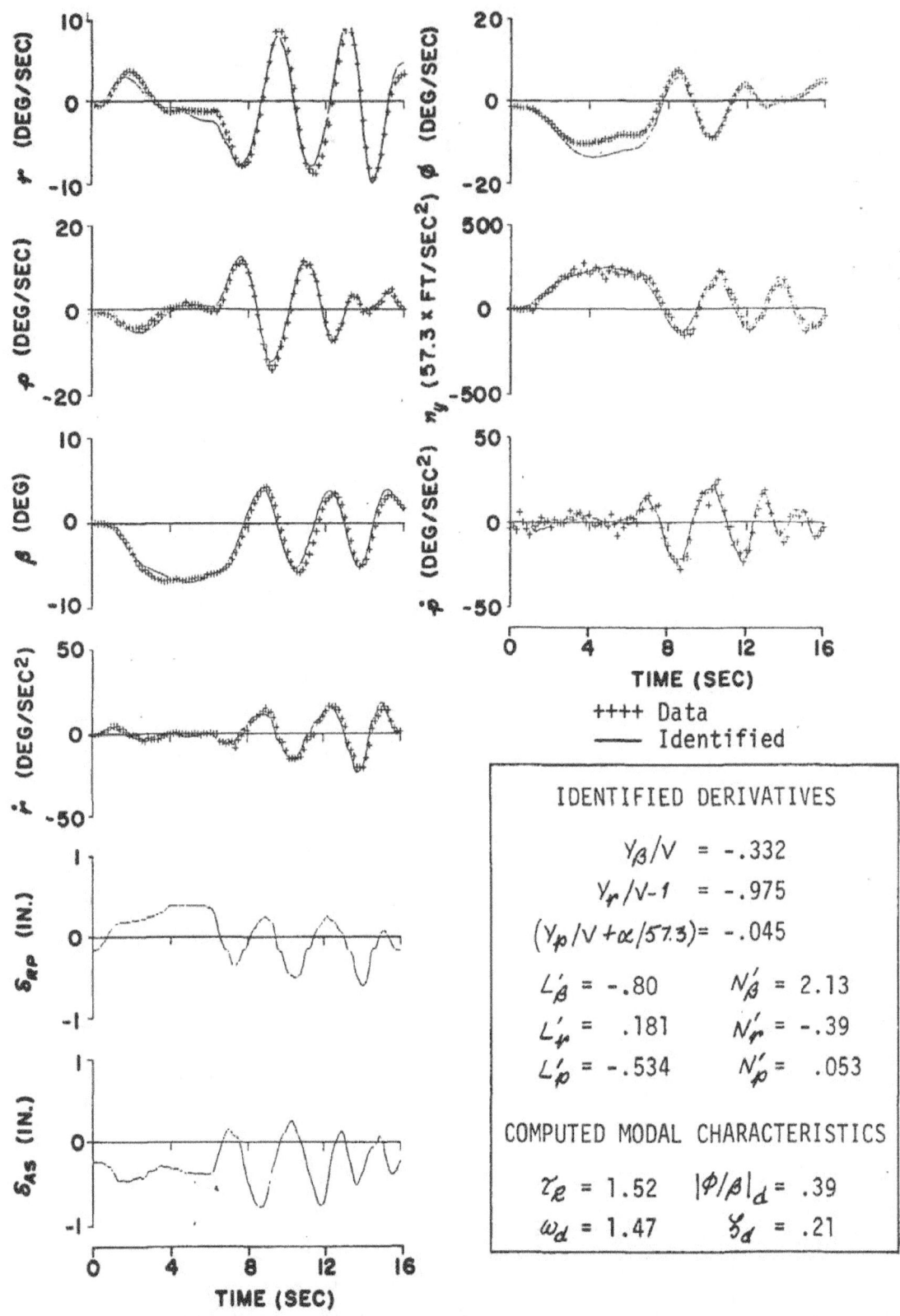

Figure 8   Identification of Configuration 6, Lateral-Directional

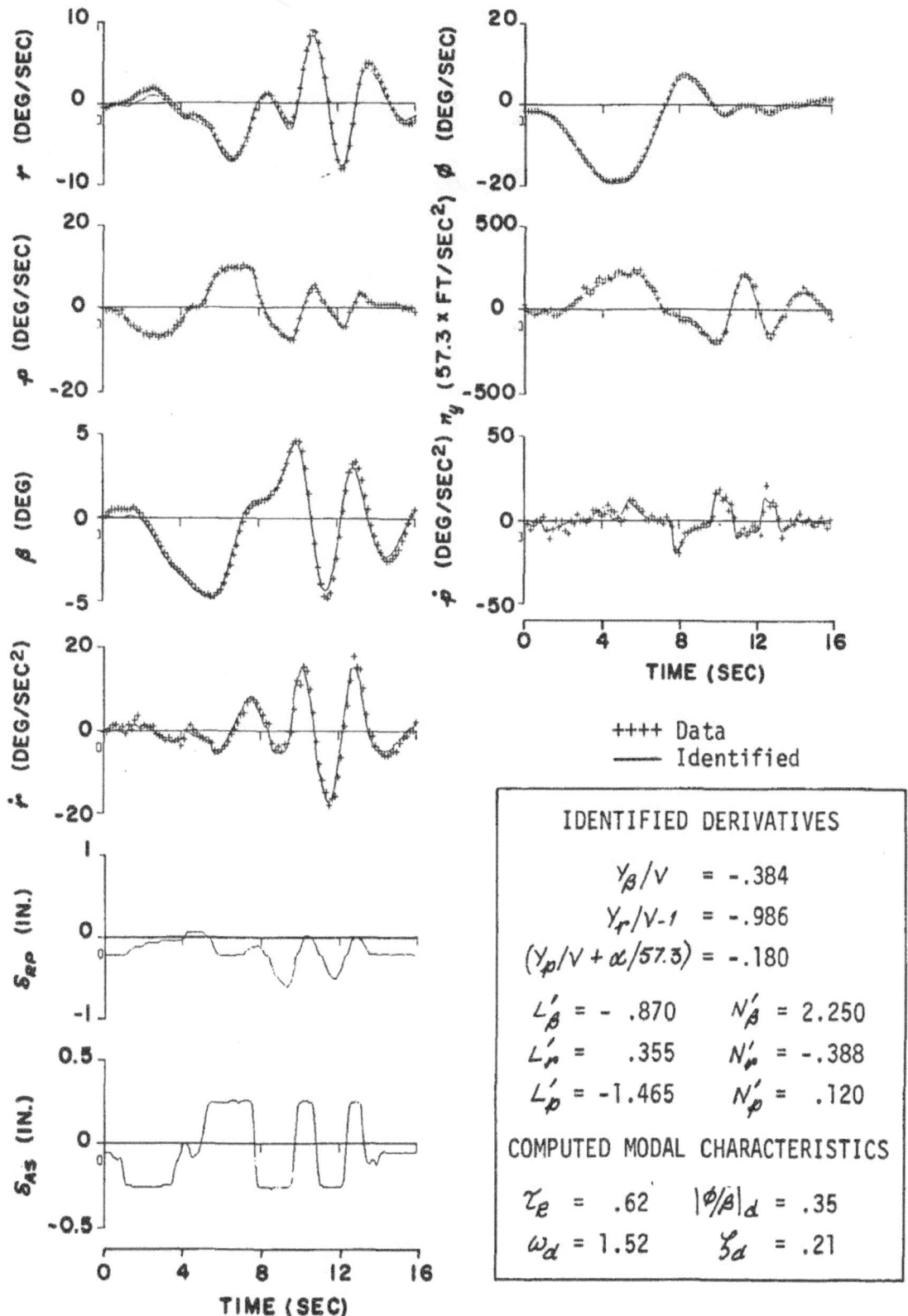

Figure 9   Identification of Configuration 4 With Initial Estimates at Zero

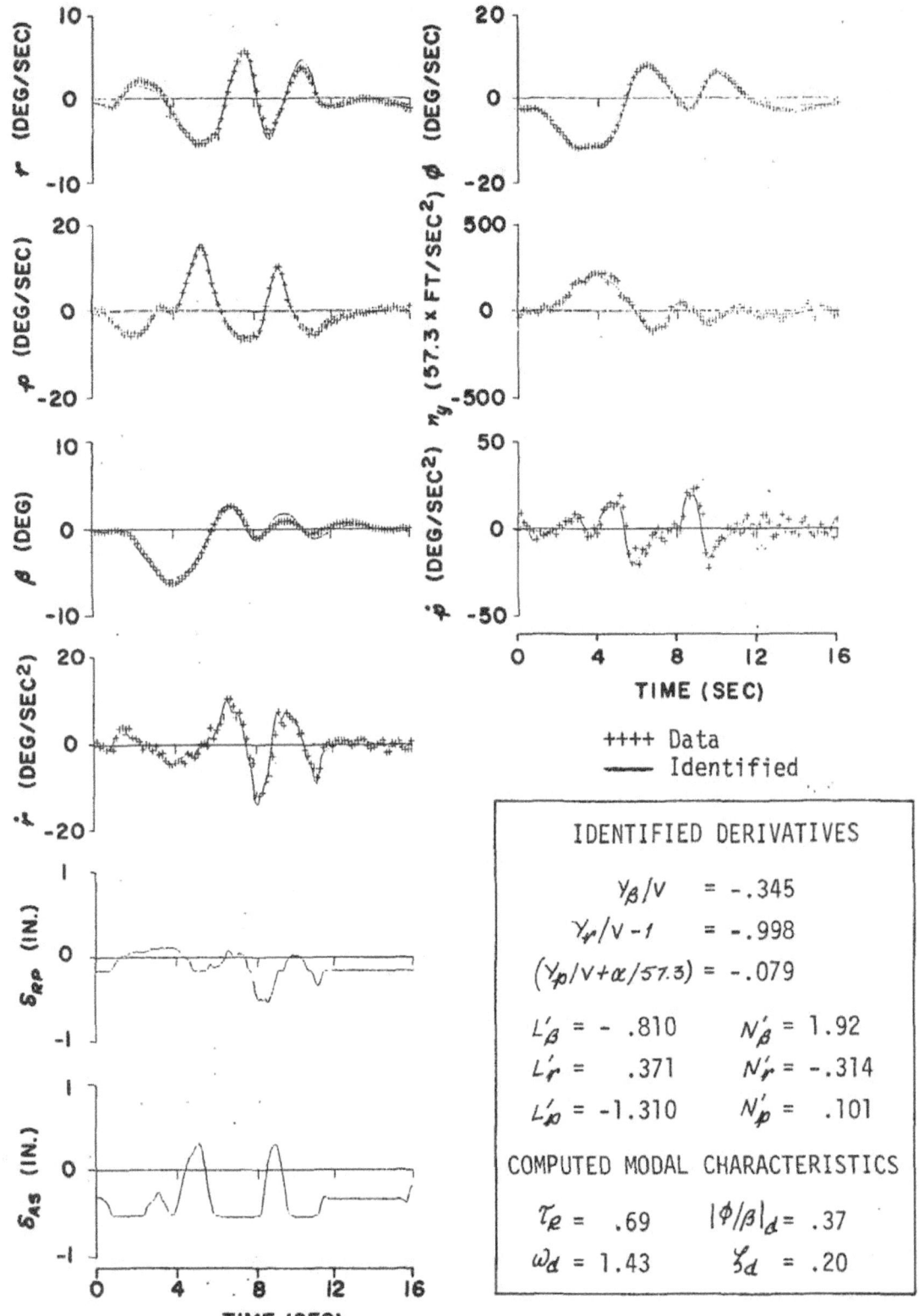

Figure 10   Identification of Configuration 4 With Different Input

174

HELICOPTER DERIVATIVE IDENTIFICATION FROM ANALYTIC

MODELS AND FLIGHT TEST DATA

By John Molusis and Stan Briczinski
Aerodynamicists, Aeromechanics Branch

Sikorsky Aircraft

Division of United Aircraft Corporation

SUMMARY

Recent results of stability derivative identification from helicopter analytic models and flight test data are presented. Six and nine degree-of-freedom (DOF) linear models are identified from an analytic nonlinear helicopter simulation using a least square technique. The identified models are compared with the conventional partial differentiation method for obtaining derivatives to form the basis for interpretation of derivatives identified from flight data. Six degree-of-freedom models are identified from CH-53A and CH-54B flight data, using an extended Kalman filter modified to process several maneuvers simultaneously. The a priori derivative estimate is obtained by optimal filtering of the data and then using a least square method.

The results demonstrate that a six DOF identified model is sufficient to determine the low frequency modes of motion, but a nine DOF rotor/body model is necessary for proper representation of short-term response.

INTRODUCTION

Because of their complexity and coupled behavior, helicopters are prime canditates for derivative identification. For example, some analytic models incorrectly predict high speed dynamics associated with articulated rotor helicopters. Another phenomenon not predicted accurately by many existing models is rotor tip path plane oscillation, which places upper limits on the feedback gains used in stability augmentation of many helicopters. Isolation of these discrepancies is among the motivating factors for pursuing derivative identification, but there has also been a general need for identification to provide correlation with stability and control prediction techniques.

Identification requirements for articulated rotor helicopters differ from fixed wing or other VTOL aircraft primarily in the following ways:

1. The helicopter has a large number of degrees of freedom (DOF), all of which are highly coupled and many of which are significant. In addition to the 6 DOF of the body, each blade of the articulated rotor has a flapping DOF and a lag DOF. (A number of bending DOF may also be used to describe the dynamics of each blade more accurately.)

175

2. Due to the rotary characteristics of the helicopter, linearization of the equations of motion results in a model with periodic coefficients.

3. The plant driving noise and measurement noise are usually larger for helicopters than for fixed wing aircraft.

4. The basic helicopter has long period instabilities and, in some cases, has unstable Dutch roll roots.

To our knowledge, the first attempt at using advanced estimation techniques to obtain derivatives from helicopter flight test data was presented in Reference 1. The particular problems related to helicopter identification (some of which are mentioned above) were discussed, and methods were proposed for solution. References 2 and 3 applied the methods proposed in Reference 1 to helicopter flight test data. In addition, Reference 4 concentrated on obtaining derivatives from an analytic computer model for a 9 DOF helicopter description.

This paper summarizes results obtained to date at Sikorsky Aircraft in helicopter identification and indicates future research areas.

SYMBOLS

L_p Rolling moment with respect to roll rate stability derivative

X_u Longitudinal force with respect to longitudinal velocity stability derivative identified from flight data

Y_u Lateral force with respect to longitudinal velocity stability derivative identified from flight data

Z_{BIS} Normal force with respect to longitudinal cyclic control stability derivative

HELICOPTER ANALYTIC MODELS

Description of Linear Models Under Investigation

Four linear models are obtained from a Sikorsky nonlinear computer helicopter simulation program called General Helicopter. This nonlinear model incorporates 6 independent blade flapping DOF in addition to the 6 body DOF. The helicopter under analytic investigation is the CH-53A at 100 kts, 33,500 lbs, with an aft c.g. location. The linear models are:

1.   6 DOF quasi-static model derived by independently perturbing each state variable associated with the 6 body DOF in the nonlinear model and then allowing the rotor to attain a new quasi-static trim condition.

2.   6 DOF identified model obtained from the nonlinear simulation by a least square method, which uses only the 6 body DOF data and lumps an average rotor contribution into the body DOF.

3.   9 DOF identified model obtained from the nonlinear simulation by the least square method. This nonlinear model also includes a Kalman estimator to resolve the flapping data into a tip path plane (for details see Reference 4), thus supplying the identification method with 3 rotor DOF data as well as 6 body DOF data.

4.   6 DOF reduced model obtained from the 9 DOF model described above, by solving a quasi-static rotor in the 9 DOF identified model and algebraically simplifying to 6 DOF.

Table I summarizes the models under investigation.

Procedure and Analysis of Techniques and Models

Stability derivatives refer to the coefficients of the variables in the Taylor series expansion representation of the aircraft equations of motion. The expansion is linearized typically by discarding all higher order terms. Therefore, the stability derivatives are equivalent to the partial derivatives of the linearized equations of motion of the aircraft. When applied to fixed wing aircraft, the system identification technique yields quantities that essentially are derivatives. Only higher order effects are handled improperly. The linearized helicopter problem is complicated by at least three nonlinear effects:

1.   Higher-order terms of the rotor and body, which are omitted from linear models.

2.   Additional rotor DOF, which are either lumped with the body in a 6 DOF linear model or approximated in a 9 DOF linear model.

3.   Periodic coefficients, which represent an exact linearization of the equations of motion, are replaced typically by a constant coefficient linear model.

The linearized helicopter equations of motion usually include the quasi-static rotor assumption, so this assumption is reflected in the conventional helicopter stability derivatives.

Figure 1 illustrates the normalized forces and moments obtained in the nonlinear General Helicopter model by perturbing one body parameter at a time by a unit value while keeping all other body parameters constant. These two examples show the variation with time of these terms as the rotor attains a new quasi-static trim condition. By definition, the final values are the coefficients of the 6 DOF quasi-static linear model. (An averaging technique is used to remove the periodic variation due to individual blade contributions about the rotor azimuth.) These values differ from the definition of helicopter stability derivatives only by the inclusion of the effects of higher-order terms. Since these effects are small, the coefficients composing the 6 DOF quasi-static model are essentially equal to the derivatives.

The perturbation technique for obtaining derivatives of a linear model can be applied only to analytic models. Identification techniques can be applied to flight data as well as to analytic models. Superimposed on the perturbation derivative of Figure 1 are the values of the corresponding coefficients of the three other linear models under study. The location of the symbols representing these values is arbitrary; the 9 DOF identified value is placed near time equal to zero in order to emphasize the somewhat instantaneous nature of this method, because it does not carry a quasi-static rotor assumption. Since the 9 DOF representation of the helicopter is different from the 6 DOF representation, the 9 DOF identified coefficients, as expected, bear little resemblance to the stability derivatives. In general, the coefficients of the 6 DOF reduced model closely approximate most of the corresponding derivatives. This technique encounters difficulty for any responses that initially have sharp spikes in the perturbation value (see Z_{B1S}). The 6 DOF identified coefficients are found to be only a fair match to the stability derivatives, due to the apparent inclusion of helicopter nonlinearities into the curve fit solution of the 6 DOF identification technique. This technique might yield better values of derivatives if shorter, more stable records of maneuvers were used in the identification process. The coefficients of all the models that depend on the identification method will include effects due to all three helicopter nonlinearities described above, whether these models are derived from analytic models or flight data. Thus, coefficients of either the 6 DOF reduced or the 6 DOF identified models at best should be considered only approximations of the helicopter quasi-static stability derivatives as defined above.

Once the four linear models are obtained, roots are collected, using an Eigenvalue program. The body roots all appear in Figure 2. Time history responses of the 6 body accelerations are then obtained for 3 of the linear models for the same longitudinal pulse control input. Figure 3 illustrates the roll acceleration from the linear models and the response from the non-linear model. With the exception of the 6 DOF identified method, the root locus plots show good correlation of all techniques. The scatter seen in the high frequency pair of roots is expected, because this pair is greatly affected by the rotor contribution, which is represented differently by each technique. It is found that both the long and short period modes due to linear effects, can be captured by the system identification method, even if it is applied only to short duration maneuvers. The 6 DOF identified model indicates, and attempts to reproduce, the non-oscillatory instability described by the nonlinear model, which is due to the increasing importance of nonlinearities as the helicopter deviates far from trim. (A pilot or feedback system normally would keep the helicopter within a stable flight regime.) Applying the 6 DOF identification to shorter or more stable maneuvers should result in making the 6 DOF identified model roots more consistent with the roots of the other methods. This was found to be true in a previous study (Reference 1).

As seen by the dynamic response example of Figure 3, the 9 DOF identified model is the only one that faithfully reproduces the short term, high frequency body accelerations due to transient rotor response. (The noisy data seen in the nonlinear time history are the effects due to individual blade dynamics.) The dynamic response obtained from the 6 DOF quasi-static model does not indicate these initial responses, nor does it even give the correct direction for the initial roll acceleration. This 6 DOF quasi-static model, which consists of the conventional helicopter stability derivatives, is therefore not necessarily the most desirable linear model for studying system dynamics.

IDENTIFICATION OF DERIVATIVES FROM FLIGHT DATA

Six degree-of-freedom derivatives are identified from CH-53A and CH-54B flight test data. The identification method used is the extended Kalman filter modified to accomodate simultaneous processing of different maneuvers. The use of simultaneous maneuvers has two main advantages. First, significant amounts of helicopter flight data are currently available that are only of 4 to 6 seconds duration. Sufficient data requirements and proper control input excitation can be obtained only by using several of these segments. Secondly, simultaneous, rather than sequential, processing of the data segments eliminates the need to start up each successive maneuver with the derivative and variance computed at the end of the previous maneuver. An extensive study was conducted (Reference 3) to investigate the best filtering method and a priori derivative estimate to be used with the extended Kalman filter identification method. The procedure found most accurate is summarized below.

1. Filter the data with an extended Kalman filter which is formulated to filter the data and determine bias error and not identify derivatives.

2. Use these optimally filtered data with a least square estimator to obtain an improved a priori derivative estimate and variance.

3. Modify the a priori derivative variance to reflect more accurately the initial uncertainty in the derivative estimate.

4. Use the least square derivative estimate and modified variance to initialize the multi-maneuver extended Kalman filter derivative identification algorithm.

A good derivative estimate is required to assure validity of the linearizations in the Kalman identification algorithm. Additionally, smaller amounts of data are needed for good derivative convergence when an accurate estimate and variance are provided. The least square method using Kalman filtered data provides an excellent estimate (Reference 2) for the derivatives; however, the variance must be modified. Multiplication of this variance by a factor of 100 or selecting a value based on engineering judgment yields good results.

Application to CH-53A

The procedure discussed above is applied to four simultaneously processed 6-second CH-53A maneuvers performed at 100 knots trim. Complete results are given in Reference 3. Only highlights are discussed here.

Of the 60 identified derivatives, Figure 4 shows one typical convergence plot. Two a priori derivative values are used to start the Kalman derivative identification method: an arbitrary estimate and the value from the least square method. Because it is more accurate with smaller uncertainty, the least square value shows fewer oscillations, particularly in the first 2 seconds of data. The arbitrary estimate indicates that, if the initial guess is bad, a considerably longer data length may be required before convergence occurs. Thus, for short data records, it is important to make a good derivative estimate and to select a variance large enough to reflect actual uncertainty, but not so large that long data records are required.

Simulation of the identified derivative model that used the least square method shows a good match with test data used in the identification and with test data not used in the identification (a more conclusive test of the accuracy of the identified model).

The method is also applied to CH-53A flight data at 150-knot trim conditions, using 4 simultaneous 5-second maneuvers. The lateral control inputs are applied at the middle of the data record and again at the end. This does not allow proper mode excitation or sufficient data lengths, as reflected in the derivative convergence plot of Figure 5. Thus, derivative convergence should be examined, since it contains information of the length of data required and whether proper control inputs were applied.

Application to CH-54B

The derivative identification method is applied to one 16-second data record of the CH-54B at 45-knot trim condition. Since the open loop helicopter is unstable in flight, this long data record is obtained with the pilot flying the vehicle, thus providing the required stabilization. Derivatives are identified for two runs using the same data record; both runs use the least square derivative estimate. The derivative variance for the first run is selected by engineering judgment, and the second run uses this variance divided by four. Figure 6 shows the various identified models resimulated against the flight data. Two conclusions are reached. First, the engineering judgment selection of the derivative variance was too large. In fact, Figure 6 shows that the least square model yields a superior match with the data. Reducing the variance to an optimum value shows the large improvement made in the time history match. Secondly, the identified linear model has unstable phugoid characteristic roots and, when simulated against the test data, the match diverges as shown in Figure 6. This is because the error variance equation (error between test data and simulated identified model) is governed by a differential equation that has unstable roots. Reference 3 discusses this problem in detail. In Figure 6, the identified linear models are reinitialized every 4 seconds because of this unstable error variance equation.

Derivative convergence is shown in Figure 7 for selected derivatives. Convergence is shown to be good, with the final derivative values quite close to the a priori least square value. Using the final derivative values as a priori estimates and rerunning the extended Kalman filter could further improve the results.

Test Results vs. Theoretical Prediction

Six DOF identified coefficients represent the lumped effect of body and rotor. Since these coefficients are obtained from transient input/output data, the rotor is being excited continually. Thus, the lumped identified coefficients represent the body plus average rotor contribution. The conventional quasi-static helicopter derivarive assumes the rotor is in a steady state condition (quasi-static rotor) and, thus, represents the body plus rotor contribution after the tip path plane reaches steady state. For this reason, 6 DOF identified coefficients cannot be correlated one-to-one with conventional quasi-static derivatives. Yet characteristic roots can provide a meaningful comparison for rigid body modes. Figure 8 shows characteristic roots obtained from the derivatives that were identified from the CH-53A flight data at 100 knots trim against the roots identified from a nonlinear simulation model. Excellent agreement is shown for the phugoid and Dutch roll roots.

Figure 9 shows a similar comparison at the 150-knot trim condition. At this speed, the unaugmented CH-53A is known to have slightly unstable Dutch roll characteristic roots, as shown by roots identified from flight test. The nonlinear simulation model incorrectly predicts stable Dutch roll roots. Examination of the identified six DOF model reveals that the unstable Dutch roll roots result from longitudinal-to-lateral coupling. While these discrepancies are revealed in 6 DOF identification, isolation of the causes requires at least a 9 DOF identification to separate rotor effects from the body.

FUTURE RESEARCH

The complexity and coupled behavior of helicopter motion provide strong motivation for applying identification methods. Six DOF identification provides quantitative information of rigid body stability. Discrepancies between actual helicopter motion and motion predicted by nonlinear models suggest identification as a means to isolate these discrepancies, but this requires identification of at least 9 DOF models and perhaps more. This can be accomplished by first using a least square method for initial estimate and then improving upon the derivatives one row (or several rows) at a time with the extended Kalman filter. A bias correction term might be added to each row that is not being updated to account for modeling errors. This approach can provide a computationally efficient means to identify large derivative arrays and, thus, permits complete correlation of helicopter derivatives obtained from test and theory.

181

CONCLUDING REMARKS

Results of helicopter analytic modeling using system identification have led to an improved linear modeling capability. The linear models investigated all yield the same long-term characteristic roots, with the exception of the 6 DOF identified model. This model yields the same roots only when the data used in the identification are representative of small perturbation response. Small perturbation response is, thus, a necessary requirement when identifying derivatives from flight test data. The 9 DOF identified model proves to be the most accurate linear model that can be obtained from another more complicated analytic model. The 9 DOF identified model duplicates the short period, high frequency response of the helicopter resulting from rotor transients.

Applications of the multi-maneuver extended Kalman filter for the identification of six DOF models from flight data provide a quantitative means of evaluating helicopter stability. Derivative convergence, control input, and a priori derivative estimate requirements are established for successful identification. Comparison of characteristic roots identified from CH-53A flight data at 100 knots trim shows excellent agreement with theoretically predicted roots. At 150 knots, the correlation is poor, indicating the need for improvement in analytic modeling.

Six DOF identification from flight data provides a practical means to correct discrepancies between theory and test. Nine DOF identification can provide even greater correlation capability, since the body and rotor effects can be isolated. A method is proposed to permit efficient identification of systems with many degrees of freedom using the extended Kalman filter, making 9 DOF identification computationally practical.

REFERENCES

1. Molusis, J. A.: "Helicopter Stability Derivative Extraction and Data Processing Using Kalman Filtering Techniques," presented at the 28th Annual National Forum of The American Helicopter Society, Washington, D. C., Preprint No. 641, May 1972.

2. Molusis, J. A.: "Helicopter Stability Derivative Extraction from Flight Data Using the Bayesian Approach to Estimation," Jour. American Helicopter Society, July 1973.

3. Molusis, J. A.: Analytical Study to Define a Helicopter Stability Derivative Extraction Method, Final Report of NASA Contract NAS1-11613, April 1973.

4. Briczinski, S. J.: Flight Investigation of Rotor/Vehicle State Feedback, Phase I and II Analytic Report of NASA Contract NAS1-11563, May 1973.

TABLE I. HELICOPTER MODELING
(Ignoring Blade Lag and Elastic D.O.F.)

Model Name	Number of DOF	Type of Model	How Obtained	Representation of Rotor
Nonlinear Helicopter Computer Model	12 (6-Body) (6-Rotor)	Nonlinear (Periodic Coeff.)	Newton's Laws	6 Individual Blades
"Exact" Linearization*	12 (6-Body) (6-Rotor)	Linear (Periodic Coeff.)	Linearization of Newton's Eqs.	6 Individual Blades
9 DOF Identified	9 (6-Body) (3-Rotor)	Linear (Constant Coeff.)	Identified **	3 DOF Tip path plane
6 DOF Reduced	6 (6-Body)	Linear (Constant Coeff.)	Algebraic Reduction of 9 DOF Identified	Quasi-static Rotor Lumped into Body
6 DOF Identified	6 (6-Body)	Linear (Constant Coeff.)	Identified**	Average Rotor Contribution Lumped into Body
6 DOF Quasi-static	6 (6-Body)	Linear (Constant Coeff.)	Perturbation of Nonlinear Model	Quasi-static Rotor Lumped into Body

*   This Model was not investigated in this study. (Exact linearization of helicopter equations of motion yield a periodic coefficient linear model. All other linear models are approximate linear representations.)

**   System identification method was used to obtain linear coefficients from input/output data from the nonlinear computer model.

183

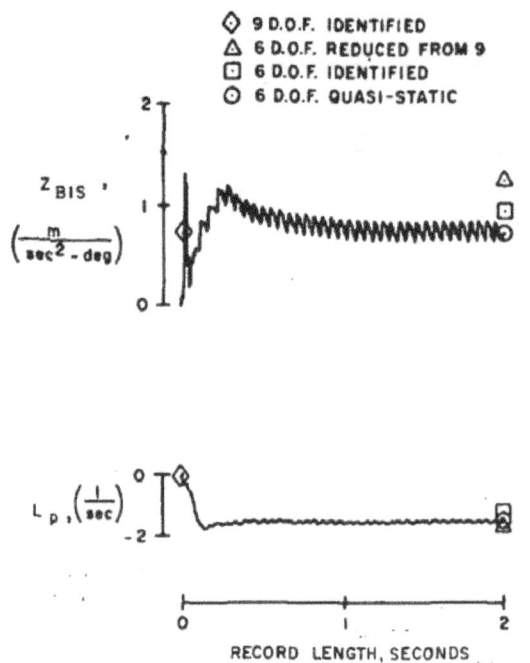

Figure 1 - Derivative Value with Time, Obtained by Perturbing the Nonlinear Model, and Comparison to Linear Model Coefficients.

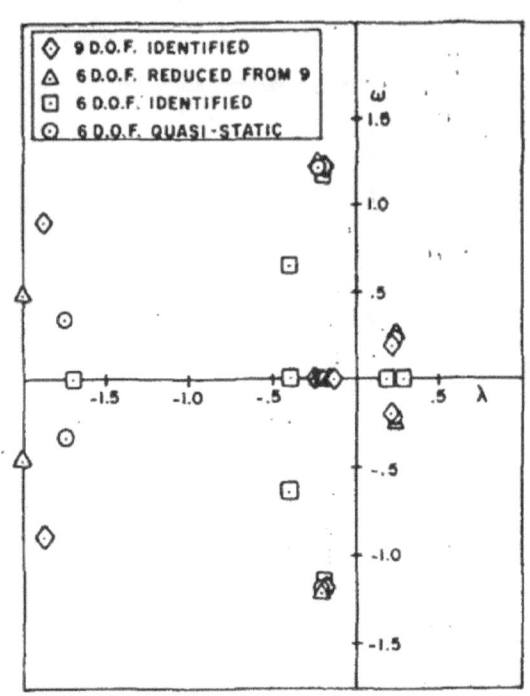

Figure 2. - Characteristic Body Roots of the Four Different Linear Models Obtained from a CH-53A Nonlinear Helicopter Computer Simulation. (100 knots, AFT. C.G., AFCS off).

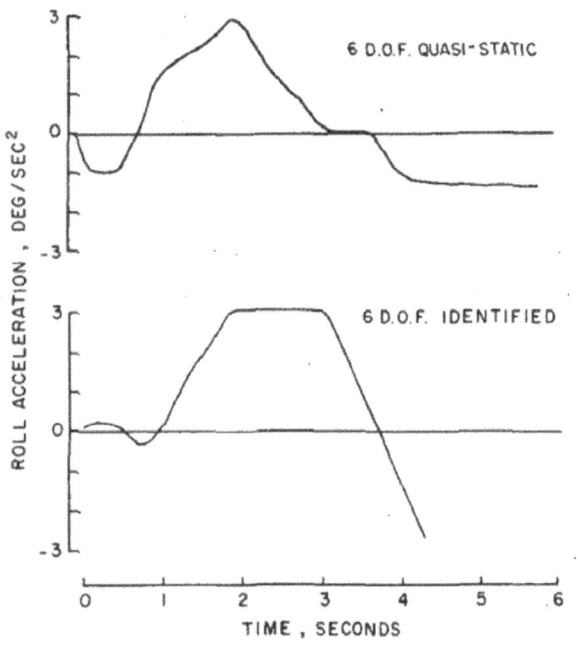

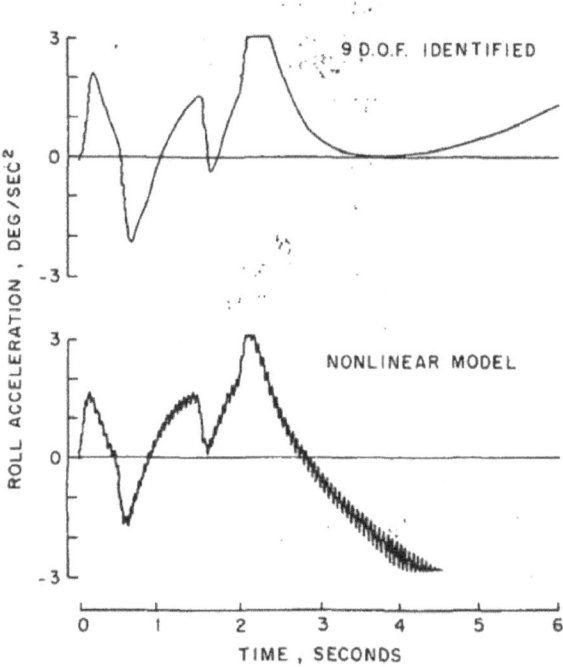

Figure 3. - Time History Comparison of Three Different Linear Models with the Nonlinear Helicopter Computer Simulation.

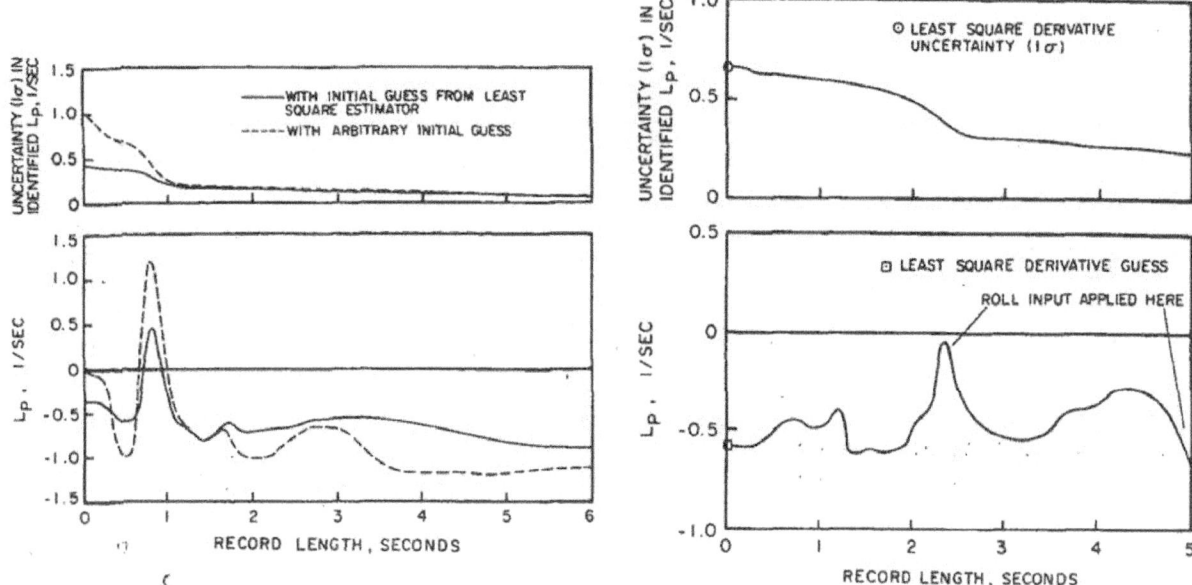

Figure 4. - Roll Damping Derivative Convergence and Uncertainty from the Multi-Maneuver Extended Kalman Filter Method Using Two Different Initial Derivative Estimates. (From CH-53A Flight Data, 100 knots Trim).

Figure 5. - Roll Damping Derivative Convergence and Uncertainty from the Multi-Maneuver Extended Kalman Filter Method Showing Effect of Control Inputs and Incomplete Convergence. (From CH-53A Flight Data, 150 knots Trim).

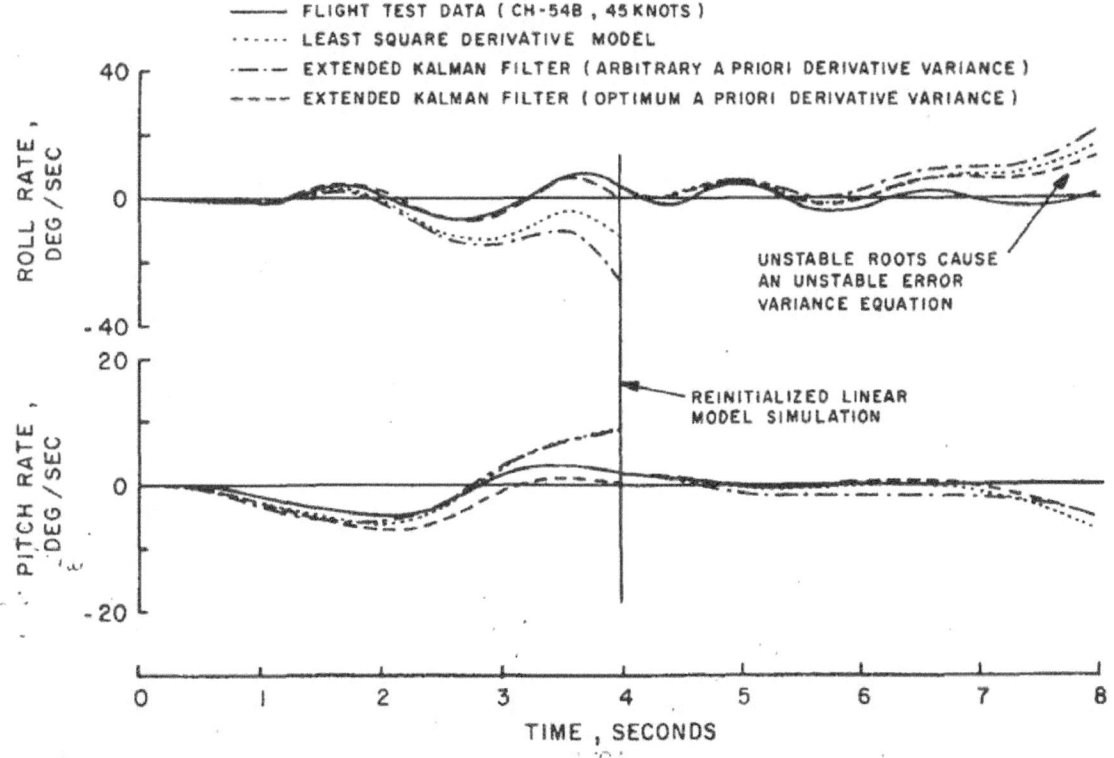

Figure 6. - Simulation of Identified Derivative Models from the Extended Kalman Filter Method Using the Least Square a Priori Derivative Estimate with Two Different a Priori Derivative Variances.

185

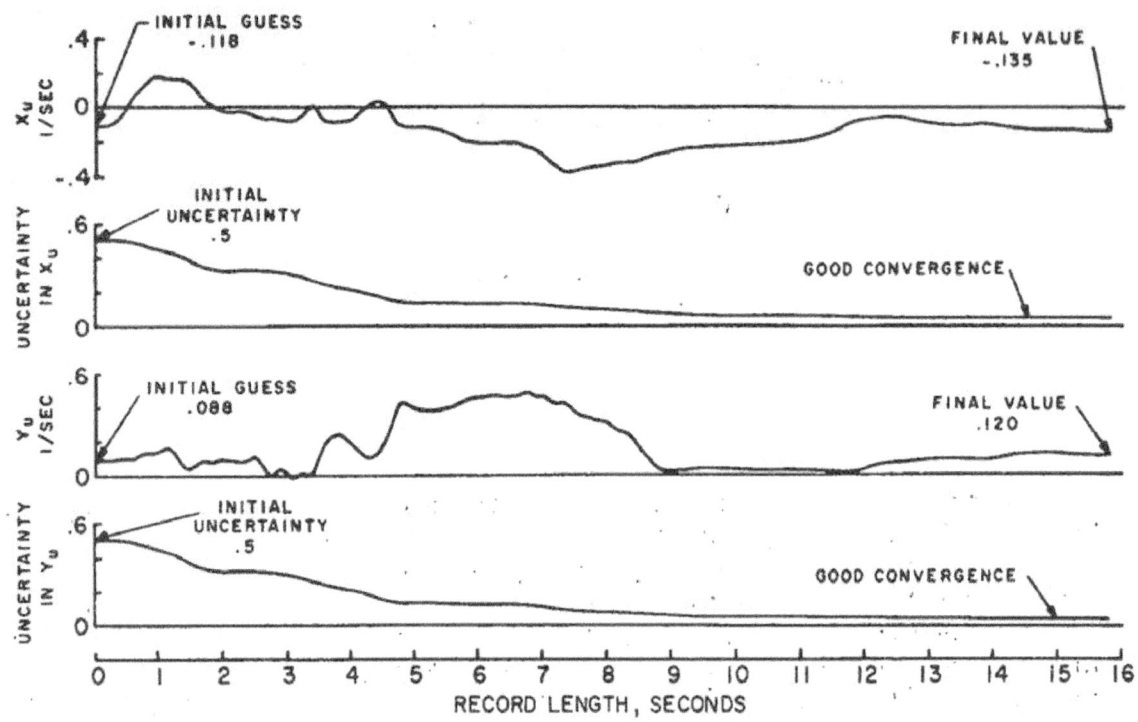

Figure 7. – Derivative Convergence and 1 sigma Uncertainty from the Extended Kalman Filter Method Using One Long Data Record (CH-54B Flight Data, 45 kts)

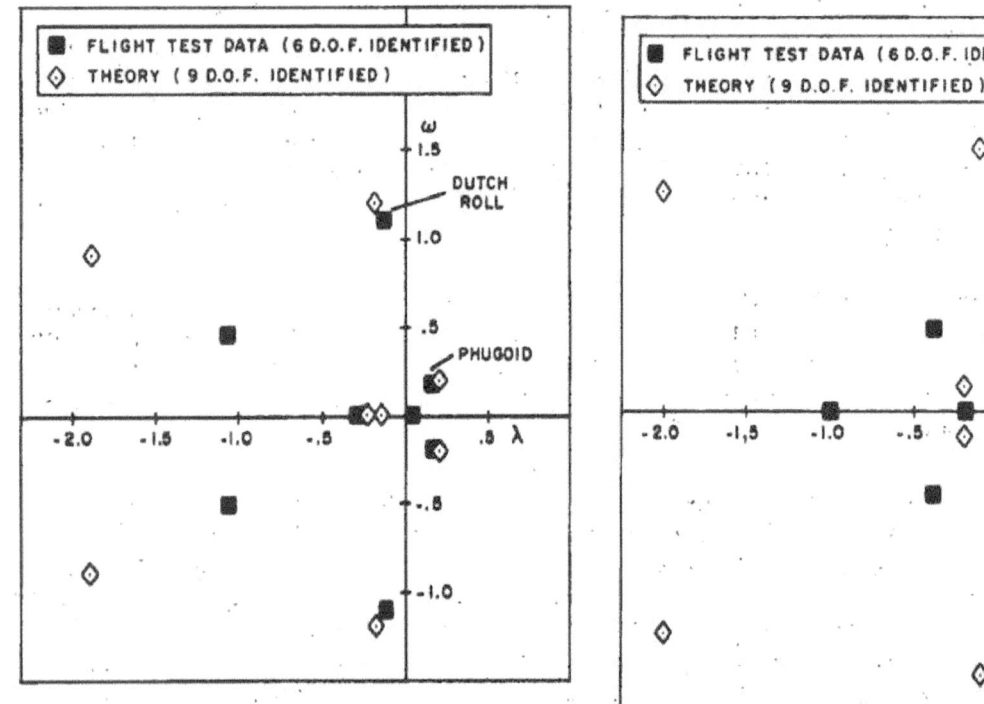

Figure 8. – Characteristic Roots Identified from Flight Data by the Kalman Filter Method vs. Analytic Model Roots (CH-53A, 100 kts, AFCS off)

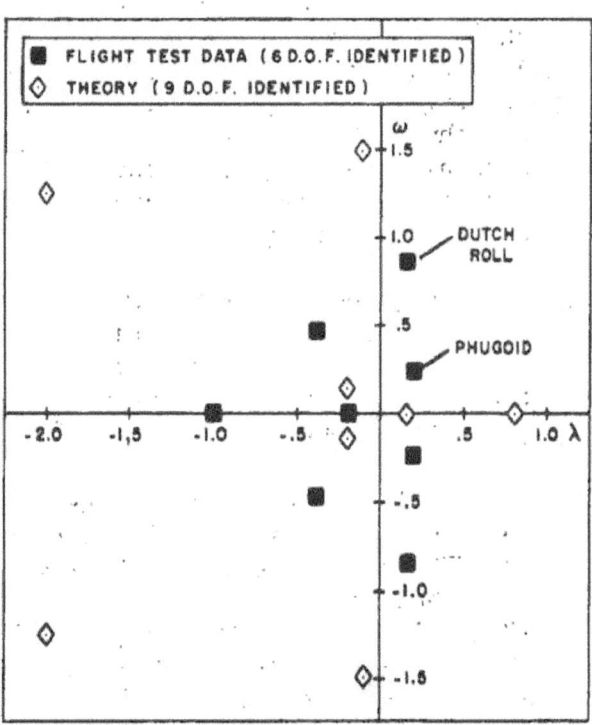

Figure 9. – Characteristic Roots Identified from Flight Data by the Kalman Filter Method vs. Analytic Model Roots (CH-53A, 150 kts, AFCS off)

IDENTIFICATION OF HIGH PERFORMANCE
AIRCRAFT STABILITY AND CONTROL COEFFICIENTS
AT HIGH ANGLES OF ATTACK

By W. E. Hall and D. E. Stepner

Systems Control, Inc.
Palo Alto, California

ABSTRACT

Aircraft parameter identification is the process of extracting numerical values for the aerodynamic stability and control coefficients, and other subsidiary parameters (wind gust statistics, sensor errors, etc.) from a set of flight test data (a time history of the flight control inputs and the resulting aircraft response variables). The key elements for achieving this goal are: (1) proper choices of model structure and identifiable parameters, (2) the identification algorithm, (3) the flight control input, and (4) the instrumentation.

A program to integrate these elements into a high angle of attack aircraft parameter identification technology is described. The basic tools of this program discussed in this paper are: (1) a nonlinear aircraft simulation, (2) a preprocessing algorithm for determining the structure and significant parameters to be identified, and (3) the maximum likelihood identification algorithm. The purpose of the simulation and the structure determination algorithm is to isolate those problems in the parameter estimation which are known to occur from over-parameterization, inadequate instrumentation, and improper control inputs.

The six degree-of-freedom nonlinear aircraft simulation includes detailed representations of the aircraft dynamics, aerodynamics, control system, and instrumentation. This simulation, correlated with observed phenomena of the high angle of attack regions, is used to generate data which is representative of that obtained from flight test.

The output data from the simulation corresponding to a known control input and known measurement statistics are used as the input to a model structure determination algorithm. This algorithm is based on an optimal subset regression technique. On the basis of an exhaustive assumed a priori model, this model regression algorithm determines which of these assumed model parameters significantly affect the aircraft response, and discards the rest. Subsequent parameter identification via the maximum likelihood method would assign values to this reduced parameter set.

IDENTIFICATION OF NONLINEAR AERODYNAMIC STABILITY AND CONTROL

PARAMETERS AT HIGH ANGLES OF ATTACK[*]

By B.J. Eulrich and E.G. Rynaski

Calspan Corporation
Buffalo, New York

ABSTRACT

This paper presents the research program plan, the techniques used and the progress to date of the identification of the aerodynamic characteristics of the post-stall gyrations of the F-4 aircraft from records taken during the Air Force acceptance tests of this high performance fighter. The major emphasis is placed on the practical considerations in using identification techniques when analyzing experimental data, such as the determination of instrumentation errors and the development and verification of the aerodynamic representation of the model.

The primary identification techniques used are a nonlinear, locally-iterated Kalman filter/fixed-point smoother algorithm and a least squares equation error method. Model form is initially determined from wind tunnel data by representing the force and moment coefficients by Taylor's series expansion for selected ranges of angle-of-attack. This leads to a large number of unknown parameters, which in many cases are not all identifiable from the flight data. The least squares method is used, along with past experience and physical reasoning, to provide the initial parameter and covariance estimates for the iterated Kalman filter and also to provide an initial indication as to parameter identifiability. However, prior to the extraction of the aerodynamic coefficients with the least squares method, state estimation and instrument error identification are performed with the Kalman filter. This increases the accuracy of the least squares results and allows easy separation of instrumentation errors from errors in the assumed form of the aerodynamic model.

To reduce the computation burden, without unduly sacrificing accuracy when employing the iterated Kalman filter, the six-degree-of-freedom equations of motion of the airplane have been separated into two systems, one for extracting the longitudinal coefficients and the other for extracting the lateral-directional coefficients. Preliminary results are presented in the form of time history matches and extracted coefficients.

---

[*] This work is being supported under Contract No. F33615-72-C-1248, Air Force Flight Dynamics Laboratory, Wright-Patterson Air Force Base, Ohio.

189

NONLINEAR PARAMETER IDENTIFICATION - BALLISTIC RANGE

EXPERIENCE APPLICABLE TO FLIGHT TESTING

By Gary Chapman and Donn Kirk

NASA Ames Research Center
Moffett Field, California

ABSTRACT

Over the past fifteen years, considerable experience in aerodynamic parameter identification has been obtained at the Ames Research Center. In particular, attention has been given to parameter identification involving nonlinear aerodynamic models and large amplitude motion. The purpose of this paper is to discuss the results of recent efforts that can be applied to aircraft flight testing.

The parameter identification scheme being used is a differential correction least squares procedure (Gauss-Newton method). The position, orientation, and derivatives of these quantities with respect to the parameters of interest (i.e., sensitivity coefficients) are determined by digital integration of the equations of motion and the parametric differential equations (ref. 1). The application of this technique to three vastly different sets of data is used to illustrate the versatility of the method and to indicate some of the problems that still remain. The first set of data are for the X-15 research aircraft. These data were originally obtained over ten years ago but could not be analyzed with the techniques in use at that time. The data to be analyzed are from a slowly rolling flight at small angles of attack. For these conditions, a conventional linear aerodynamic model is applicable and the fits to the $\alpha - \beta$ motion obtained with the linear model were very good (see fig. 1). The aerodynamic parameters determined (not shown) by this technique were found to agree well with flight and wind tunnel values.

The second example deals with motions of an axisymmetric vehicle trimmed near 90° angle of attack. This type of motion (see figs. 2a and b) is very similar to an aircraft in a flat spin with some residual oscillations in angle of attack. The residual oscillations here were sufficiently large to require using a nonlinear force and moment system. Both a conventional force and moment representation and that of Tobak, Schiff, Peterson and Levy (refs. 2 and 3) (TSPL) were utilized. The equations of motion were more easily handled in an Eulerian system (resultant angle and its orientation) than in a conventional modified Eulerian system (yaw, pitch and roll). Past experience in nonlinear parameter identification (ref. 4) has shown that the angular motion waveform is not sensitive to nonlinearities (see fig. 3); therefore, simultaneous reduction of several sets of data (four, in this case) with different amplitudes was required to define the nonlinearities. Long data samples, if available, will often fulfill the same requirement. The static forces and moments determined from these free flight data obtained at M ~ 14 are in relatively good agreement with some low Mach number (M = 3.2) wind tunnel data (see figs. 4 a, b, and c). It was not possible with these data to ascertain whether there is an advantage to the TSPL force and moment representation because the pitch and spin damping (where the major difference between modeling systems appear) are both so small as to make good identification difficult.

The last case to be presented is again for an axisymmetric body trimmed near zero degrees with an amplitude near 20 degrees. The test was a free flight wind tunnel test (ref. 5). This case represents an as yet unsolved modeling problem and illustrates some of the things that can be tried and the implications involved. The motion was first analyzed with the conventional linear static moment representation and constant damping. The fit that was obtained (not shown) resulted in a standard deviation of $1.1°$. A cubic term was then added to the static moment; this addition resulted in a significant improvement in the standard deviation of fit $(0.73°)$. This fit, shown in figure 5, is still not nearly as good as the scatter in the smooth data suggests it should be. In addition, the differences between the predicted curve (fit) and the smoothed data were not distributed uniformly about the predicted curve. In an attempt to improve the fit, additional nonlinear static moment terms were tried with no significant improvement. Finally, the nonlinear spin damping $(N_{\dot{\theta}})$ of the TSPL moment representation was tried, but no improvement was noted. At the present time it is not understood why the modeling attempts do not produce better fits to the data; however, the discrepancy may be associated with nonconformities in the wind tunnel flow.

An important question, for which as yet there is no answer, is whether the addition of the cubic term as described above makes any sense when the residuals are strongly correlated, even though it did produce a significant reduction in the standard deviation. Attempts to improve the fit and understand the modeling problem are continuing.

In summary, the parameter identification technique in use on ballistic range data has been successful in handling some nonlinearities, but modeling remains a problem in many cases, as our one example showed. The differences between parameter identification in ballistic ranges and in aircraft flight testing are more of degree than of type, with the possible exception of turbulence; however, much flight testing is not strongly influenced by turbulence and in this respect, the ballistic range experience can be of use.

REFERENCES

1. Chapman, Gary T., and Kirk, Donn B.: A Method for Extracting Aerodynamic Coefficients from Free-Flight Data, AIAA Jour., vol. 8, no. 1, April 1970.
2. Tobak, Murray, Schiff, Lewis, and Peterson, Victor L.: Aerodynamics of Bodies of Revolution in Coning Motion. AIAA Jour., vol. 7, no. 1, Jan. 1969.
3. Levy Jr., L. L., and Tobak, M.: Nonlinear Aerodynamics of Bodies of Revolution in Free Flight. AIAA Jour., vol. 8, no. 12, Dec. 1970.
4. Chapman, Gary T.: Aerodynamic Parameter Identification Ballistic Range Tests, Proceedings of the 1972 Army Numerical Analysis Conference, Edgewood Arsenal, Edgewood, Md., April 1972.
5. Jaffe, Peter: Non-Planar Tests Using the Wind Tunnel Free-Flight Technique, AIAA 2nd Atmospheric Flight Mechanics Conference, Palo Alto, Calif., Paper No. 72-983, Sept. 11-13, 1972.

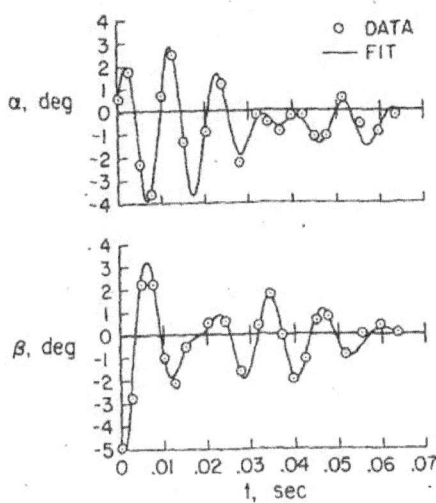

Figure 1.- Curve fit to X-15 airplane model motion obtained in a ballistic range.

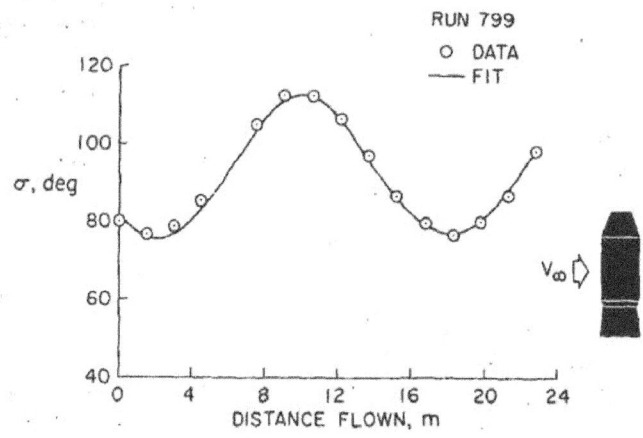

(a) Resultant angle of attack versus distance flown.

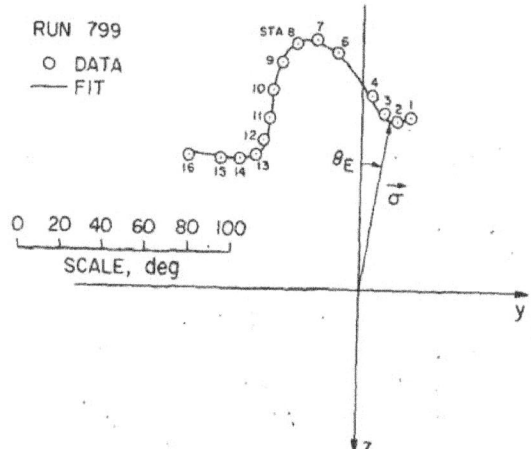

(b) Large amplitude motion about a near 90° trim point.

Figure 2.- The motion of an axisymmetric body near a 90° trim point.

193

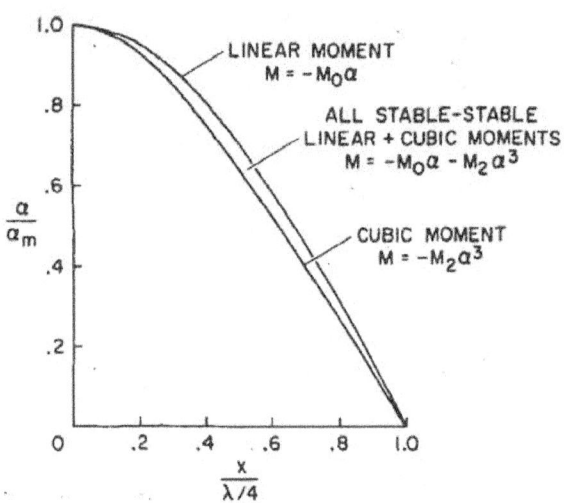

Figure 3.- Effect of nonlinear pitching moment on waveform of oscillation.

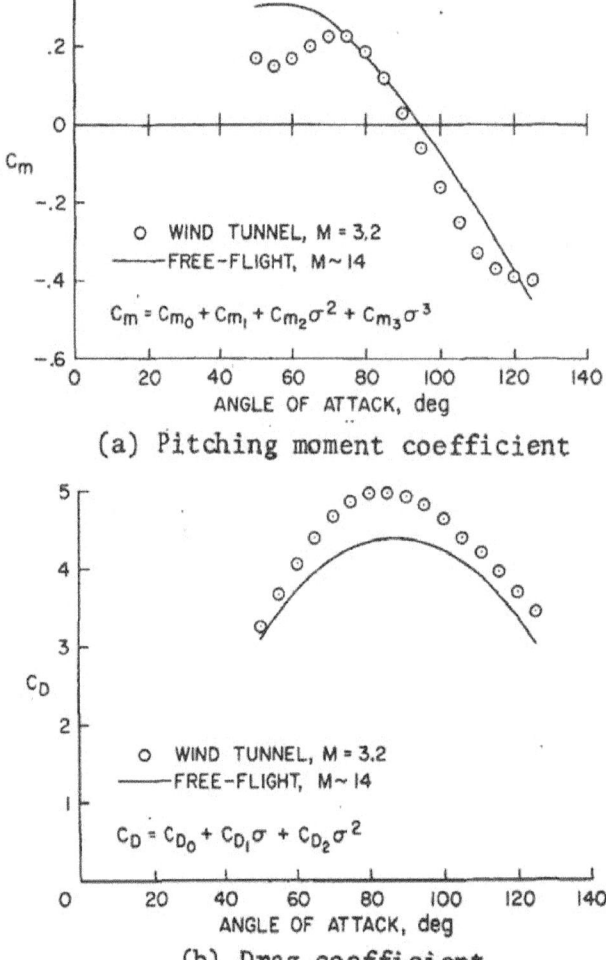

(a) Pitching moment coefficient

(b) Drag coefficient

Figure 4.- The aerodynamic characteristics of an axisymmetric body trimmed near 90° angle of attack.

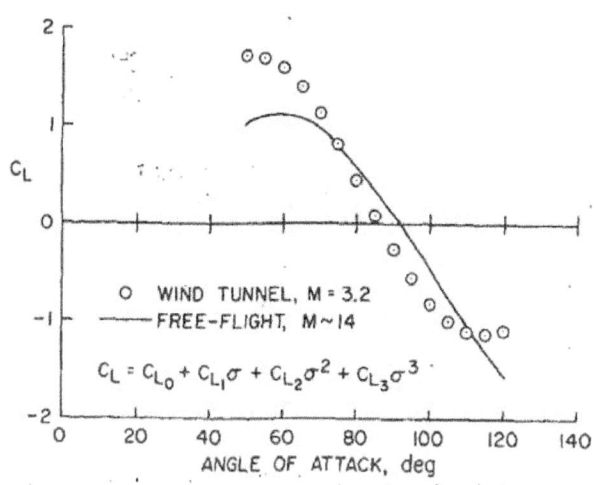

(c) Lift coefficient

Figure 4.- Concluded.

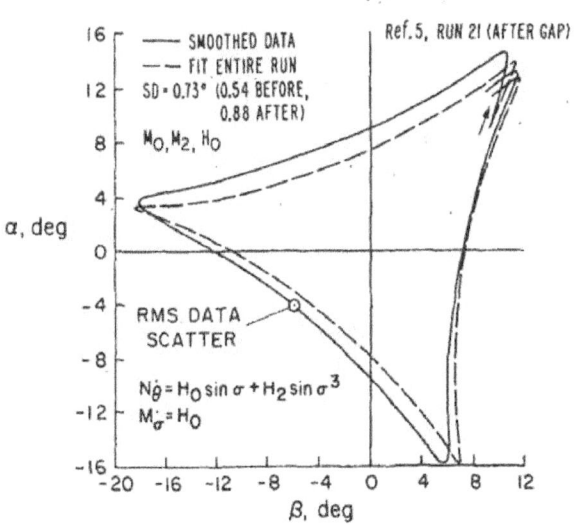

Figure 5.- A modeling problem.

Application of the Maximum Likelihood

Method to the Identification of Aircraft Parameters

at High Angles-of-Attack

By Dr. W. R. Wells and J. B. Callahan, Capt., USAF

Air Force Flight Dynamics Laboratory
Wright-Patterson Air Force Base, Ohio

ABSTRACT

An application of the maximum likelihood algorithm to the high angle-of-attack stability derivative estimation problem as applied to a fighter type aircraft is described. Used in the study are the full six degree of freedom non-linear equations of motion with the inclusion of selected terms from a Taylor's series third-order expansion of the aerodynamic force and moment coefficients. This procedure introduces into the problem higher order cross-coupling of the angles-of-attack and sideslip effects through coefficients whose values are to be determined. The algorithm uses as start-up values the linear coefficients determined by the program when the higher order coefficients are dropped, leaving the same coefficients as Groves, et. al. in a NASA-LRC study. The higher order derivatives are then considered in selected groups until all coefficients are found. Preliminary results indicate numerical problems associated with the inversion of ill-conditioned matrices. A convergent solution is highly dependent upon the "goodness" of the start-up values. Since it is necessary that the coefficient considered must make a change in the aircraft response, further difficulties arise from the uncertainty that the parameter in question is, in fact, reflected in the flight data.

IDENTIFICATION OF AIRCRAFT STABILITY AND CONTROL PARAMETERS

USING MULTILEVEL, HIERARCHICAL ESTIMATION*

By C. M. Fry and A. P. Sage

Information and Control Sciences Center
SMU Institute of Technology
Dallas, Texas 75222

SUMMARY

Previous attempts to identify aircraft stability and control derivatives from flight test data, using three-degrees-of-freedom (3-DOF) longitudinal or lateral-directional perturbation-equations-of-motion models, suffer from the disadvantage that the coupling between the longitudinal and lateral-directional dynamics has been ignored. In this paper the identification of aircraft stability parameters is accomplished using a more accurate 6-DOF model which includes this coupling. Hierarchical system identification theory is used to reduce the computational effort involved. The 6-DOF system of equations is decomposed into two 3-DOF subsystems, one for the longitudinal dynamics and the other for the lateral-directional dynamics. The two subsystem parameter identification processes are then coordinated in such a way that the overall system parameter identification problem is solved.

1. INTRODUCTION

The desirability of obtaining accurate estimates of aircraft stability and control derivatives from flight test data has been recognized for some time. The identification of aircraft derivatives from flight test data is a problem in parameter identification. Parameter identification problems have been extensively investigated. References [1-2] are survey papers and references [4] and [5] are books on system identification. There has also been a good deal of effort devoted to the application of system identification techniques to the problem of aircraft parameter identification. Mehra [6,7] has made perhaps the most successful contribution thus far to the solution of this problem. Reference [7] contains a summary of various methods of parameter identification and also contains numerous references to other contributions to the aircraft parameter identification problem.

Most previous attempts to identify aircraft stability and control derivatives from flight test data have been hindered by the fact that the airframe was modeled using three-degrees-of-freedom (3-DOF) linear perturbation equations of motion. The longitudinal and lateral-directional 3-DOF models ignore the effects of dynamic coupling between the longitudinal and lateral-directional motion. This dynamic coupling is often significant in flight test manuevers. Any attempt to identify parameters in a 3-DOF model using flight test data in which the coupling is significant will lead to error in the parameter estimates. This is due to the fact that the identification process essentially adjusts the parameters of the model so that the model time histories match, as well as possible, the aircraft flight test time histories. Any error due to use of an overly simplified model will show up as error in the parameter estimates.

This research was supported by the National Science Foundation under Grant GK 33348.

The present work is an attempt to at least partially avoid errors in the parameter estimates caused by inaccurate modeling. Six-DOF nonlinear perturbation equations of motion, which include dynamic coupling between the longitudinal and lateral-directional modes, are used as the airframe model. Unfortunately, the fact that the 6-DOF model is nonlinear and is higher-dimensional than a 3-DOF model makes the solution of the parameter identification problem much more demanding computationally. This difficulty can be overcome to a great extent by using hierarchical estimation techniques to reduce the computational effort involved. In this paper the 6-DOF system of equations is decomposed into two 3-DOF subsystems, one for the longitudinal dynamics and one for the lateral-directional dynamics. The two subsystem parameter identification processes are then coordinated in such a way that the overall system parameter identification problem is solved.

2. SYSTEM IDENTIFICATION USING HIERARCHICAL SYSTEM THEORY

The basic idea of hierarchical system theory is to decompose a large unwieldy system into several smaller subsystems which are more easily analyzed. The analyses of the subsystems are then coordinated in such a way that the overall system problem is solved. Smith and Sage [8] have presented a very readable introduction to hierarchical system theory as applied to the dynamic optimization problem. A more general discussion of the dynamic optimization problem is given by Pearson [9] while a very general development of hierarchical system theory has been accomplished by Mesarovic, Macko and Takahara [10]. Smith and Sage [11] have applied hierarchical system theory to the problem of system identification. A summary of the development of their identification algorithms will be presented to serve as an introduction to hierarchical system identification.

Continuous-Time MAP Estimation

The parameter identification problem can be effectively approached through the use of state estimation techniques. The maximum-a-posteriori (MAP) method of state estimation will now be presented to serve as a tool for the identification of system parameters. Consider the nth order nonlinear continuous-time system given by

$$\dot{x}(t) = \overline{f}[x(t),t] + w(t) \qquad (2.1)$$

and

$$z(t) = h[x(t),t] + v(t) \qquad (2.2)$$

with $E\{x(t_0)\} = \mu_{xo}$ and $\text{var}\{x(t_0)\} = V_{xo}$, where $\overline{f}$ denotes the original system function and where $x(t_0)$ is assumed to be gaussian and uncorrelated with the plant and measurement noise processes; thus,

$$\text{cov}\{x(t_0),w(t)\} = 0 = \text{cov}\{x(t_0),v(t)\}$$

for all $t > t_0$ where we define $\text{cov}\{a,b\} \triangleq E\{[a-\mu_a][b-\mu_b]^T\}$.

200

Equation (2.1) represents the plant state variable dynamics and equation (2.2) represents the state observations where

> $x(t)$ is the n-dimensional state vector,
>
> $\overline{f}[x(t),t]$ is the n-dimensional vector-valued nonlinear function describing the plant structure and includes any known inputs,
>
> $w(t)$ is the n-dimensional plant noise vector,
>
> $z(t)$ is the r-dimensional observation vector,
>
> $h[x(t),t]$ is the r-dimensional vector-valued nonlinear function describing the relation between the system states and observations, and
>
> $v(t)$ is the r-dimensional measurement noise vector.

It is assumed that the plant and measurement noises are uncorrelated, zero-mean, white gaussian processes with covariances

$$\text{cov}\{w(t),w(\tau)\} = \Psi_w(t)\delta_D(t-\tau)$$

$$\text{cov}\{v(t),v(\tau)\} = \Psi_v(t)\delta_D(t-\tau)$$

and

$$\text{cov}\{w(t),v(\tau)\} = 0$$

The (nxn) matrix $\Psi_w(t)$ and the (rxr) matrix $\Psi_v(t)$ are assumed to be positive definite symmetric covariance matrices.

Now let $X(t_f)$ and $Z(t_f)$ be defined by the sets

$$X(t_f) \overset{\Delta}{=} \{x(t) : t_0 \le t \le t_f\}$$

$$Z(t_f) \overset{\Delta}{=} \{z(t) : \ _0 \ \ \ \ \le t_f\}$$

Then $p[X(t_f)|Z(t_f)]$ represents the conditional probability density of $X(t_f)$ given the observations $Z(t_f)$. The maximum-a-posteriori (MAP) estimate of $x(t)$ over the interval $[t_0 \le t \le t_f]$ is that estimate obtained by maximizing the density $p[X(t_f)|Z(t_f)]$ with respect to $X(t_f)$. It is shown in references [4] and [12] that maximizing $p[X(t_f)|Z(t_f)]$ with respect to $X(t_f)$ is equivalent to minimizing with respect to $w(t)$ the functional

$$\overline{J} = \frac{1}{2}\|x(t_0)-\mu_x(t_0)\|^2_{V^{-1}_{xo}} + \frac{1}{2}\int_{t_0}^{t_f}\left\{\|z(t)-h[x(t),t]\|^2_{\Psi_v^{-1}(t)} \right.$$

$$\left. + \|w(t)\|^2_{\Psi_w^{-1}(t)}\right\} dt \tag{2.3}$$

where t_f is fixed, subject to the dynamic equality constraint given by equation (2.1)

$$\dot{x}(t) = \overline{f}[x(t),t] + w(t)$$

Thus the MAP estimate of the system state $X(t_f)$ given the observations $Z(t_f)$ can be obtained by solving the above dynamic optimization problem.

Thus the MAP estimate of the system state $X(t_f)$ given the observations $Z(t_f)$ can be obtained by solving the above dynamic optimization problem.

The optimization problem defined by equations (2.1) and (2.3) can be solved by applying the Pontryagin Maximum Principle [13]. Forming the Hamiltonian, we have

$$\overline{H}[x(t),w(t),\Lambda(t),t] = \frac{1}{2}\left\| z(t)-h[x(t),t]\right\|^2_{\Psi_v^{-1}(t)} + \frac{1}{2}\left\| w(t)\right\|^2_{\Psi_w^{-1}(t)}$$

$$+ \Lambda^T(t)\{\overline{f}[x(t),t] + w(t)\} \tag{2.4}$$

where $\Lambda(t)$ is an n-dimensional vector Lagrange multiplier. The necessary conditions for a minimum of $\overline{J}$ are that

$$\frac{\partial \overline{H}}{\partial \hat{w}} = 0 = \Psi_w^{-1}(t)\hat{w}(t) + \Lambda(t)$$

so that

$$\hat{w}(t) = -\Psi_w(t)\Lambda(t) \tag{2.5}$$

and

$$\frac{\partial \overline{H}}{\partial \Lambda} = \dot{\hat{x}}(t) = \overline{f}[\hat{x}(t),t] + \hat{w}(t) \tag{2.6}$$

$$-\frac{\partial \overline{H}}{\partial \hat{x}} = \dot{\Lambda}(t) = \frac{\partial h^T[\hat{x}(t),t]}{\partial \hat{x}(t)}\Psi_v^{-1}(t)\{z(t)-h[\hat{x}(t),t]\} - \frac{\partial \overline{f}^T[\hat{x}(t),t]}{\partial \hat{x}(t)}\Lambda(t) \tag{2.7}$$

with boundary conditions

$$\Lambda(t_0) = -V_{xo}^{-1}[\hat{x}(t_0)-\mu_{xo}] \tag{2.8}$$

and

$$\Lambda(t_f) = 0 \tag{2.9}$$

where $\hat{x}(t)$ and $\hat{w}(t)$ represent optimum values. Equations (2.5) through (2.9) represent a two-point boundary-value problem (TPBVP) whose solution will yield the fixed-interval smoothing estimate of $x(t)$, denoted $\hat{x}(t|t_f)$, which is the estimate of $x(t)$, $t\varepsilon[t_0,t_f]$, based upon the data $Z(t_f)$ which represents all of the observations $z(t)$, $t\varepsilon[t_0,t_f]$. Thus, strictly speaking, the solution of the TPBVP must be performed off-line after all of the observation data are available. However, if the TPBVP is solved by means of invariant imbedding, an algorithm is obtained which yields the filter estimate of $x(t)$, denoted $\hat{x}(t|t)$, which is based only on the observation set $\{z(\tau) : t_0 \leq \tau \leq t\}$. This is a desirable solution since it allows sequential processing of the observation data.

Parameter Identification Using State Estimation

If there are unknown parameters in the plant or observation models, state estimation can be used to identify them. Suppose the system plant and observation equations are written as

$$\dot{y} = \overline{f}'[y(t),p(t),t] + w'(t) \tag{2.10}$$

and

$$z(t) = h'[y(t),p(t),t] + v(t) \tag{2.11}$$

where $y(t)$ is the state vector and $p(t)$ is the vector of unknown parameters.

202

Let the unknown parameter vector be modeled by

$$\dot{p}(t) = Ap(t) + w''(t) \tag{2.12}$$

where $w''(t)$ is a zero-mean gaussian white noise vector. If $p(t)$ is a constant vector, it is often adequate to model it by letting the A matrix in (2.12) be the null matrix so that

$$\dot{p}(t) = w''(t) \tag{2.13}$$

The unknown parameter vector $p(t)$ can be adjoined to the state vector $y(t)$ to form an augmented state vector $x(t)$ defined by

$$x(t) \triangleq \begin{bmatrix} y(t) \\ p(t) \end{bmatrix}$$

Now equations (2.10)-(2.12) can be combined and rewritten in the form of equations (2.1) and (2.2) by defining

$$\bar{f}[x(t),t] \triangleq \begin{bmatrix} \bar{f}'[y(t),p(t),t] \\ Ap(t) \end{bmatrix} \quad ; \quad w(t) \triangleq \begin{bmatrix} w'(t) \\ w''(t) \end{bmatrix}$$

Thus, the MAP estimate of the state of the new system will yield estimates of both the states and unknown parameters of the original system. The noise term $w''(t)$ in the parameter model is needed to prevent the phenomena of data saturation and divergence of the estimation algorithm. Reference [4] contains considerable discussion of this approach to system identification.

Hierarchical System Identification

It is an unfortunate consequence of the state augmentation approach to parameter identification that the dimension of the state vector is increased. This fact can often make the direct solution of MAP estimation TPBVP computationally prohibitive. Hierarchical system theory can be applied to the solution of the MAP minimization problem to reduce the computational effort. This is accomplished by decomposing the original minimization problem into several more easily solved subproblems. These subproblems are coordinated in such a way that their composite solution results in the solution of the overall problem.

Assume that the original system described by equations (2.1) and (2.2) can be be decomposed into N subsystems where the ith subsystem state vector has dimension n_i and the ith subsystem observation vector has dimension r_i so that

$$\sum_{i=1}^{N} n_i = n \quad \text{and} \quad \sum_{i=1}^{N} r_i = r$$

where the state equation for the ith subsystem is given by

$$\dot{x}_i(t) = f_i[x_i(t),\pi_i(t),t] + w_i(t) \tag{2.14}$$

and where the ith subsystem observation is given by

$$z_i(t) = h_i[x_i(t),t] + v_i(t)$$

The vector $\pi_i(t)$ represents the coupling into the ith subsystem from all of the other subsystems. Thus, $\pi_i(t)$ is a function only of states other than those in $x_i(t)$ so that

$$\pi_i(t) = g_i[x_j(t)] \quad , \quad j \neq i \tag{2.15}$$

The equations (2.15) are called interconnection constraints. Now assume that $\Psi_w(t)$ and $\Psi_v(t)$ are block diagonal matrices so that

$$\Psi_w = \begin{bmatrix} \Psi_{w_1} & & 0 \\ & \Psi_{w_2} \cdot & \\ & & \cdot \\ 0 & & \Psi_{w_N} \end{bmatrix} \quad , \quad \Psi_v = \begin{bmatrix} \Psi_{v_1} & & 0 \\ & \Psi_{v_2} \cdot & \\ & & \cdot \\ 0 & & \Psi_{v_N} \end{bmatrix}$$

where Ψ_{w_i} is an $(n_i \times n_i)$ dimensional matrix and Ψ_{v_i} is an $(r_i \times r_i)$ dimensional matrix. This assumption implies that there is no plant noise correlation and no measurement noise correlation among the subsystems. If such correlation does exist, Ψ_w and Ψ_v can be made block diagonal by applying standard matrix techniques given in reference [11]. Thus, with Ψ_w and Ψ_v block diagonal, the performance functional $\overline{J}$ given by equation (2.3) can be decomposed into the additive form

$$\overline{J} = \sum_{i=1}^{N} \overline{J}_i \tag{2.16}$$

where

$$\overline{J}_i = \frac{1}{2} \| x_i(t_0) - \mu_{x_{io}} \|^2_{V_{x_{io}}^{-1}} + \frac{1}{2} \int_{t_0}^{t_f} \left\{ \| z_i - h_i[x_i, t] \|^2_{\Psi_{v_i}^{-1}} + \| w_i \|^2_{\Psi_{w_i}^{-1}} \right\} dt \tag{2.17}$$

Thus, the overall minimization problem can be solved by solving the N minimization problems defined by minimizing $\overline{J}_i$ with respect to $w_i(t)$ subject to the constraints given by equations (2.14) and (2.15).

The overall cost functional may be rewritten as

$$J = \sum_{i=1}^{N} \left\{ \overline{J}_i + \int_{t_0}^{t_f} \beta_i^T(t) [\pi_i(t) - g_i(x_j)] dt \right\} \tag{2.18}$$

where equations (2.15) have been adjoined to assess a penalty for failing to satisfy the interconnection constraints. We will assume that the system decomposition can be carried out so that

$$g_i(x_j) = \sum_{\substack{j=1 \\ j \neq i}}^{N} g_{ij}(x_j)$$

By algebraic manipulation it can be shown that

$$\sum_{i=1}^{N} \beta_i^T \sum_{\substack{j \neq i}}^{N} g_{ij}(x_j) = \sum_{i=1}^{N} \sum_{\substack{j \neq i}}^{N} \beta_j^T g_{ji}(x_i) \tag{2.19}$$

Then the overall cost J may be written as

$$J = \sum_{i=1}^{N} \left\{ \overline{J}_i + \int_{t_0}^{t_f} \left[\beta_i^T \pi_i - \sum_{\substack{j \neq i}}^{N} \beta_j^T g_{ji}(x_i) \right] dt \right\} \tag{2.20}$$

204

or

$$J = \sum_{i=1}^{N} J_i \tag{2.21}$$

where

$$J_i = \frac{1}{2}\|x_i(t_0) - \mu_{x_{io}}\|^2_{V^{-1}_{x_{io}}} + \int_{t_0}^{t_f} \left\{ \frac{1}{2}\|z_i - h_i(x_i,t)\|^2_{\Psi^{-1}_{v_i}} + \frac{1}{2}\|w_i(t)\|^2_{\Psi^{-1}_{w_i}} \right.$$

$$\left. + \beta_i^T \pi_i - \sum_{j \neq i}^{N} \beta_j^T g_{ji}(x_i) \right\} dt \tag{2.22}$$

Thus, the minimization problem associated with the $\underline{i}$th subsystem is to minimize equation (2.22) with respect to $w_i(t)$ subject to equation (2.14) given by

$$\dot{x}_i(t) = f_i[x_i, \pi_i, t] + w_i(t) \tag{2.23}$$

Rather than write necessary conditions for $\beta_i(t)$ and $\pi_i(t)$, we will use the Prediction Principle of Mesarovic, $\underline{et\ al.}$ [10]. Using the Prediction Principle, a supremal controller or supremal unit predicts values for the variables $\pi(t)$ and $\beta(t)$ and supplies these values to the subsystem minimization processes. The minimization problem associated with the $\underline{i}$th subsystem will be hereafter referred to as the $\underline{i}$th infimal unit. The infimal units then solve their problems using the values of $\pi(t)$ and $\beta(t)$ supplied by the supremal. The supremal unit then uses the results of the infimal solutions to predict new values for $\pi(t)$ and $\beta(t)$. This process is repeated until $\pi(t)$ and $\beta(t)$ are correctly predicted so that the interconnection constraints are satisfied. Because the function of the supremal unit is to coordinate the infimal units, it is sometimes referred to as the "coordinator" and $\pi(t)$ and $\beta(t)$ are called "coordination variables." Since $\pi(t)$ is associated with the subsystem interconnections, it is called the "model coordination variable". Since $\beta(t)$ is more closely associated with the infimal performance functions, it is referred to as the "goal coordination variable".

The infimal minimization problems can be solved using the Maximum Principle. The $\underline{i}$th Hamiltonian may be written as

$$H_i = \frac{1}{2}\|z_i - h_i(x_i,t)\|^2_{\Psi^{-1}_{v_i}} + \frac{1}{2}\|w_i\|^2_{\Psi^{-1}_{w_i}} + \lambda_i^T[f_i(x_i,\pi_i,t) + w_i]$$

$$+ \beta_i^T \pi_i - \sum_{j \neq i}^{N} \beta_j^T g_{ji}(x_i) \tag{2.24}$$

Necessary conditions for a minimum of J_i are

$$\frac{\partial H_i}{\partial \hat{w}_i} = 0 = \Psi^{-1}_{w_i} \hat{w}_i + \lambda_i$$

so that

$$\hat{w}_i = -\Psi_{w_i} \lambda_i \tag{2.25}$$

and

$$\frac{\partial H_i}{\partial \lambda_i} = \dot{\hat{x}}_i = f_i(\hat{x}_i, \pi_i, t) + \hat{w}_i \qquad (2.26)$$

$$-\frac{\partial H_i}{\partial \hat{x}_i} = \dot{\lambda}_i = \frac{\partial h_i^T(\hat{x}_i, t)}{\partial \hat{x}_i} \Psi_{v_i}^{-1}[z_i - h_i(\hat{x}_i, t)] - \frac{\partial f_i^T(\hat{x}_i, \pi_i, t)}{\partial \hat{x}_i} \lambda_i$$

$$+ \frac{\partial}{\partial \hat{x}_i} \left[\sum_{\substack{j \neq i}}^{N} \beta_j^T g_{ji}(\hat{x}_i) \right] \qquad (2.27)$$

with

$$\lambda_i(t_0) = -V_{x_{io}}^{-1}[\hat{x}_i(t_0) - \mu_{x_{io}}] \qquad (2.28)$$

and

$$\lambda_i(t_f) = 0 \qquad (2.29)$$

The TPBVP defined by equations (2.25)-(2.29) can be solved using the method of continuous invariant imbedding [4,11-13] to obtain the state estimation algorithm given in Table 1. It is the nature of the invariant imbedding solution that the filter estimate $\hat{x}_i(t|t)$ is obtained. Thus the algorithm of Table 1 is processed sequentially.

Coordination Procedures

Thus far in the present development, the exact procedure the supremal unit uses to predict new values of the coordination variables has not been discussed. There are actually many ways to accomplish coordination of the infimal units. The particular method used depends upon the nature of both the problem treated and the desired solution. The one overriding requirement that the coordination procedure must meet is that it must, of course, result in a convergent algorithm for the overall problem solution. One coordination procedure using the Prediction Principle that has shown good convergence properties is the equality method of Guinzy and Sage [14]. This procedure is developed by comparing the TPBVP for the coordinated system with the TPBVP for the uncoordinated system. For the uncoordinated system, the Hamiltonian is, from equations (2.14)-(2.17)

$$\bar{H}_u = \sum_{i=1}^{N} \left\{ \frac{1}{2} \|z_i - h_i(x_i, t)\|_{\Psi_{v_i}^{-1}}^2 + \frac{1}{2} \|w_i\|_{\Psi_{w_i}^{-1}}^2 + \lambda_i^T \left[f_i(x_i, g_i(x_j), t) + w_i \right] \right\} \qquad (2.30)$$

If we compare the canonical equations for the coordinated and uncoordinated ith infimal unit, it becomes evident that for the TPBVP's to be equivalent, it is required that

$$\pi_i(t) = g_i[\hat{x}_j(t)] \qquad (2.31)$$

and

$$\sum_{\substack{j \neq i}}^{N} \frac{\partial}{\partial \hat{x}_i} [g_{ji}^T(\hat{x}_i)] \beta_j = -\sum_{\substack{j \neq i}}^{N} \frac{\partial}{\partial \hat{x}_i} f_j^T[\hat{x}_j, g_j(\hat{x}_i), t] \lambda_j \qquad (2.32)$$

for $i=1,2,\ldots,N$. Unfortunately, this coordination procedure, as stated above, is nonsequential. The supremal unit must supply the infimal units with

206

coordination variables $\beta(t)$ and $\pi(t)$ for all $t\varepsilon[t_0,t_f]$. Then the infimal units must solve their problems based on these predictions for all $t\varepsilon[t_0,t_f]$. Only then can the supremal predict new values of $\beta(t)$ and $\pi(t)$. Thus, we see that even though the infimal estimation algorithm is sequential, the overall system algorithm is iterative.

One method for obtaining an overall-sequential algorithm is to use the predictor-corrector coordination method of Smith and Sage [15]. The procedure developed there is to use approximations to the derivatives of the coordination variables given by

$$\overset{\bullet}{\beta}(t) = \frac{1}{\Delta t}\,[\beta(t)-\beta(t-\Delta t)] \qquad (2.33)$$

and

$$\overset{\bullet}{\pi}(t) = \frac{1}{\Delta t}\,[\pi(t)-\pi(t-\Delta t)] \qquad (2.34)$$

to predict new values for the coordination variables according to

$$\beta_p(t+\Delta t) = \beta(t) + \overset{\bullet}{\beta}(t)\Delta t \qquad (2.35)$$

and

$$\pi_p(t+\Delta t) = \pi(t) + \overset{\bullet}{\pi}(t)\Delta t \qquad (2.36)$$

where the subscript p denotes predicted value. The supremal then corrects these predicted values using the relations

$$\beta(t) = \beta_p(t) + K_\beta\,\frac{\partial H(t)}{\partial\beta(t)} \qquad (2.37)$$

and

$$\pi(t) = \frac{1}{2}\,\{\pi_p(t)+g[x(t)]\} \qquad (2.38)$$

where $\beta(t)$ is corrected to minimize the Hamiltonian using a one-step gradient technique while $\pi(t)$ is corrected using an averaging process.

3. AIRCRAFT PARAMETER IDENTIFICATION

Equations of Motion of the Aircraft

The 6-DOF airframe perturbation equations of motion used for this study were developed following reference [16]. The development of these equations will not be given, but the assumptions used in the derivation will now be listed:

1. The airframe is assumed to be a rigid body.
2. The earth is assumed to be fixed in space and the atmosphere is assumed to be fixed with respect to the earth.
3. The mass of the airplane is assumed constant over the duration of any manuever.
4. The xz plane is assumed to be a plane of symmetry.
5. Initially the airplane is assumed to be in steady flight with wings level and with all components of velocity, including angular velocity, zero except for horizontal velocity U_0 and vertical velocity W_0.
6. The air flow is assumed to be quasi-steady; that is, as the airplane changes its orientation with respect to its flight path, the air flow is assumed to change instantaneously to a steady-state flow pattern.

7. The angular <u>deflections</u> are assumed small enough to allow setting the sines of the angles equal to the angles and the cosines equal to one. Assume products of the angular deflections to be small enough to neglect.

The Eulerian axis system used will be the principal axis system since for this axis system the moment of inertia I_{xz} is zero, and there is a resulting simplification of the equations. Stability derivatives identified with respect to the principal axis system can be transformed to correspond to a different axis system by a coordinate transformation. The model of the airframe to be used here is given by:

Longitudinal Equations:

$$\dot{u} = -W_0 q - wq + vr - q\theta \cos \theta_0 + X_u u + X_q q + X_w w + X_{\delta_e} \delta_e$$

$$\dot{w} = -pv + U_0 q + qu - g\theta \sin \theta_0 + Z_u u + Z_q q + Z_w w + Z_{\delta_e} \delta_e \qquad (3.1)$$

$$\dot{q} = \left(\frac{I_{zz} - I_{xx}}{I_{yy}} \right) pr + M_u u + M_q q + M_w w + M_{\dot{w}} \dot{w} + M_{\delta_e} \delta_e$$

Lateral-Directional Equations:

$$\dot{v} = -U_0 r - ru + W_0 p + wp + g\psi \sin \theta_0 + q\phi \cos \theta_0 + Y_v v + Y_r r + Y_p p + Y_{\delta_a} \delta_a + Y_{\delta_r} \delta_r$$

$$\dot{r} = -\left(\frac{I_{yy} - I_{xx}}{I_{zz}} \right) pq + N_v v + N_r r + N_p p + N_{\delta_a} \delta_a + N_{\delta_r} \delta_r \qquad (3.2)$$

$$\dot{p} = -\left(\frac{I_{zz} - I_{yy}}{I_{xx}} \right) qr + L_v v + L_r r + L_p p + L_{\delta_a} \delta_a + L_{\delta_r} \delta_r$$

where the stability derivatives are dimensional derivatives defined by expressions of the form, reference [16],

$$X_u = \frac{1}{m} \frac{\partial X}{\partial u} \qquad \text{and} \qquad L_r = \frac{1}{I_{xx}} \frac{\partial L}{\partial r}$$

Note that terms involving products containing angular velocity perturbations have been retained. It is through these terms that the dynamic coupling between the longitudinal and lateral-directional modes is represented. The magnitude of these terms can easily become significant even for small angular deflections since the angular rates of change may become sizeable even for small angular deflections.

Identification Procedures

The method of hierarchical system parameter identification will be applied to the identification of unknown aircraft stability and control derivatives. The airframe equations of motion, equations (3.1) and (3.2), and the state observations can readily be put in the form of equations (2.1) and (2.2). Any unknown stability derivatives can be adjoined to the state vector as additional states, but the dimension of the resulting state vector will be high. The airframe equations of motion rather naturally decompose into a longitudinal subsystem and a lateral-directional subsystem since the coupling between the modes is light. A schematic of the aircraft hierarchical parameter identification process is shown in Fig. 1.

It will be assumed that atmospheric turbulence can be adequately modeled by first-order dynamical models driven by white gaussian noise. If the turbulence is zero-mean and the noise covariance terms are known, or if there is no turbulence, the results of Section 2 can be used directly to identify the unknown system parameters. However, if turbulence is present, it is unlikely that the covariance of the turbulence will be known exactly. Smith and Sage [15] have developed an adaptive estimation algorithm for hierarchical systems which can be used when the plant and observation noise moments are unknown. The procedure used there is to alter the sequential algorithm of Table 1 so as to allow their use in situations where the noises are assumed to be white and gaussian but with unknown means and variances. This is done by incorporating the adaptive estimation algorithms of Sage and Husa [17] and Sage and Wakefield [18] into the hierarchical identification algorithms. Algorithms from reference [17] are used to identify unknown measurement noise moments and algorithms from reference [18] are used to treat the case of unknown plant noise covariances. The reader is referred to the referenced papers and to the companion paper for further details.

It will be assumed that only measurements of the system states are available in the linear form

$$z(t) = Hx(t) + v(t) \tag{3.3}$$

Thus, either the linear and angular accelerations are either not available or available but not used. If measurements of these accelerations were to be used, then these observations could be adjoined to the state observation vector. Thus, suppose we have the observations:

Observation of x : $z_1 = H_1 x(t) + v_1(t)$

Observation of $\dot{x}$: $z_2 = H_2 \dot{x}(t) + v_2(t) = H_2 \bar{f}[x(t),t] + H_2 w(t) + v_2(t)$

These can be combined to form

$$z(t) = \begin{bmatrix} z_1(t) \\ z_2(t) \end{bmatrix} = \begin{bmatrix} H_1 x(t) \\ H_2 \bar{f}[x(t),t] \end{bmatrix} + \begin{bmatrix} v_1(t) \\ H_2 w(t)+v_2(t) \end{bmatrix}$$

Now, the observation equation is a nonlinear function of the system state and the measurement noise is now correlated with the plant noise. The case of correlated plant and measurement noise can be treated by writing equation (2.2) as

$$z(t) - h[x(t),t] - v(t) = 0$$

Equation (2.1) can be written, by adding "nothing" to it, as

$$\dot{x}(t) = \bar{f}[x(t),t]+w(t)+\Xi\{z(t)-h[x(t),t]-v(t)\} \tag{3.4}$$

or

$$\dot{x}(t) = \theta[x(t),t] + w(t) \tag{3.5}$$

where, since $z(t)$ is a known function of t,

$$\theta[x(t),t] = \bar{f}[x(t),t] - \Xi h[x(t),t] + \Xi z(t) \tag{3.6}$$

$$\omega(t) = w(t) - \Xi v(t) \tag{3.7}$$

Now, it is desired to adjust Ξ such that the plant noise $\omega(t)$ and the measurement noise $v(t)$ are uncorrelated. Now, postmultiplication of equation (3.7) by $v^T(\tau)$ gives

$$\omega(t)v^T(\tau) = w(t)v^T(\tau) - \Xi v(t)v^T(\tau)$$

Taking the expected value of both sides of this equation, we have that

$$V_{\omega v} - V_{wv} - \Xi V_v = 0$$

where we have enforced the requirement that $V_{\omega v} = 0$. Thus, for $\omega(t)$ and $v(t)$ to be uncorrelated, Ξ must be given by

$$\Xi = V_{wv} V_v^{-1}$$

Thus, for the case of plant and measurement noise correlation, the plant equation (2.1) can be replaced by equation (3.5); and the results obtained previously may be used directly with appropriate substitutions.

It is of interest to point out that if the atmospheric turbulence inputs were somehow observed during the flight test manuevers, then the plant noise moments would not need to be identified by adaptive estimation methods. This would be a problem in system identification with a noise-corrupted plant noise observation

$$z_2(t) = H_2 w(t) + v_2(t)$$

in addition to the state observation

$$z_1(t) = H_1 x(t) + v_1(t)$$

The observation vectors can be combined to form

$$z(t) = \begin{bmatrix} z_1(t) \\ z_2(t) \end{bmatrix} = \begin{bmatrix} H_1 \\ 0 \end{bmatrix} x(t) + \begin{bmatrix} v_1(t) \\ H_2 w(t) + v_2(t) \end{bmatrix}$$

Thus, once again the noises are correlated, and this situation can be treated just as described before. This problem is given more detailed treatment in Sage and Wakefield [19].

210

Solution Procedure

The method of parameter identification using hierarchical estimation has been applied to the problem of the identification of unknown aircraft stability derivatives. The development of the solution procedure used is now presented. The system model used in the identification is given by equations (3.1) and (3.2). It is assumed here that the plant and measurement noise covariance matrices Ψ_w and Ψ_v are known and constant. This allows the use of the estimation algorithms of Table 1. Suggestions for dealing with the case of unknown noise moments will be given later in the paper.

It is the nature of the 6-DOF airframe perturbation equations of motion that the longitudinal equations and the lateral-directional equations are rather weakly coupled. This fact can be used to advantage by employing the method of Section 2 to hierarchically structure the identification problem. The airframe equations of motion can be rewritten in state variable form by defining state variables $y_i(t)$, $i = 1,\dots,n$ as

$$
\begin{aligned}
y_1 &= u & y_5 &= u_g & y_9 &= \psi \\
y_2 &= w & y_6 &= w_g & y_{10} &= p \\
y_3 &= q & y_7 &= v & y_{11} &= \phi \\
y_4 &= \theta & y_8 &= r & y_{12} &= v_g
\end{aligned}
$$

where u_g, w_g, and v_g are the x, y, and z components of atmospheric turbulence, respectively. With the state variables so defined, equations (3.1) and (3.2) become

$$
\dot{y}_1 = -W_o y_3 - y_2 y_3 + y_7 y_8 - (g \cos \theta_o) y_4 + X_u y_1 + X_w y_2 + X_q y_3 + X_\delta \delta_e
$$

$$
- y_3 y_6 + y_8 y_{12} + X_u y_5 + X_w y_6
$$

$$
\dot{y}_2 = -y_7 y_{10} + U_o y_3 + y_1 y_3 - (g \sin \theta_o) y_4 + Z_u y_1 + Z_w y_2 + Z_q y_3 + Z_\delta \delta_e
$$

$$
- y_{10} y_{12} + y_3 y_5 + Z_u y_5 + Z_w y_6
$$

$$\dot{y}_3 = \left(\frac{I_{zz}-I_{xx}}{I_{yy}}\right) y_8 y_{10} + M_u y_1 + M_w y_2 + M_q y_3 + M_{\delta_e}\delta_e + M_u y_5 + M_w y_6$$

$$\dot{y}_4 = y_3$$

$$\dot{y}_5 = -\frac{1}{\tau_{ug}} y_5 + w_5$$

$$\dot{y}_6 = -\frac{1}{\tau_{wg}} y_6 + w_6$$

$$\dot{y}_7 = -U_o y_8 - y_1 y_8 + W_o y_{10} + y_2 y_{10} + (g\sin\theta_o)y_9 + (g\cos\theta_o)y_{11} + Y_v y_7$$
$$+ Y_r y_8 + Y_p y_{10} + Y_{\delta_a}\delta_a + Y_{\delta_r}\delta_r - y_5 y_8 + y_6 y_{10} + Y_v y_{12}$$

$$\dot{y}_8 = -\left(\frac{I_{yy}-I_{xx}}{I_{zz}}\right) y_3 y_{10} + N_v y_7 + N_r y_8 + N_p y_{10} + N_{\delta_a}\delta_a + N_{\delta_r}\delta_r + N_v y_{12}$$

$$\dot{y}_9 = y_8$$

$$\dot{y}_{10} = -\left(\frac{I_{zz}-I_{yy}}{I_{xx}}\right) y_3 y_8 + L_v y_7 + L_r y_8 + L_p y_{10} + L_{\delta_a}\delta_a + L_{\delta_r}\delta_r + L_v y_{12}$$

$$\dot{y}_{11} = y_{10}$$

$$\dot{y}_{12} = -\frac{1}{\tau_{vg}} y_{12} + w_{12}$$

with the observation set assumed to be given by

$$z_1 = y_1 + y_5 + v_1 \qquad z_5 = y_7 + y_{12} + v_5$$
$$z_2 = y_2 + y_6 + v_2 \qquad z_6 = y_8 + v_6$$
$$z_3 = y_3 + v_3 \qquad z_7 = y_9 + v_7$$
$$z_4 = y_4 + v_4 \qquad z_8 = y_{10} + v_8$$
$$\qquad\qquad\qquad\qquad z_9 = y_{11} + v_9$$

It has been assumed in the equations above that $M_{\dot{w}} = 0$ since this results in considerable simplification of the equations, and is usually a good assumption anyway. The observation noises v_i, $i = 1,\ldots,9$ are assumed to be zero-mean and uncorrelated with each other. The three components of atmospheric turbulence, u_g, w_g, and v_g, are also assumed to be zero-mean and mutually uncorrelated and uncorrelated with the observation noises.

Any unknown stability derivatives may be adjoined as additional states. For the problem discussed here it is assumed that all of the system parameters are known except for the parameter representing the change in pitching moment due to pitch rate, M_q. This assumption was made based on two considerations. First, the effort involved in deriving the estimation algorithms was considerable even for the case of one unknown parameter; and second, the procedure for the case of many unknown parameters is identical to the procedure for the case of one unknown parameter. Thus, since the primary purpose here is to exemplify the method, the simpler problem is solved.

The first step in the derivation of the hierarchical estimation algorithms is to decompose the system into two subsystems by replacing the terms coupling the longitudinal and lateral-directional equations by model coordination variables $\pi_1(t)$ and $\pi_2(t)$. Thus, with M_q adjoined as an additional state, define the subsystem state vectors as

$$x_1 = \begin{bmatrix} x_{1,1} \\ x_{1,2} \\ x_{1,3} \\ x_{1,4} \\ x_{1,5} \\ x_{1,6} \\ x_{1,7} \end{bmatrix} \triangleq \begin{bmatrix} y_1 \\ y_2 \\ y_3 \\ y_4 \\ y_5 \\ y_6 \\ y_{13} \end{bmatrix} = \begin{bmatrix} u \\ w \\ q \\ \theta \\ u_g \\ w_g \\ M_q \end{bmatrix} \quad \text{and} \quad x_2 = \begin{bmatrix} x_{2,1} \\ x_{2,2} \\ x_{2,3} \\ x_{2,4} \\ x_{2,5} \\ x_{2,6} \end{bmatrix} \triangleq \begin{bmatrix} y_7 \\ y_8 \\ y_9 \\ y_{10} \\ y_{11} \\ y_{12} \end{bmatrix} = \begin{bmatrix} v \\ r \\ \psi \\ p \\ \phi \\ v_g \end{bmatrix}$$

and let the coordination vectors be given by

$$\pi_1 = \begin{bmatrix} \pi_{1,1} \\ \pi_{1,2} \\ \pi_{1,3} \\ \pi_{1,4} \end{bmatrix} = \begin{bmatrix} g_{1,1}(x_2) \\ g_{1,2}(x_2) \\ g_{1,3}(x_2) \\ g_{1,4}(x_2) \end{bmatrix} \triangleq \begin{bmatrix} x_{2,1} \\ x_{2,2} \\ x_{2,4} \\ x_{2,6} \end{bmatrix}$$

and

$$\pi_2 = \begin{bmatrix} \pi_{2,1} \\ \pi_{2,2} \\ \pi_{2,3} \\ \pi_{2,4} \\ \pi_{2,5} \end{bmatrix} = \begin{bmatrix} g_{2,1}(x_1) \\ g_{2,2}(x_1) \\ g_{2,3}(x_1) \\ g_{2,4}(x_1) \\ g_{2,5}(x_1) \end{bmatrix} \triangleq \begin{bmatrix} x_{1,1} \\ x_{1,2} \\ x_{1,5} \\ x_{1,6} \\ x_{1,3} \end{bmatrix}$$

We will follow the convention that the subscript in front of the comma in the above expressions refers to the subsystem while the subscript following the comma refers to the component of the vector. This has proven to be a very convenient notation for high-dimensional problems. With the subsystem state vectors and coordination vectors thus defined, the subsystem equations can now be written.

<u>Subsystem 1:</u>

$$\dot{x}_{1,1} = -W_o x_{1,3} - x_{1,2} x_{1,3} + \pi_{1,1} \pi_{1,2} - (g \cos \theta_o) x_{1,4} + X_u x_{1,1} + X_w x_{1,2}$$

$$+ X_q x_{1,3} - x_{1,3} x_{1,6} + \pi_{1,2} \pi_{1,4} + X_u x_{1,5} + X_w x_{1,6} + X_{\delta_e} \delta_e$$

$$\dot{x}_{1,2} = -\pi_{1,1} \pi_{1,3} + U_o x_{1,3} + x_{1,1} x_{1,3} - (g \sin \theta_o) x_{1,4} + Z_u x_{1,1} + Z_w x_{1,2}$$

$$+ Z_q x_{1,3} - \pi_{1,3} \pi_{1,4} + x_{1,3} x_{1,5} + Z_u x_{1,5} + Z_w x_{1,6} + Z_{\delta_e} \delta_e$$

$$\dot{x}_{1,3} = \left(\frac{I_{zz} - I_{xx}}{I_{yy}} \right) \pi_{1,2} \pi_{1,3} + M_u x_{1,1} + M_w x_{1,2} + x_{1,3} x_{1,7} + M_u x_{1,5}$$

$$+ M_w x_{1,6} + M_{\delta_e} \delta_e$$

$$\dot{x}_{1,4} = x_{1,3}$$

$$\dot{x}_{1,5} = - \frac{1}{\tau_{u_g}} x_{1,5} + w_{1,5}$$

$$\dot{x}_{1,6} = - \frac{1}{\tau_{w_g}} x_{1,6} + w_{1,6}$$

$$\dot{x}_{1,7} = w_{1,7}$$

with

$$z_{1,1} = x_{1,1} + x_{1,5} + v_{1,1}$$

$$z_{1,2} = x_{1,2} + x_{1,6} + v_{1,2}$$

$$z_{1,3} = x_{1,3} + v_{1,3}$$

$$z_{1,4} = x_{1,4} + v_{1,4}$$

<u>Subsystem 2:</u>

$$\dot{x}_{2,1} = -U_o x_{2,2} - \pi_{2,1} x_{2,2} + W_o x_{2,4} + \pi_{2,2} x_{2,4} + (g \sin \theta_o) x_{2,3}$$

$$+ (g \cos \theta_o) x_{2,5} + Y_v x_{2,1} + Y_r x_{2,2} + Y_p x_{2,4} - \pi_{2,3} x_{2,2} + \pi_{2,4} x_{2,4}$$

$$+ Y_v x_{2,6} + Y_{\delta_a} \delta_a + Y_{\delta_r} \delta_r$$

214

$$\dot{x}_{2,2} = -\left(\frac{I_{yy}-I_{xx}}{I_{zz}}\right)\pi_{2,5}x_{2,4} + N_v x_{2,1} + N_r x_{2,2} + N_p x_{2,4} + N_v x_{2,6}$$
$$+ N_\delta{}_a \delta_a + N_\delta{}_r \delta_r$$

$$\dot{x}_{2,3} = x_{2,2}$$

$$\dot{x}_{2,4} = -\left(\frac{I_{zz}-I_{yy}}{I_{xx}}\right)\pi_{2,5}x_{2,2} + L_v x_{2,1} + L_r x_{2,2} + L_p x_{2,4}$$
$$+ L_v x_{2,6} + L_\delta{}_a \delta_a + L_\delta{}_r \delta_r$$

$$\dot{x}_{2,5} = x_{2,4}$$

$$\dot{x}_{2,6} = -\frac{1}{\tau_{v_g}} x_{2,6} + w_{2,6}$$

with

$$z_{2,1} = x_{2,1} + x_{2,6} + v_{2,1}$$
$$z_{2,2} = x_{2,2} + v_{2,2}$$
$$z_{2,3} = x_{2,3} + v_{2,3}$$
$$z_{2,4} = x_{2,4} + v_{2,4}$$
$$z_{2,5} = x_{2,5} + v_{2,5}$$

Now that the subsystem models have been defined, the algorithms of Table 1 may be used to solve the infimal unit estimation problems. Note that the coordination vector in Subsystem 1 was defined so that each of its components is equal to a state in Subsystem 2. This resulted in products of coordination variables appearing in Subsystem 1. By observing the original system state equations, it is seen that different coordination variables could have been defined for Subsystem 1 as products of the states of Subsystem 2. There is considerable practical motivation in this problem for using the former approach, since if all the coordination variables are each defined in terms of a single state, the last term in the error variance algorithm of Table 1 vanishes, and results in a reduction of effort involved in writing down the individual error variance equations.

It is worthwhile pointing out that the primary motivation for using hierarchical techniques is to reduce computational effort. This reduction is realized by a decrease in the number of error variance equations which must be solved. If the problem solved here had been solved without hierarchical structuring, there would have been $(13 \times 14)/2 = 91$ error variance equations. For the solution presented here there are $(7 \times 8)/2 + (6 \times 7)/2 = 49$ error variance equations. If more unknown parameters had been adjoined as additional states, the savings would have been more dramatic. For example, if it were assumed that all of the static derivatives could be satisfactorily obtained from wind tunnel tests and only the rate derivatives were to be identified from flight test data, there would have been nine additional states instead of only one, and the original system identification algorithm would have required $(21 \times 22)/2 = 231$ error variance equations. By comparison, the two-subsystem hierarchical structure would require the solution of $(9 \times 10)/2 + (12 \times 13)/2 = 123$ error variance equations. This is a considerable reduction, but it is evident that even the decomposed problem could be difficult to solve. Thus, there is good reason for decomposing the original system into more than two subsystems. If the system with nine additional states were decomposed into six subsystems with state vectors given by

$$x_1 = \begin{bmatrix} x_{1,1} \\ x_{1,2} \\ x_{1,3} \end{bmatrix} = \begin{bmatrix} u \\ u_g \\ X_q \end{bmatrix} \qquad x_2 = \begin{bmatrix} x_{2,1} \\ x_{2,2} \\ x_{2,3} \end{bmatrix} = \begin{bmatrix} w \\ w_g \\ Z_q \end{bmatrix}$$

$$x_3 = \begin{bmatrix} x_{3,1} \\ x_{3,2} \\ x_{3,3} \end{bmatrix} = \begin{bmatrix} q \\ \theta \\ M_q \end{bmatrix} \qquad x_4 = \begin{bmatrix} x_{4,1} \\ x_{4,2} \\ x_{4,3} \\ x_{4,4} \end{bmatrix} = \begin{bmatrix} v \\ v_g \\ Y_r \\ Y_p \end{bmatrix}$$

$$x_5 = \begin{bmatrix} x_{5,1} \\ x_{5,2} \\ x_{5,3} \\ x_{5,4} \end{bmatrix} = \begin{bmatrix} r \\ \psi \\ N_r \\ N_p \end{bmatrix} \qquad x_6 = \begin{bmatrix} x_{6,1} \\ x_{6,2} \\ x_{6,3} \\ x_{6,4} \end{bmatrix} = \begin{bmatrix} p \\ \psi \\ L_r \\ L_p \end{bmatrix}$$

then the identification algorithms would have $3(3 \times 4)/2 + 3(4 \times 5)/2 = 48$ error variance equations instead of 231. However, there are now many more coordination constraints which must be satisfied; thus, the infimal unit identification processes may be difficult to coordinate. No computational experience has been acquired with this six-subsystem decomposition. It is interesting to point out that if a system can be satisfactorily decomposed into a number of subsystems equal to the number of system states, then the number of error variance equations equals the number of states, and the maximum reduction in error variance equations occurs.

If the covariance of the turbulence is unknown, and it is desired to use the algorithms of Smith and Sage [15] to adaptively estimate the system states, there is again considerable motivation for a maximal decomposition. These algorithms are most efficient for the case of a scalar observation. But if the system were decomposed into several subsystems, each with a scalar observation, these algorithms could be used effectively. Of course, other adaptive estimation algorithms could also be incorporated to obtain the subsystem state estimates.

216

Computational Results

The algorithms of Table 1 have been applied to the sybsystem equations written above to obtain the optimum estimates of the system states and thus the parameter M_q. To coordinate the infimal units a sequential coordination technique was used which is based on a modification of equations (2.33) - (2.38). Rather than use equations (2.33) - (2.36) to predict the new values of the coordination variables, equations (2.37) and (2.38) were modified to the form

$$\beta_i(t+\Delta t) = \beta_i(t) + K_{\beta_i} \frac{\partial H_i(t)}{\partial \beta_i(t)} \tag{3.9}$$

and

$$\pi_i(t+\Delta t) = g_i[\hat{x}_j(t)] \tag{3.10}$$

Thus, the updating of β_i is accomplished by correcting the old value of β_i with the gradient of H_i with respect to β_i. The technique for updating π_i was motivated by the equality updating method equation (2.31). Use of a sequential updating technique such as that represented by equations (2.33) - (2.38) or equations (3.9) and (3.10) makes the overall estimation algorithm sequential thereby avoiding the increased computation and memory requirements of an iterative technique.

The flight test data used in this study was generated from a simulation of a more general set of 6-DOF equations than those given by equations (3.1) and (3.2) in that the assumption that the angular deflections are small (Assumption 7) was not made. This was done in an attempt to make the observations more realistic than they would be if equation (3.1) and (3.2) had been used to generate the observations. Atmospheric turbulence was simulated to give standard deviations of 2.5 ft/sec for the horizontal turbulence components u_g and v_g and 1.5 ft/sec for the vertical component. This turbulence level is fairly light, but is not considered unrealistic since a concerted effort is usually made to obtain flight test data in calm air. The stability and control parameters are very close to being those published for the A-7 Corsair II flying in the cruise configuration at Mach 0.6 at sea level. Equations (3.1) and (3.2) are written with respect to the principal axis system. Further simplification was achieved for the problem solved here by assuming that in the reference flight condition the principal x-axis is in the direction of the relative wind; that is, in this reference flight condition, the principal axis system is the body-stability axis system. This results in W_0 being zero for the reference flight condition. It is also assumed that the airplane has wings level and is neither climbing nor diving in the reference flight condition so that the principal x-axis is not inclined to the horizontal and thus $\theta_0 = 0$. It should be mentioned here that none of the above assumptions about the choice of axis-system and the reference flight condition are necessary to solve the problem. They were made only for convenience in completing this investigation with an amount of effort commensurate with the purpose of the paper, namely, to demonstrate the use of hierarchical estimation in the context of the aircraft parameter identification problem.

The aerodynamic data used in the modeling of the airplane appear in Figure 2. Atmospheric turbulence was modeled by a first order filter driven by gaussian white noise. A time constant of 0.15 sec was used for each shaping filter so that

217

$$\tau_{u_g} = \tau_{w_g} = \tau_{v_g} = 0.15$$

The control inputs applied to the airplane in the flight test simulation are shown in Figure 3. The control surface deflection sign conventions follow reference [16]. The conventions are that positive δ_e is trailing edge up, positive δ_a is right wing aileron up, and positive δ_r is trailing edge left. The observation noise standard deviations used are given in Figure 2.

The observation data obtained from the flight test simulation was used with the algorithms of Table 1 to identify the unknown parameter M_q. The identification of M_q was accomplished using two different values of the initial guess of the value of M_q. For an initial guess of $\hat{x}_{1,7}(0) = -0.500$, the identification yielded a value of -0.88 as compared to a true value of -0.865. For $\hat{x}_{1,7}(0) = -1.500$, M_q was identified as being -0.93. Neither of these estimates is a bad estimate of M_q, but they do show that the estimate of M_q is somewhat sensitive to the initial guess. Estimates of M_q were also obtained with the infimals decoupled (uncoordinated) by setting $\pi_i = 0$ and $\beta_i = 0$ for all time. This is exactly equivalent to identifying M_q using the longitudinal equations of motion and ignoring the lateral-directional dynamics and observations. The estimates of M_q obtained by the decoupled solution were -0.93 for $\hat{x}_{1,7}(0) = -0.500$ and -1.02 for $\hat{x}_{1,7}(0) = -1.500$. It is seen that these estimates are less accurate than those obtained from the 6-DOF solution. Considering that the control inputs used in this study were probably not optimal identification inputs, the quality of the estimates obtained by hierarchical estimation was good. Mehra [20] has shown that for time-invariant linear systems the optimal inputs for parameter identification are sums of sine and cosine functions at appropriate frequencies. It is likely that the results of that work could be used to improve the parameter estimates obtained from the present study.

4. CONCLUSIONS

It has been shown that this method has good potential for use in the identification of aircraft stability and control parameters. There are, however, many problems involved in using actual flight test data that were not confronted in this paper. The authors are presently studying the case where the original subsystem is decomposed into more than two subsystems and the case of unknown plant and measurement noise moments. It is recommended that hierarcical identification techniques for 6-DOF models be given careful consideration for use in future aircraft parameter identification research.

5. REFERENCES

[1]   Cuenod, M. and Sage, A. P.: "Comparison of Some Methods Used for Process Identification." Automatica, Vol. 4, no. 4, May 1968.

[2]   Balakrishnan, A. V. and Peterka, V.: "Identification in Automatic Control Systems." Fourth Congress IFAC, Warsaw, 1969.

[3]   Astrom, K.J. and Eykhoff, P.: "System Identification." IFAC Symposium on Identification and Process Parameter Estimation, Prague, June 1970; also, Automatica, Vol. 7, no. 2, March 1971.

[4]   Sage, A. P. and Melsa, J. L.: System Identification, Academic Press, 1971.

[5]   Graupe, Daniel: Identification of Systems, Van Nostrand, 1972.

[6]   Mehra, R. K.: "Maximum Likelihood Identification of Aircraft Parameters." Proc. JACC, pp. 442-444, 1970.

[7]   Mehra, R. K., Stepner, D. E., and Tyler, J. S.: "A Generalized Method for the Identification of Aircraft Stability and Control Derivatives from Flight Test Data." Proc. JACC, 1972.

[8]   Smith, N. J. and Sage, A. P.: "An Introduction to Hierarchical Systems Theory." Computers and Electrical Engineering, to appear, 1973.

[9]   Pearson, J. D., in Wismer, D. A. (Ed.): Optimization Methods for Large-Scale Systems, McGraw-Hill, 1971.

[10]   Mesarovic, M. D., Macko, D., and Takahara, Y.: Theory of Hierarchical, Multilevel, Systems, Academic Press, 1970.

[11]   Smith, N. J. and Sage, A. P.: "Hierarchical Structuring for System Identification." Information Sciences, to appear, 1973.

[12]   Sage, A. P. and Melsa, J. L.: Estimation Theory with Applications to Communications and Control, McGraw-Hill, 1971.

[13]   Sage, A. P.: Optimum Systems Control, Prentice Hall, 1968.

[14]   Guinzy, N. J. and Sage, A. P.: "System Identification in Large-Scale Systems with Hierarchical Structures." Computers and Electrical Engineering, to appear, 1973.

[15]   Smith, N. J. and Sage, A. P.: "A Sequential Method for System Identification in Hierarchical Structures." Proceedings of IFAC, 3rd Symposium on System Identification, Netherlands, June 1973; also, Automatica, November 1973.

[16]   McRuer, D. T., Bates, C. L. and Ashkenas, I. L.: Dynamics of the Airframe, Navy Bureau of Aeronautics Report AE-61-4II, 1952.

[17]   Sage, A. P. and Husa, G. W.: "Adaptive Filtering with Unknown Prior Statistics." Proc. JACC, pp. 760-769, 1969.

[18]   Sage, A. P. and Wakefield, C. D.: "Maximum Likelihood Identification of Time-Varying and Random System Parameters." Int. Journal of Control, pp. 81-100, July 1972.

[19]   Sage, A. P. and Wakefield, C. D.: "System Identification with Noise Corrupted Input Observation." Proc. SWIEEECO, pp. 333-337, Dallas, April 1970.

[20]   Mehra, R. K., "Optimal Inputs for Linear System Identification, Proc. JACC, Stanford, August 1972.

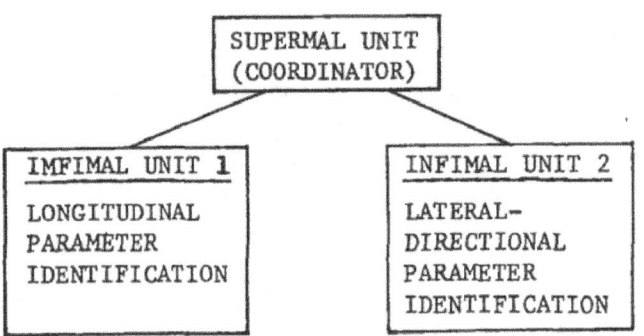

Figure 1.   Schematic of Hierarchical System Identification

Table 1

Hierarchical Invariant Imbedding Identification
Algorithms for Subsystem i

System Model	$\dot{x}_i(t) = f_i[x_i,\pi_i,t]+w_i(t)$
Observation Model	$z_i(t) = h_i[x_i,t]+v_i(t)$
Prior Statistics	$E\{x_i(t_o)\} = \mu_{x_{io}}$, $\mathrm{Var}\{x_i(t_o)\} = V_{x_{io}}$ $E\{w_i(t)\} = 0 = E\{v_i(t)\}$ $\mathrm{Cov}\{w_i(t),w_i(\tau)\} = \Psi_{w_i}(t)\delta_D(t-\tau)$ $\mathrm{Cov}\{v_i(t),v_i(\tau)\} = \Psi_{v_i}(t)\delta_D(t-\tau)$
Innovation	$\tilde{z}_i(t) = z_i(t)-h_i[\hat{x}_i,t]$
Definitions	$\beta(t)$ and $\pi(t)$ are coordination variables. $F_i(t) \overset{\Delta}{=} \dfrac{\partial f_i[\hat{x}_i,\pi_i,t]}{\partial \hat{x}_i(t)}$, $H_i(t) = \dfrac{\partial h_i[\hat{x}_i,t]}{\partial \hat{x}_i(t)}$
Filter Algorithm	$\dot{\hat{x}}_i(t) = f_i[\hat{x}_i,\pi_i,t]+P_i(t)H_i^T(t)\Psi_{v_i}^{-1}(t)\tilde{z}_i(t)+P_i(t)\dfrac{\partial}{\partial \hat{x}_i}\{\sum_{j\neq i}^{N}\beta_j^T(t)g_{ji}(\hat{x}_i)\}$
Error Variance Algorithm	$\dot{P}_i(t) = P_i(t)F_i^T(t)+F_i(t)P_i(t)+\Psi_{w_i}(t)+P_i(t)\dfrac{\partial}{\partial \hat{x}_i}\{H_i^T(t)\Psi_{v_i}^{-1}(t)\tilde{z}_i(t)\}P_i(t)$ $+ P_i(t)\dfrac{\partial}{\partial \hat{x}_i}\{\dfrac{\partial}{\partial \hat{x}_i}\sum_{j\neq i}^{N}\beta_j^T(t)g_{ji}(\hat{x}_i)\}P_i(t)$
Initial Conditions	$\hat{x}_i(t_o) = \mu_{x_{io}}$, $P_i(t_o) = V_{x_{io}}$

220

AERODYNAMIC DATA

$U_o = 670.0$ ft/sec

$I_{xx} = 15,324$ slug-ft^2

$I_{yy} = 69,528$ slug-ft^2

$I_{zz} = 79,046$ slug-ft^2

$X_u = -0.01169 \; \dfrac{1}{\text{sec}}$

$X_w = 0.0234 \; \dfrac{1}{\text{sec}}$

$X_q = 0.0 \; \dfrac{\text{ft}}{\text{sec-rad}}$

$X_{\delta_e} = 0.0 \; \dfrac{\text{ft}}{\text{sec}^2\text{-deg}}$

$Z_u = -0.1122 \; \dfrac{1}{\text{sec}}$

$Z_w = -1.66 \; \dfrac{1}{\text{sec}}$

$Z_q = 0.0 \; \dfrac{\text{ft}}{\text{sec-rad}}$

$Z_{\delta_e} = 2.49 \; \dfrac{\text{ft}}{\text{sec}^2\text{-deg}}$

$M_u = 0.000290 \; \dfrac{1}{\text{sec-ft}}$

$M_w = -0.0203 \; \dfrac{1}{\text{sec-ft}}$

$M_{\dot{w}} = 0.0 \; \dfrac{1}{\text{ft}}$

$M_q = -0.865 \; \dfrac{1}{\text{sec-rad}}$

$M_{\delta_e} = 0.456 \; \dfrac{1}{\text{sec}^2\text{-deg}}$

Observation Noise Standard Deviations:

$\sigma_{v_1} = \sigma_{v_2} = \sigma_{v_5} = 0.1$ ft/sec

$\sigma_{v_3} = \sigma_{v_6} = \sigma_{v_8} = 0.1$ deg/sec

$\sigma_{v_4} = \sigma_{v_7} = \sigma_{v_9} = 0.1$ deg

$Y_v = -0.272 \; \dfrac{1}{\text{sec}}$

$Y_r = 2.07 \; \dfrac{\text{ft}}{\text{sec-rad}}$

$Y_p = 0.725 \; \dfrac{\text{ft}}{\text{sec-rad}}$

$Y_{\delta_a} = -0.204 \; \dfrac{\text{ft}}{\text{sec}^2\text{-deg}}$

$Y_{\delta_r} = 0.788 \; \dfrac{\text{ft}}{\text{sec}^2\text{-deg}}$

$N_v = 0.01033 \; \dfrac{1}{\text{ft-sec}}$

$N_r = -0.837 \; \dfrac{1}{\text{sec-rad}}$

$N_p = 0.01226 \; \dfrac{1}{\text{sec-rad}}$

$N_{\delta_a} = 0.0230 \; \dfrac{1}{\text{sec}^2\text{-deg}}$

$N_{\delta_r} = -0.1326 \; \dfrac{1}{\text{sec}^2\text{-deg}}$

$L_v = -0.0601 \; \dfrac{1}{\text{ft-sec}}$

$L_r = 1.241 \; \dfrac{1}{\text{sec-rad}}$

$L_p = -4.79 \; \dfrac{1}{\text{sec-rad}}$

$L_{\delta_a} = 0.935 \; \dfrac{1}{\text{sec}^2\text{-deg}}$

$L_{\delta_r} = 0.1607 \; \dfrac{1}{\text{sec}^2\text{-deg}}$

Figure 2

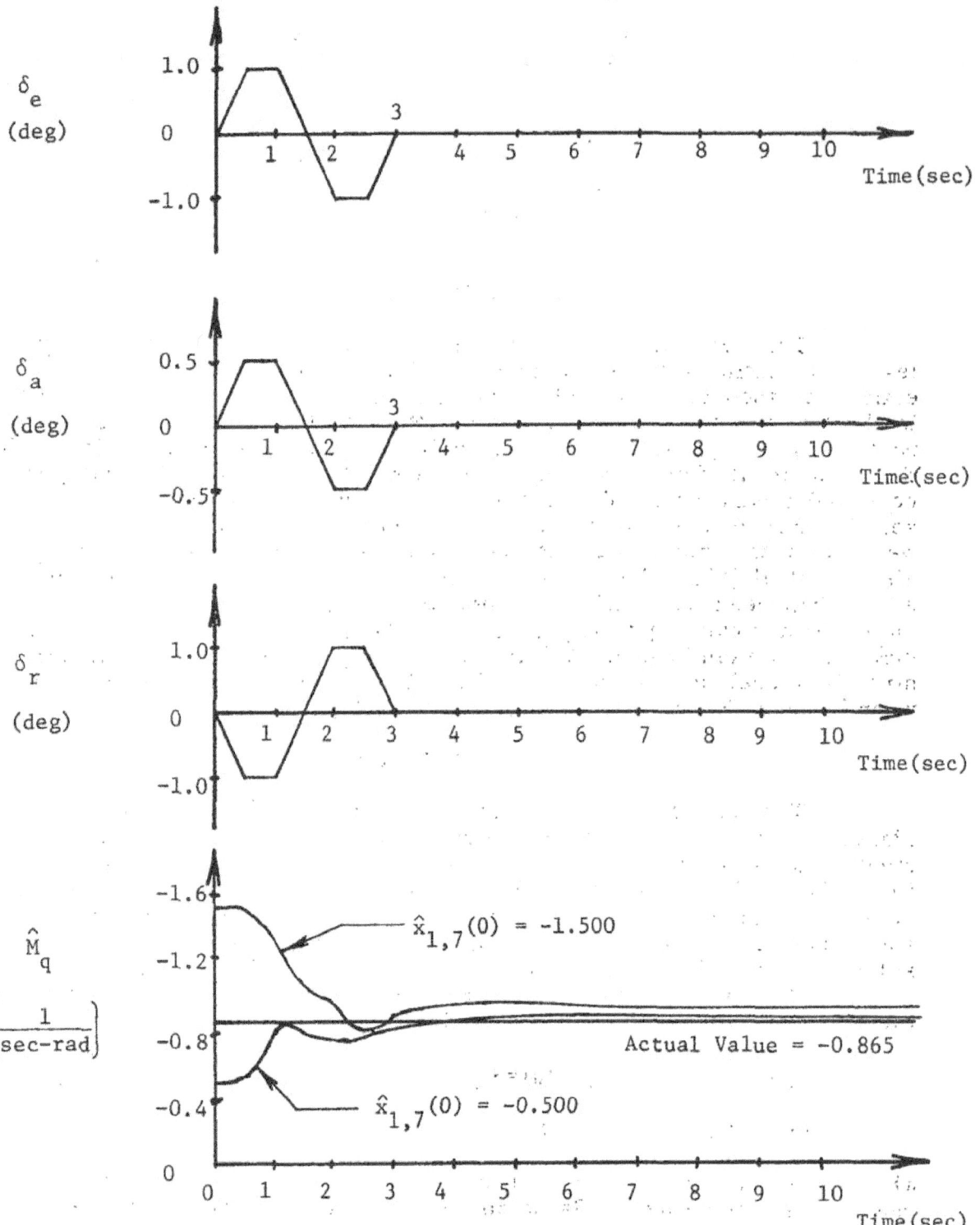

Figure 3

222

Parameter Estimation Using
an A Posteriori Criterion

Ralph E. Bach, Jr.[*]
Electrical Engineering Dept.
Northeastern University
Boston, Mass. 02115

INTRODUCTION

The determination of aircraft stability derivatives from flight-test data continues to be an important application of parameter-estimation theory. It is not surprising that many recent developments in the theory and practice have been reported by workers in the aeronautics field. [1-8]. The identification problem, of course, is much the same in all fields, although few physical models with typical process and measurement disturbances are as well understood as the aerodynamic model. A procedure for estimating noise statistics as well as system model parameters has been described recently by Mehra, et al [6]. The method, based on a maximum likelihood criterion, has been successfully applied to a number of aircraft parameter-identification problems. In this investigation, an a posteriori criterion is examined for use in such an application, when it is desired to model process noise. The criteria lead to identical results when there is no process noise.

MAP CRITERION

The use of an a posteriori, or "smoothing" criterion for identification of unknown parameters, including process and measurement noise statistics, was suggested by Bryson and Frazier in 1962 [7]. An implementation was described by Tyler, et al [8], in which state and parameter estimates were combined to form an augmented state vector. In this presentation, however, the estimates are considered separately in order to develop an optimization procedure.

Here we assume, for simplicity of notation, a linear, time-invariant, discrete message-observation model

$$x(i+1) = \varphi x(i) + w(i), \quad x(0) = x_o ; \tag{1}$$

$$z(i+1) = Hx(i+1) + v(i+1), \tag{2}$$

where x is an n-vector and z is an m-vector. The noise sources are independent, zero-mean, Gaussian "white" sequences with covariances

[*] The research reported here was supported in part by NASA, under Contract No. NAS2-7397 (Ames Research Center).

$$E[w(i)w^T(j)] = Q \delta_K(i-j) ; \tag{3}$$

$$E[v(i)v^T(j)] = R \delta_K(i-j). \tag{4}$$

We have available N measurements

$$Z(N) = \{z(1), z(2), \ldots z(N)\} , \tag{5}$$

and wish to estimate the initial condition x_o, the system parameter vector p (unknown elements of φ), and covariance vectors q and r (unknown elements of Q and R). One procedure is to formulate an a posteriori probability density function

$$p[X(N), \theta/Z(N)], \tag{6}$$

where

$$X(N) = \{x(1), x(2), \ldots, x(N)\} ; \tag{7}$$

$$\theta = (x_o^T, p^T, q^T, r^T)^T, \tag{8}$$

and to maximize it with respect to θ. Hence the term "maximum a posteriori" (MAP) estimation.

An expression for the a posteriori density function of (6) can be shown to be [9]

$$A \cdot \exp(-B/2) \cdot \left\{ p(\theta)/p[Z(N)] \right\} , \tag{9}$$

where

$$A = 1/[(2\pi)^{N(n+m)/2} \cdot |Q|^{N/2} \cdot |R|^{N/2}]; \tag{10}$$

$$B = \sum_{i=0}^{N-1} [w^T(i)Q^{-1}w(i) + v^T(i+1)R^{-1}v(i+1)]. \tag{11}$$

Note that $p[Z(N)]$ in (9) is just a constant. In the absence of a priori statistical information regarding θ, a useful performance measure for maximizing the a posteriori density function* is

$$J = (1/2) \sum_{i=0}^{N-1} [w^T(i)Q^{-1}w(i) + v^T(i+1)R^{-1}v(i+1)]$$

$$+ (N/2)\ln(|Q| \cdot |R|). \tag{12}$$

If there exists some a priori information about θ, we just add to J the term

$$(1/2)[\theta - \hat{\theta}]^T S^{-1}[\theta - \hat{\theta}] , \tag{13}$$

*J is to be minimized.

where $\hat{\theta}$, S are the a priori mean and variance, respectively, and $p(\theta)$ is assumed to be Gaussian.

GRADIENT DETERMINATION

The estimation problem may be solved by applying a standard variational procedure [10] to obtain a first variation of the performance measure (12), using the resulting gradient information to implement a "down-hill" numerical method. As usual, the dynamic constraints (1) are adjoined to (12) with a Lagrange multiplier to form

$$\bar{J} = J + \sum_{i=0}^{N-1} \lambda^T(i+1)[\varphi x(i) + w(i) - x(i+1)]. \tag{14}$$

Note that the last term of (14) can be written as

$$\lambda^T(0)x(0) - \lambda^T(N)x(N) - \sum_{i=0}^{N-1} \lambda^T(i)x(i). \tag{15}$$

Now, if we make differential changes in $x(i)$, $w(i)$ and θ, we obtain the expression

$$\delta\bar{J} = (\partial J/\partial q)\delta q + (\partial J/\partial r)\delta r + \lambda^T(0)\delta x(0) - \lambda^T(N)\delta x(N)$$

$$+ \sum_{i=0}^{N-1} \left\{ (\partial H/\partial p)\delta p + [\partial H/\partial w(i)]\delta w(i) + [\partial H/\partial x(i) + \lambda^T(i)]\delta x(i) \right\}, \tag{16}$$

where

$$H = (1/2)[w^T(i)Q^{-1}w(i) + v^T(i+1)R^{-1}v(i+1)]$$

$$+ \lambda^T(i+1)[\varphi x(i) + w(i)] \tag{17}$$

A necessary condition for a minimum is that $\delta\bar{J} = 0$. We can design our numerical procedure so that $\delta\bar{J}$ approaches zero with successive choices of θ. First, choose $\lambda(i)$ such that

$$\lambda(i) = [\partial H/\partial x(i)]^T, \quad \lambda(N) = 0; \tag{18}$$

$$\partial H/\partial w(i) = 0. \tag{19}$$

Expansion of (18) and (19) results in

$$\lambda(i) = \varphi^T\lambda(i+1) - \varphi^T H^T R^{-1}[z(i+1) - Hx(i+1)], \quad \lambda(N) = 0; \tag{20}$$

$$w(i) = -Q\varphi^{-T}\lambda(i). \tag{21}$$

The linear, two-point boundary problem of equations (1), (20), and (21) may be solved by a "sweep" method [10], by assuming a solution

$$x(i) = \hat{x}(i) - P(i)\lambda(i). \tag{22}$$

Note that for i=0, (22) must reduce to the initial condition given in (1). The resulting "smoothed" solution x(i) is obtained by first solving the discrete Kalman filter equations

$$\hat{x}(i+1) = \varphi\hat{x}(i) + P(i+1)H^T R^{-1}[z(i+1) - H\varphi\hat{x}(i)], \quad \hat{x}(0) = x_o; \tag{23}$$

$$P(i+1) = [(\varphi P(i)\varphi^T + Q)^{-1} + H^T R^{-1} H]^{-1}, \quad P(0) = 0, \tag{24}$$

forward in time, and then solving (20), (21) and (22) backward in time. Note that $x(N) = \hat{x}(N)$.

With (18) and (19) satisfied for any choice of θ, the variation δJ of (16) reduces to

$$\delta\bar{J} = (\partial J/\partial q)\delta q + (\partial J/\partial r)\delta r + \lambda(0)\delta x(0) + \sum_{i=0}^{N-1} (\partial H/\partial p)\delta p. \tag{25}$$

The gradients $\partial J/\partial q$, $\partial J/\partial r$ may be calculated from

$$(\partial J/\partial q_j) = N \cdot \mathrm{trace}\, Q^{-1}(\partial Q/\partial q_j)Q^{-1}\left\{Q - \sum_{i=0}^{N-1}[w(i)w^T(i)]/N\right\}; \tag{26}$$

$$(\partial J/\partial r_j) = N \cdot \mathrm{trace}\, R^{-1}(\partial R/\partial r_j)R^{-1}\left\{R - \sum_{i=0}^{N-1}[v(i+1)v^T(i+1)]/N\right\}. \tag{27}$$

It can be seen, however, that (26) and (27) vanish identically if at each iteration

$$Q = (1/N)\sum_{i=0}^{N-1}[w(i)w^T(i)]; \tag{28}$$

$$R = (1/N)\sum_{i=0}^{N-1}[v(i+1)v^T(i+1)]. \tag{29}$$

It is possible that numerical instability will result if (28) and (29) are used, especially if θ is some distance from its minimizing value. If (28) and (29) are used, the performance measure (12) simplifies to

$$J = (N/2)[(n+m) + \ln(|Q| \cdot |R|)]. \tag{30}$$

The usual smoothing solution, when p, q, and r are known, requires only adjustment of x_o to minimize J. In this case, however, we need to implement a "downhill" method, using gradients from (25), where, in addition to (26) and (27), we have

$$\partial\bar{J}/\partial x_o = \lambda^T(0); \tag{31}$$

$$\partial\bar{J}/\partial p = \sum_{i=0}^{N-1} (\partial H/\partial p). \tag{32}$$

If (28) and (29) are used, the variation $\delta\bar{J}$ becomes

$$\delta\bar{J} = (\partial\bar{J}/\partial x_0)\delta x_0 + (\partial\bar{J}/\partial p)\delta p. \tag{33}$$

Implementation of a gradient method may result in difficulties with step-size determination and slow convergence in the neighborhood of a minimum. An alternate strategy having better convergence properties is the method of quasilinearization, which has been successfully utilized in the maximum-likelihood applications described in References [1-6]. An extension of the quasilinearization method for minimizing the a posteriori criterion (12) is considered in the next section.

QUASILINEARIZATION

First, recall that if the parameter set θ is known, the a posteriori criterion is minimized by the "smoothed" solution, generated by computing

$$\hat{x}(i+1) = \varphi\hat{x}(i) + P(i+1)H^T R^{-1}[z(i+1) - H\varphi\hat{x}(i)]; \tag{34}$$

$$P(i+1) = [(\varphi P(i)\varphi^T + Q)^{-1} + H^T R^{-1} H]^{-1}, \tag{35}$$

forward in time, with $\hat{x}(0) = x_0$ and $P(0) = 0$, and then computing

$$x(i+1) = \hat{x}(i+1) - P(i+1)\lambda(i+1); \tag{36}$$

$$\lambda(i) = \varphi^T \lambda(i+1) - \varphi^T H^T R^{-1}[z(i+1) - Hx(i+1)]; \tag{37}$$

$$w(i) = -Q\varphi^{-T}\lambda(i), \tag{38}$$

backward in time, with $\lambda(N) = 0$, where

$$Q = (1/N) \sum_{i=0}^{N-1} [w(i)w^T(i)]; \tag{39}$$

$$R = (1/N) \sum_{i=0}^{N-1} [v(i+1)v^T(i+1)], \tag{40}$$

Now, suppose that we wish to improve our estimate of x_0 and p. Let

$$\alpha = [x_0^T, p^T]^T. \tag{41}$$

The method of quasilinearization provides an explicit expression for a parameter step $\delta\alpha$ that reduces the corresponding variation of J to zero. Actually, the procedure is equivalent to expanding J in a Taylor series about a solution (in this case, the "smoothed" solution) and keeping one of the second-order terms.* The minimizing parameter

---

* See Denery [2].

change is given by

$$\delta\alpha = -M^{-1}(\partial J/\partial\alpha)^T, \tag{42}$$

where, for the a posteriori criterion of (12), it is easily shown that

$$\partial J/\partial\alpha = \sum_{i=0}^{N-1} [w^T(i)Q^{-1}\partial w(i)/\partial\alpha + v^T(i+1)R^{-1}\partial v(i+1)/\partial\alpha]; \tag{43}$$

$$M = \sum_{i=0}^{N-1} [(\partial w(i)/\partial\alpha)^T Q^{-1}\partial w(i)/\partial\alpha + (\partial v(i+1)/\partial\alpha)^T R^{-1}\partial v(i+1)/\partial\alpha], \tag{44}$$

where, from (38) we have

$$\partial w(i)/\partial\alpha = -Q\varphi^{-T}[-(\partial\varphi/\partial\alpha)^T\varphi_{\lambda(i)}^{-T} + \partial\lambda(i)/\partial\alpha], \tag{45}$$

and from (2)

$$\partial v(i+1)/\partial\alpha = -H\partial x(i+1)/\partial\alpha. \tag{46}$$

Evaluation of (43)-(46) requires computation of the "sensitivity" functions $\partial\hat{x}(i+1)/\partial\alpha$, $\partial P(i+1)/\partial\alpha$, $\partial\lambda(i)/\partial\alpha$, and $\partial x(i+1)/\partial\alpha$. These are determined from (34)-(37) to be

$$\partial\hat{x}(i+1)/\partial\alpha_j = [I - K(i+1)H][\varphi\partial\hat{x}(i)/\partial\alpha_j + (\partial\varphi/\partial\alpha_j)\hat{x}(i)]$$

$$+ (\partial P(i+1)/\partial\alpha_j)H^T R^{-1}[z(i+1) - H\varphi\hat{x}(i)], \tag{47}$$

with $\quad \partial\hat{x}(0)/\partial p = 0$ and $\partial\hat{x}(0)/\partial x_o = I$;

$$\partial P(i+1)/\partial\alpha_j = [I - K(i+1)H][(\partial\varphi/\partial\alpha_j)P(i)\varphi^T + \varphi(\partial P(i)/\partial\alpha_j)\varphi^T$$

$$+ \varphi P(i)(\partial\varphi/\partial\alpha_j)^T][I - K(i+1)H]^T, \tag{48}$$

with $\quad \partial P(0)/\partial\alpha_j = 0$, and $K(i+1) = P(i+1)H^T R^{-1}$;

$$\partial\lambda(i)/\partial\alpha_j = (\partial\varphi/\partial\alpha_j)^T[\lambda(i+1) - H^T R^{-1}(z(i+1) - Hx(i+1))]$$

$$+ \varphi^T H^T R^{-1}H\partial x(i+1)/\partial\alpha_j, \tag{49}$$

where $\quad \partial\lambda(N)/\partial\alpha_j = 0$, and

$$\partial x(i+1)/\partial\alpha_j = \partial\hat{x}(i+1)/\partial\alpha_j - P(i+1)\partial\lambda(i+1)/\partial\alpha_j - [\partial P(i+1)/\partial\alpha_j]\lambda(i+1). \tag{50}$$

We note from (48) that $\partial P(i+1)/\partial x_0 = 0$.

ALGORITHM

A procedure for identification of the initial state, system parameters, and noise statistics utilizing quasilinearization to minimize the a posteriori criterion would be as follows:

(1) Choose θ and solve the Kalman-filter equations (34), (35) and the sensitivity equations (47), (48) forward in time.

(2) Solve the smoothing equations (36), (37), (38) and the sensitivity equations (49), (50) backward in time.

(3) Evaluate M, $\partial J/\partial \alpha$ during the backward run and compute the revised θ from (39), (40) and (42).

(4) Continue until $\delta J = 0$ and a minimum of J is obtained.

COMMENTS

If the system model is without process noise, i.e., $w(i) = 0$ for all i, there is no need to smooth the data, and from (35) we see that $P(i) = 0$ for all i. In this case, the expressions for $\partial J/\partial \alpha$ in (43) and M in (44) reduce exactly to those given by Grove [5] and Mehra [6] for parameter estimation using a maximum-likelihood criterion (no process noise) and quasilinearization.

The number of sensitivity equations to be solved during each iteration using quasilinearization for minimization of the MAP criterion is given by

$$N_{map} = [2n + n(n+1)/2]1 - n^2(n+1)/2 , \tag{51}$$

where 1 is the number of parameters in α and n is the order of the model. For a typical lateral-stability identification problem with $1 = 23$, $n = 4$, we find that $N_{map} = 374$. The corresponding number for the maximum=likelihood criterion, where the performance measure is expanded about the <u>filtered</u> solution [11], is given by

$$N_{ml} = [n + n(n+1)/2]1 - n^2(n+1)/2, {}^* \tag{52}$$

which yields $N_{ml} = 282$ for the example considered. For the case of no process noise, the two criteria yield identical results, with the number of sensitivity equations given by

$$N = nl. \tag{53}$$

*Mehra's procedure reduces this number[6].

Finally, it should be emphasized that application of the MAP criterion requires further processing of the data than the ML criterion. In the context of the presentation here, it can be seen that the MAP estimation procedure is an extension of the ML procedure. A basic question remains: how do the results obtained compare in practice? At this writing only laboratory experiments on first-order systems have been performed and no comparisons with maximum-likelihood results have yet been made.

REFERENCES

[1] Mehra, R.K., "Maximum-Likelihood Identification of Aircraft Parameters", JACC, June 1970.

[2] Denery, D.G., "Identification of System Parameters from Input-Output Data with Application to Air Vehicles", NASA TN D-6468, August 1971.

[3] Steinmetz, G.G., et al, "Longitudinal Stability and Control Derivatives of a Jet Fighter Airplane Extracted from Flight-Test Data by Utilizing Maximum-Likelihood Estimation", NASA TN D-6532, March 1972.

[4] Iliff, K.W.; Taylor, L.W., Jr., "Determination of Stability Derivatives from Flight Data Using a Newton-Raphson Minimization Technique", NASA TN D-6579, March 1972.

[5] Grove, R.D., et al, "A Procedure for Estimating Stability and Control Parameters from Flight Test Data by Using Maximum Likelihood Methods Employing a Real-Time Digital System", NASA TN D-6735, May 1972.

[6] Mehra, R.K.; Stepner, D.E.; Tyler, J.S., "A Generalized Method for the Identification of Aircraft Stability and Control Derivatives from Flight-Test Data", JACC, August 1972.

[7] Bryson, A.E., Jr.; Frazier, M., "Smoothing for Linear and Nonlinear Dynamic Systems", ASD-TDR-63-119, February 1963.

[8] Tyler, J.S.; Powell, J.D.; Mehra, R.K., "The Use of Smoothing and Other Advanced Techniques for VTOL Aircraft Parameter Identification", Final Report, Contract No. N000 19-69-C-0534, Systems Control, Inc., June 1970.

[9] Sage, A.P.; Melsa, J.L., System Identification, Ch. 3, Academic Press, 1971.

[10] Bryson, A.E., Jr.; Ho, Y.C., Applied Optimal Control, Ch. 2, Ch. 13, Blaisdell, 1969.

[11] Mayne, D.Q., "Parameter Estimation", Automatica, Vol. 3, pp. 245-255, Pergamon Press, 1966.

A SQUARE ROOT FORMULATION FOR THE COMBINED STATE-PARAMETER ESTIMATOR WITH APPLICATION TO THE IDENTIFICATION OF SAILPLANE PERFORMANCE

By Michel R. Froidevaux and Charles E. Hutchinson

Electrical and Computer Engineering Department
University of Massachusetts, Amherst, Massachusetts 01002

SUMMARY

A square root formulation is presented for the discrete combined state-parameter estimation problem with linear plant dynamics, Gaussian random disturbances and constant but uncertain parameters. The estimator is a combination of the classical Kalman filter and a maximum likelihood algorithm which maximizes the parameter log-likelihood function using a first-order search routine.

INTRODUCTION

This paper offers a new way to propagate the gradient of the log-likelihood function, extending the concept of the "popular" square root filtering technique (ref. 1), to the combined estimator, thus preserving the outstanding properties of the square root filter algorithm.

The derivation of the complete recursive equation for the log-likelihood gradient in its "square root matrix" form is given for successive scalar measurement updates as is current practice in many implementations, since it improves both the numerical characteristics and the efficiency of the estimation procedure.

Two ($n \times n$) complexified Cholesky square root decompositions and two ($2n \times n$) triangularization algorithms are shown to be necessary for producing the needed updates for the square root covariance gradient matrices.

COMBINED STATE-PARAMETER ESTIMATOR

The following notation is used to describe the discrete time linear system:

$$\underline{x}_{n+1} = \phi_n(\underline{\alpha})\underline{x}_n + G_n(\underline{\alpha})\underline{v}_n \tag{1}$$

$$\underline{z}_n = H_n(\underline{\alpha})\underline{x}_n + \underline{w}_n \tag{2}$$

* The research reported herein was partially supported by the Air Force Office of Scientific Research, Air Force Systems Command, USAF, under Grant No. AFOSR-73-2443, and by the Measurement Systems Laboratory of the Massachusetts Institute of Technology.

$\underline{x}_n$: state vector, dimension N_x

$\underline{z}_n$: measurement vector, dimension N_z

$\underline{\alpha}$: uncertain parameter vector, dimension p

$\underline{w}_n$, $\underline{v}_n$: zero mean, independent, Gaussian sequences, dimension N_z and N_v, with covariance matrices R and Q, respectively.

If we let $P_n(\underline{\alpha})$ be the state error covariance matrix at stage n, then one formulation of the discrete square-root Kalman filter is as follows (ref. 1):

$$P_n(\underline{\alpha}) \overset{\Delta}{=} P_n = S_n S_n^T \tag{3}$$

$$Q = UU^T \tag{4}$$

$$R = WW^T \tag{5}$$

Where S_n, U, W are the uniquely defined "Cholesky" square-root matrices for P_n, Q and R, as discussed in ref. 1.

Time update:

$$\hat{\underline{x}}'_{n+1} = \Phi_n(\underline{\alpha}) \hat{\underline{x}}_n \tag{6}$$

$$\begin{bmatrix} S^{T'}_{n+1} \\ \hline 0 \end{bmatrix} = T \begin{bmatrix} S_n^T \Phi_n^T \\ \hline U_n^T G_n^T \end{bmatrix} \tag{7}$$

where T is a suitable orthogonal matrix (which need not be calculated in practice).

Measurement update: Use N_z scalar measurements (assume R is diagonal)

$$\hat{\underline{x}}_n(k) = \hat{\underline{x}}_n(k-1) + \underline{K}_n(k) [z_n(k) - \underline{h}_n^T(k) \hat{\underline{x}}_n(k-1)] \tag{8}$$

where $\underline{h}_n^T(k)$ is the kth row of the $H_n(\underline{\alpha})$ matrix and for k = 0, $\hat{\underline{x}}_n(0) = \hat{\underline{x}}'_n$, $S_n(0) = S'_n$ and we have:

$$\underline{K}_n(k) = a_n(k) S_n(k) S_n^T(k) \underline{h}_n(k) \tag{9}$$

$$a_n(k)^{-1} = \underline{h}_n^T(k) S_n(k) S_n^T(k) \underline{h}_n(k) + R(k) \tag{10}$$

$$S_n(k) = S_n(k-1) - \gamma_n(k) \underset{n}{\underline{K}}(k) \underset{n}{\underline{h}^T}(k) S_n(k-1) \tag{11}$$

$$\gamma_n(k)^{-1} = 1 + [a_n(k) R(k)]^{1/2} \tag{12}$$

where $k = 1, 2, \ldots, N_z$,

$n = 0, 1, 2, \ldots, N$, and $R(k)$ is the appropriate element of R.
For the purposes of initialization $\underset{0}{\underline{x}}$ and P_0 are assumed known.

Now, if we define $\underset{N}{\underline{Z}}$ as the $(N \cdot N_z)$ vector formed with "all" the successive measurement vectors, then we can define the likelihood function of the parameter $\underline{\alpha}$ as follows. Let $p(\underset{N}{\underline{Z}}/\underline{\alpha})$ be the conditional probability density of $\underset{N}{\underline{Z}}$ given $\underline{\alpha}$. Thus, $\hat{\underline{\alpha}}$, the maximum likelihood estimate of the parameter $\underline{\alpha}$, is the value of $\underline{\alpha}$ which maximizes $p(\underset{N}{\underline{Z}}/\underline{\alpha})$ for a particular $\underset{N}{\underline{Z}}$. The log-likelihood function is defined as

$$\xi_N(\underline{\alpha}) = \ln [p(\underset{N}{\underline{Z}}/\underline{\alpha})] \tag{13}$$

For the general linear system (1) and (2) with Gaussian noise, the log-likelihood function can be shown (ref. 3) to be the sum of a "bias" term, independent of the measurements, and an "observation" term which depends on $\underset{N}{\underline{Z}}$ as follows:

$$2\xi_N(\underline{\alpha}) = \xi_{N,bias}(\underline{\alpha}) + \xi_{N,observation}(\underline{\alpha}) \tag{14}$$

where

$$\xi_{N,bias}(\underline{\alpha}) = -NN_z \ln(2\pi) - \sum_{n=1}^{N} \ln \left| \Delta_n^{\prime}(\underline{\alpha}) \right| \tag{15}$$

$$\xi_{N,observation}(\underline{\alpha}) = - \sum_{n=1}^{N} \underset{n}{\underline{\delta}^T}(\underline{\alpha}) \Delta_n^{\prime}(\underline{\alpha})^{-1} \underset{n}{\underline{\delta}}(\underline{\alpha}) \tag{16}$$

and

$$\underset{n}{\underline{\delta}}(\underline{\alpha}) = \underset{n}{\underline{z}} - H_n \hat{\underline{x}}_n^{\prime} \tag{17}$$

$$\Delta_n^{\prime}(\underline{\alpha}) = E \{ \underset{n}{\underline{\delta}}(\underline{\alpha}) \underset{n}{\underline{\delta}^T}(\underline{\alpha}) \} = H_n P_n^{\prime}(\underline{\alpha}) H_n^T + R \tag{18}$$

($\underset{n}{\underline{\delta}}(\underline{\alpha})$ is sometimes referred to as the innovation process and must be white for satisfactory performance).

If the $\underset{n}{\underline{z}}$ vector is processed one component at a time, the covariance of

233

$k\underline{^{th}}$ innovation (scalar) is simply:

$$\Delta_n^-(k) = \underline{h}_n^T(k) \ S_n^-(k) \ S_n^{-T}(k) \ \underline{h}_n(k) + R(k) = a_n^{-1}(k) \tag{19}$$

so that we obtain the desired result:

$$2\xi_N(\underline{\alpha}) = -NN_z \ \ln(2\pi) - \sum_{n=1}^{N} \sum_{k=1}^{N_z} [\ln a_n^{-1}(k) + a_n(k) \ \delta_n^2(k)] \tag{20}$$

where $\delta_n(k)$ is now the scalar innovation sequence:

$$\delta_n(k) = z_n(k) - \underline{h}_n^T(k) \ \underline{\hat{x}}_n(k-1) \tag{21}$$

We note that equation (20) is very easily implemented if one has already coded the square-root filter equations (6) to (12); only the maintenance of running sums is required to calculate $\xi_N(\underline{\alpha})$.

GRADIENT OF LOG LIKELIHOOD FUNCTION

We shall derive an expression for each component of the gradient for the log-likelihood

$$\frac{\partial \xi_N(\underline{\alpha})}{\partial \alpha_1}, \ 1 = 1, \ \dots, \ p,$$

by postulating a complexified Cholesky square-root form for the gradient of the covariance matrix. (Indeed, the gradient of the covariance matrix is again symmetric, but not positive semi-definite in general). That is, defining

$$M_{n,1} \ M_{n,1}^T = \frac{\partial}{\partial \alpha_1} [P_n(\underline{\alpha})] = \frac{\partial}{\partial \alpha_1} [S_n S_n^T] \tag{22}$$

only M_n need be propagated, M_n being complex in general.

The general expression for the $\ell\underline{^{th}}$ component of the gradient of the log-likelihood function is given by

$$2\frac{\partial \xi_N(\underline{\alpha})}{\partial \alpha_1} = - \sum_{n=1}^{N} \sum_{k=1}^{N_z} \left\{ \frac{\partial a_n(k)}{\partial \alpha_1} [\delta_n^2(k) - a_n^{-1}(k)] + 2a_n(k) \ \delta_n(k) \ \frac{\partial \delta_n(k)}{\partial \alpha_1} \right\} \tag{23}$$

where

$$\frac{\partial a_n}{\partial \alpha_1} = -a_n^2 \left\{ 2\frac{\partial \underline{h}_n^T}{\partial \alpha_1} S_n^- S_n^{-T} \underline{h}_n + \underline{h}_n^T M_{n,1}^- M_{n,1}^{-T} \underline{h}_n \right\} \tag{24}$$

and,

$$\frac{\partial \delta_n(k)}{\partial \alpha_1} = - \frac{\partial \underline{h}_n^T}{\partial \alpha_1} \underline{\hat{x}}_n(k-1) - \underline{h}_n^T \frac{\partial \underline{\hat{x}}_n(k-1)}{\partial \alpha_1} \tag{25}$$

234

Only the propagation equations for M_n and the gradient of $\hat{x}$ remain to be obtained. These quantities are provided as follows. First for the state gradient:

Time update: Use a single vector update

$$\frac{\partial \hat{\underline{x}}'_{n+1}}{\partial \alpha_1} = \phi_n \frac{\partial \hat{\underline{x}}_n}{\partial \alpha_1} + \frac{\partial \phi_n}{\partial \alpha_1} \hat{\underline{x}}_n \tag{26}$$

since $\underline{x}$ is known (fixed), then we have necessarily ≈ 0.

$$\frac{\partial \hat{\underline{x}}_0}{\partial \alpha_1} = \underline{0}$$

Measurement update: Use N_z scalar measurements

$$\frac{\partial \hat{\underline{x}}_n(k)}{\partial \alpha_1} = \frac{\partial \hat{\underline{x}}_n(k-1)}{\partial \alpha_1} + \frac{\partial \underline{K}_n(k)}{\partial \alpha_1} \delta_n(k) + \underline{K}_n(k) \frac{\partial \delta_n(k)}{\partial \alpha_1} \tag{27}$$

substituting (25) into the above yields

$$\frac{\partial \hat{\underline{x}}_n(k)}{\partial \alpha_1} \approx \frac{\partial \hat{\underline{x}}_n(k-1)}{\partial \alpha_1} [1 - \underline{K}_n^T(k) \underline{h}_n(k)] + \frac{\partial \underline{K}_n(k)}{\partial \alpha_1} \delta_n(k) - \underline{K}_n(k) \frac{\partial \underline{h}_n^T(k)}{\partial \alpha_1} \hat{\underline{x}}_n(k-1) \tag{28}$$

where $\dfrac{\partial \underline{K}_n(k)}{\partial \alpha_1}$ is obtained similarly by differentiating (9):

$$\frac{\partial \underline{K}_n}{\partial \alpha_1} = S_n'S_n'^T \left[\frac{\partial a_n}{\partial \alpha_1} + a_n \frac{\partial \underline{h}_n}{\partial \alpha_1}\right] + a_n M_{n,1}' M_{n,1}'^T \underline{h}_n \tag{29}$$

and $\dfrac{\partial a_n}{\partial \alpha_1}$ has already been calculated in (24).

Now, the square-root covariance gradient may be calculated according to:

Time update: Differentiating the basic covariance time update equation:

$$S_{n+1}' S_{n+1}'^T = \phi_n S_n S_n^T \phi_n^T + G_n Q G_n^T \tag{30}$$

and recalling our definition of M_n:

$$M_{n+1,\ell}' M_{n+1,\ell}'^T = \frac{\partial \phi_n}{\partial \alpha_1} S_n S_n^T \phi_n^T + \phi_n M_{n,1} M_{n,1}^T \phi_n^T + \phi_n S_n S_n^T \frac{\partial \phi_n^T}{\partial \alpha_1} + \frac{\partial G_n}{\partial \alpha_1} Q G_n^T + G_n Q \frac{\partial G_n^T}{\partial \alpha_1} \tag{31}$$

235

Letting

$$E_n + \left\{ \frac{\partial \Phi_n}{\partial \alpha_1} S_n S_n^T \phi_n^T + \frac{\partial G_n}{\partial \alpha_1} Q \, G_n^T \right\} \tag{32}$$

and

$$J_n J_n^T = E_n + E_n^T \tag{33}$$

where J_n is obtained by a complexified Cholesky decomposition, then we have:

$$\begin{bmatrix} M_{n+1,\ell}^T \\ - - - - \\ 0 \end{bmatrix} = T_1 \begin{bmatrix} J_n^T \\ \bar{M}_{n,1}^T \bar{\phi}_n^T \end{bmatrix} \tag{34}$$

with T_1 being a suitable orthogonal transformation. As for the state equation, since $P_0 = S_0 S_0^T$ is fixed, then $M_{0,1} = 0$.

Measurement update: Use N_z scalar measurements.

$$S_n(k) S_n^T(k) = S_n(k-1) S_n^T(k-1) - a_n(k) S_n(k-1) S_n^T(k-1) \underline{h}(k) \underline{h}^T(k) S_n(k-1) S_n^T(k-1) \tag{35}$$

To obtain (dropping the k index for clarity):

$$M_{n,1} M_{n,1}^T = \bar{M}_{n,1} \bar{M}_{n,1}^T - \bar{M}_{n,1} \bar{M}_{n,1}^T \underline{h}_n \frac{\partial a_n}{\partial \alpha_1} \underline{h}_n^T \bar{M}_{n,1} \bar{M}_{n,1}^T$$

$$- a_n \left\{ \bar{M}_{n,1} \bar{M}_{n,1}^T \underline{h}_n \underline{h}_n^T \bar{S}_n \bar{S}_n^T + \bar{S}_n \bar{S}_n^T \underline{h}_n \underline{h}_n^T \bar{M}_{n,1} \bar{M}_{n,1}^T \right.$$

$$\left. + \bar{S}_n \bar{S}_n^T [\underline{h}_n \frac{\partial \underline{h}_n^T}{\partial \alpha_1} + \frac{\partial \underline{h}_n}{\partial \alpha_1} \underline{h}_n^T] \bar{S}_n \bar{S}_n^T \right\} \tag{36}$$

If we let

$$\bar{\psi}_{n,1} = [\bar{M}_{n,1} \bar{M}_{n,1}^T \underline{h}_n \underline{h}_n^T + \bar{S}_n \bar{S}_n^T \underline{h}_n \frac{\partial \underline{h}_n^T}{\partial \alpha_1}] \bar{S}_n \bar{S}_n^T \tag{37}$$

(Note: $\bar{\psi}_{n,1}$ is real)

Then (36) becomes

$$M_{n,1} M_{n,1}^T = \bar{M}_{n,1} [I - \bar{M}_{n,1}^T \underline{h}_n \frac{\partial a_n}{\partial \alpha_1} \underline{h}_n^T \bar{M}_{n,1}] \bar{M}_{n,1}^T - a_n \bar{Y}_{n,1} \bar{Y}_{n,1}^T \tag{38}$$

where

$$Y_{n,1}^{\cdot} \, Y_{n,1}^{\cdot T} = \psi_{n,1}^{\cdot} + \psi_{n,1}^{\cdot T} \qquad (39)$$

Now, the bracketed term in (38) can be decomposed as

$$[\cdot] = [I - \beta_{n,1} \frac{\partial a_n}{\partial \alpha_1} M_{n,1}^{\cdot T} \frac{h\, h^T}{n\, n} M_{n,1}^{\cdot}][I - \beta_{n,1} \frac{\partial a_n}{\partial \alpha_1} M_{n,1}^{\cdot T} \frac{h\, h^T}{n\, n} M_{n,1}^{\cdot}] \qquad (40)$$

with $\beta_{n,1}$ given by:

$$\beta_{n,1}^{-1} = 1 \pm (1 - \frac{\partial a_n}{\partial \alpha_1} h_n^T M_{n,1}^{\cdot} M_{n,1}^{\cdot T} h_n)^{1/2}. \qquad (41)$$

Using this result, we rewrite (38) as

$$M_{n,1} M_{n,1}^T = N_{n,1}^{\cdot} N_{n,1}^{\cdot T} - a_n Y_{n,1}^{\cdot} Y_{n,1}^{\cdot T} \qquad (42)$$

where $N_{n,1}$ is complex in general and is now obtained by substituting (40) into (38) as

$$N_{n,1}^{\cdot} = M_{n,1}^{\cdot} - \frac{\partial a_n}{\partial \alpha_1} \beta_{n,1}^{\cdot} M_{n,1}^{\cdot} M_{n,1}^{\cdot T} \frac{h\, h^T}{n\, n} M_{n,1}^{\cdot} \qquad (43)$$

and $Y_{n,1}$ is computed via the complexified Cholesky decomposition algorithm from the expression (39) in much the same way as matrix J_n for the time update in the previous development. (Both decompositions are $^nN_x \cdot N_x$.)

Finally, the measurement update equation for the square-root covariance gradient is given by:

$$\begin{bmatrix} M_{n,1}^T(k) \\ ---- \\ 0 \end{bmatrix} = T_2 \begin{bmatrix} N_{n,1}^T(k-1) \\ ----------- \\ j \ \sqrt{a_n}(k-1) \ Y_{n,1}^{\cdot T}(k-1) \end{bmatrix}_{k=1, \ldots, N_z} \qquad (44)$$

where $j = \sqrt{-1}$ and T_2 is again a suitable orthogonal transformation which need not be further defined. Note that the update equations (39) and (44) may be implemented by using a triangularization algorithm in a complex space.

IMPLEMENTATION

A few observations can be made when attempting to reduce computation time while performing a square-root combined estimation. When the dimension of $\underline{x}$ is small, a fair amount of computer time can be saved if one substitutes a "parametric propagation" in lieu of the exact propagation equation by functionally curve fitting certain matrices such as S_n, $S_n^{\cdot}$, M_n, $M_n^{\cdot}$, or the gain vectors K_n, using a set of "relevant" values for the parameter $\underline{\alpha}$ over an assumed domain of variation (steady-state filter gains as a function of $\underline{\alpha}$ may also suffice for some specific applications).

A first-order "fast update" has also been suggested (ref. 2), computing for state estimates over N data points with a new value for the parameter estimate $\hat{\underline{\alpha}}$: use the newly evaluated Φ_n, $\underline{h}_n$ and G_n matrices to propagate $\hat{\underline{x}}$ and $\frac{\partial \hat{\underline{x}}}{\partial \alpha_1}$, but retain the same second-order statistics, S_n and M_n, as used previously, unless the parameter estimate has changed by more than some prespecified amount (the state estimate incorporates the newest $\hat{\underline{\alpha}}$, except that the gains $\underline{K}_n$ are generally based on older estimates. In general, any higher order terms may be neglected by a proper sensitivity analysis performed with extensive simulation trials, or better, using the concept of "ambiguity functions" as discussed in (ref. 3).

Although "Quasi-Newton" methods have been traditionally used in the past for finding the unconstrained maximum of a scalar function $\xi_N(\underline{\alpha})$, where $\underline{\alpha}$ is a p vector (e.g. Davidon, Fletcher and Powell, Broyden, Murtagh and Sargent, of which Fletcher and Powell's method is probably the most widely used), it is suggested here that a more recent algorithm be used, that of Jacobson and Oksman (ref. 4), which is based upon homogenous functions rather than quadratic models. The major advantages of the latter algorithm is that it does not require that a minimum along a line be found, does not require evaluation or estimation of second derivatives and is superior to Fletcher and Powell's algorithm on the classical test functions.

SAILPLANE MODEL

This section contains a practical, simplified model of an instrumented sailplane, carrying an inertial measurement unit, an altimeter, an airspeed indicator, and a device measuring the body pitch angle. The wind is assumed to be horizontal, a function of altitude and time only, and directed along the "x" axis. The inertial coordinate system and body orientation used in this example are illustrated in Figure 1.

The sailplane mass and wing area are denoted by M and S respectively. The lift and drag forces are given by

$$L = 1/2 \, \rho \, C_L \, V^2 \, S \tag{45}$$

$$D = 1/2 \, \rho \, C_D \, V^2 \, S \tag{46}$$

In the equation above ρ is the air density. For simplicity it will be assumed that C_L is given and C_D is to be identified.

Using the orientation of Figure 1 and classical techniques the dynamical equations for the sailplane performances can be expressed in the form:

$$\dot{z} = V\theta$$

$$M\dot{V} = -M\dot{W}\cos\psi - D - Mg\theta$$

$$MV\dot{\theta} = M\dot{W}\theta\cos\psi + L\cos\mu - Mg \tag{47}$$

$$MV\dot{\psi} = M\dot{W}\sin\psi - L\sin\mu$$

For the purposes of this study a realistic wind model would be of the form:

$$W(z,t) = nz[y_1(t) + y_2(t)] \tag{48}$$

where $y_1(t)$ is a random walk and $y_2(t)$ is first order markovian random process. That is

$$\dot{y}_1 = w_1(t)$$
$$\dot{y}_2 = -\beta y_2 + w_2(t) \tag{49}$$

where w_1 and w_2 are white noise processes.

Combining equations (47), (48), and (49) yields, after some rearranging a six dimensional state variable model given by:

$$\dot{x}_1 = \dot{z} = V\theta$$

$$\dot{x}_2 = \dot{V} = -\{V\theta n(y_1 + y_2) - n\beta z y_2\}\cos\psi - AC_D V^2 - g\theta$$

$$\dot{x}_3 = \dot{\theta} = -\{\theta n(y_1 + y_2) - \frac{n\beta z}{V} y_2\}\theta\cos\psi + AC_L V\cos\mu - \frac{g}{V}$$

$$\dot{x}_4 = \dot{\psi} = \{\theta n(y_1 + y_2) - \frac{n\beta z}{V} y_2\}\sin\psi - AC_L V\sin\mu \tag{50}$$

$$\dot{x}_5 = \dot{y}_1 = w_1$$

$$\dot{x}_6 = \dot{y}_2 = -\beta y_2 + w_2$$

There are assumed to be four measurements in the measurement vector: x_1, x_2, x_3, and x_4.

The numerical quantities appropriate here are:

$$C_L = \text{constant} = 1.0$$
$$C_D = C_{Do} + \frac{KC_L^2}{\pi AR}$$

$$C_{Do} = .0130 \text{ (for simulation purposes)}$$

$K = 1.22$ (for simulation purposes) $M = 28.3$ slug

AR = aspect ratio = 25.8 $\eta = \frac{1}{2000}$ ft^{-1}

$A = \frac{\rho S}{2M}$ $\beta = \frac{1}{20}$ sec^{-1}

$\rho = .00238$ slug/ft^3 $E[w_1^2] = 10$ kts^2/hr

$S = 140$ ft^2 $E[w_2^2] = 2\beta$ kts^2

SIMULATION RESULTS

Given the time and budget limitations relevant to this project, only a limited number of specific test runs were made for this problem (no Monte Carlo run series were made), none including a full square-root gradient search. Instead, a comparison between an extended Kalman filter and a maximum likelihood algorithm using a pattern search minimization was undertaken, leading to some interesting observations.

As expected from earlier experience with uncertain plant parameters, the sailplane drag parameter identification problem is quite insensitive to K. In all cases, a very "flat" behavior was observed for the log-likelihood function throughout the minimization search. The extended Kalman filter had great difficulty in getting a meaningful estimate of the drag parameter, and quite often diverged totally from the correct answer.

The measurement accuracy and measurement sampling rate which were chosen purposely within "practical" ranges (10 ft. error on altitude, 0.01 degree on pitch angle and 0.5 ft/sec accuracy on the velocity measurement with 0.5 sec. sampling rate) was seen to be "marginal" with respect to the extended Kalman filtering assumption, i.e., trajectory tracking by the filter was not quite good enough to allow good performance from local linearization. An actual flight test may be implemented in the future by the MIT measurement systems laboratory.

No attempt was made to use smooth-type sailplane trajectories, which could have led to different results. The dynamic model was simply a "fixed-stick" (uncontrolled) sailplane originating at some specified altitude with a given velocity, pitch and bank angle (generally different from the smooth steady-state flight conditions) in a disturbed atmosphere.

It is believed that the more sensitive algorithm (with double precision) using gradient information will be valuable here and will be tried for this particular problem. Also, it is clear that a very precise instrumentation is needed for trajectory tracking, with possibly a higher measurement sampling rate.

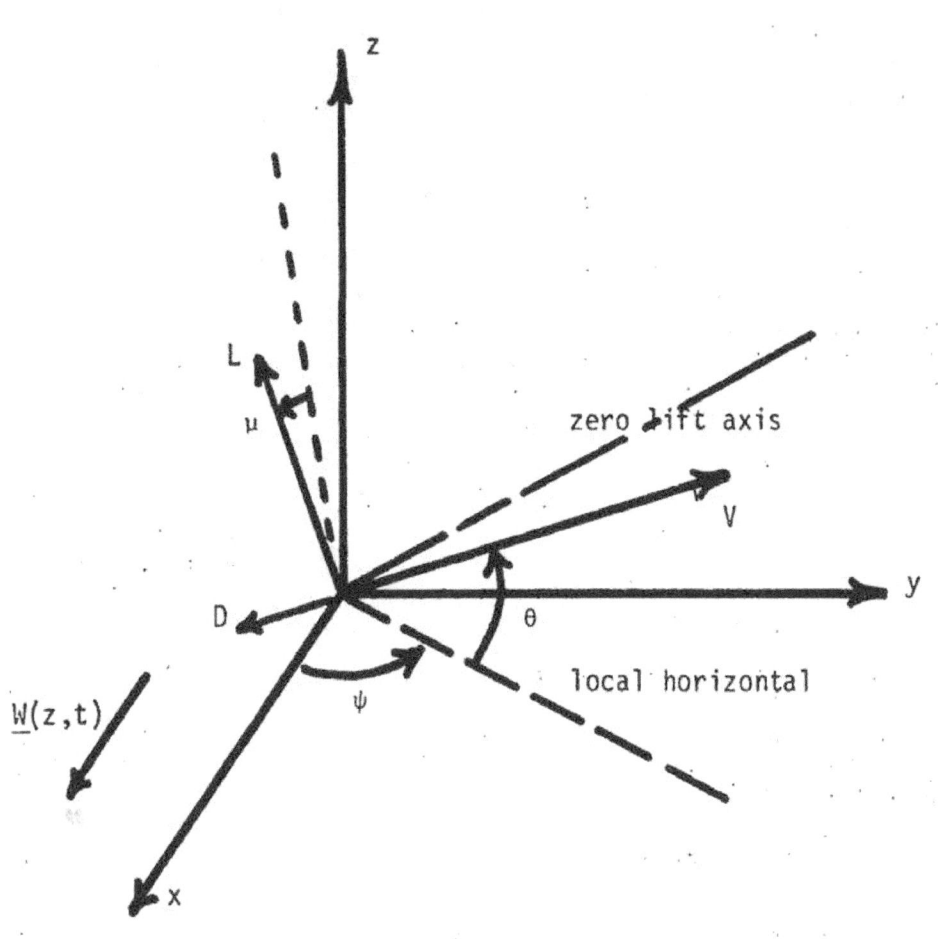

Fig. 1.   Coordinate System.

REFERENCES

1. Kaminski, P.G., Bryson, E.,Jr., Schmidt, S.T., "Discrete Square Root Filtering: A Survey of Current Techniques," IEEE Transactions on Automatic Control, Vol. AC-16, No. 6, December 1971.

2. Maybeck, P.S., "Combined State and Parameter Estimation for On-Line Applications," Ph.D. Thesis, M.I.T., Department of Aeronautics and Astronautics, C.S. Draper Laboratory Report T-557, Cambridge, Massachusetts, February 1972.

3. Schweppe, F., Uncertain Dynamic Systems, Prentice Hall, New York, 1972.

4. Jacobson, D.H. and Oksman, W., "An Algorithm that Minimizes Homogenous Functions of N Variables in N+2 Iterations and Rapidly Minimizes General Functions," ONR Technical Report No. 618, Division of Engineering and Applied Physics, Harvard University, October 1970.

A UNIFIED APPROACH TO AIRCRAFT PARAMETER IDENTIFICATION*

By David E. Stepner and John A. Sorensen

Systems Control, Inc.
260 Sheridan Ave.
Palo Alto, Ca. 94306

SUMMARY

The most accurate identification results are obtained when all three elements of the identification process - the identification algorithm, the control input and the instrumentation system - are considered in a unified approach. This paper discusses this type of approach for the design of optimal control inputs and for determining the effect of the instrumentation system, in each case with respect to the identification process. A new approach for the design of control inputs which optimize the sensitivity of the system output to the unknown parameters is given. Results using these inputs in an extensive simulation of the identification process indicate they perform measurably better than doublet type inputs. A new technique is then presented for specifying an optimal instrumentation system or for determining the effect the instrumentation system has on the accuracy of the parameter estimates. Results with this technique prove that unmodeled instrument errors can increase the uncertainty in the values of the identified parameters.

INTRODUCTION

It is generally the parameter identification algorithm that is considered as the key element in obtaining stability and control derivatives from flight test data. However, the control input and the measurement system are equally important, and the accuracy of the identified aircraft parameters depend strongly on the design of the control input and selection of the measurement system. This interrelationship is shown in Figure 1. As will be discussed in the succeeding sections, the control input and the measurement system must be designed with this interrelationship in mind, and the objective in their design must be to increase the accuracy of the identified parameter estimates. For this reason, evaluation of the techniques discussed in this paper is done via Monte Carlo (or ensemble) simulations of the entire identification process.

I. OPTIMAL INPUT DESIGN

The problem of specifying flight control inputs that will enhance the identification of parameters is one of insuring that all the response modes for which the parameters are to be identified are adequately excited. The required excitation is dependent on the modes which are to be identified, the instrumentation system that has been specified and the length of time that is available for recording the system response. For example, the short period response mode

---

*
This work was supported by NASA under Contracts NAS1-10700 and NAS1-1079
and by AFOSR under Contract F44620-71-C-0077.

requires different types of inputs than the phugoid mode. Although adequate excitation of the aircraft modes is obtained using standard inputs (step functions, doublets), the real purpose of addressing this problem is to specify quantitatively what the "optimal" inputs would be for a given aircraft, the specific stability and control derivatives to be identified and the specified instrumentation system.

The complete theoretical development for the design of optimal control inputs for identification is given in References 1-3 and only an extremely brief statement of the problem is given. The performance criterion used to compute the input is the sensitivity of the system response to the unknown parameters. It is also assumed that an output error or maximum likelihood method which can handle arbitrary inputs is used for the derivative extraction.

For the linearized aircraft equations of motion

$$\dot{x} = Fx + Gu$$

where u is the control vector, and measurements given by

$$y = Hx + n$$

where n is a vector of zero mean white noise with covariance R, the information matrix M for the unknown parameters p is written as

$$M = \int_0^T \left(\frac{\partial x}{\partial p}\right)^T H^T R^{-1} H\left(\frac{\partial x}{\partial p}\right) dt.$$

Allowing for the possibility of weighting the sensitivities, thereby reflecting the relative importance of the parameters, the scalar criterion used was

$$\text{maximize } \text{tr}\{WM\}$$

with W a diagonal weighting matrix. The class of inputs considered for this optimization problem were those with specified energy.

DESIGN OF MONTE CARLO SIMULATION

In order to provide a realistic evaluation of the advantages of the optimal input over a sub-optimal doublet input of equal energy, a Monte Carlo simulation of the identification process was performed. Both the phugoid and short period modes of the longitudinal equations of motion were used in generating the simulated flight test data, although only the short period derivatives were to be identified. In addition, the short period parameters of the 4 state model used to generate the optimal input were changed by approximately 50% from the equivalent parameters of the model used to generate the data. This was to simulate the situation of designing the input based on wind estimates.

The four state longitudinal equations of motion for the C-8 aircraft which were used for computing the optimal input are

$$\begin{bmatrix} \dot{u} \\ \dot{\theta} \\ \dot{q} \\ \dot{\alpha} \end{bmatrix} = \begin{bmatrix} -.02 & -32.2 & 0 & 33.74 \\ 0 & 0 & 1 & 0 \\ .003 & 0 & \underline{-1.588} & \underline{-.562} \\ -.004 & 0 & 1 & \underline{-.737} \end{bmatrix} \begin{bmatrix} u \\ \theta \\ q \\ \alpha \end{bmatrix} + \begin{bmatrix} 0 \\ 0 \\ \underline{-1.658} \\ \underline{.005} \end{bmatrix} \delta_e$$

$$\begin{bmatrix} y_{1_k} \\ y_{2_k} \\ y_{3_k} \\ y_{4_k} \end{bmatrix} = \begin{bmatrix} u_k \\ \theta_k \\ q_k \\ \alpha_k \end{bmatrix} + \begin{bmatrix} n_u \\ n_\theta \\ n_q \\ n_\alpha \end{bmatrix}$$

where   u is longitudinal velocity (ft/sec)

θ is pitch angle (°)

q is pitch rate (°/sec)

α is angle of attack (°)

δ_e is elevator deflection (°)

and

$$R = E\{n_i n_j\} = \begin{bmatrix} .125 & & & \\ & .125 & & \\ & & .125 & \\ & & & .125 \end{bmatrix} \delta_{ij}$$

is the discrete noise covariance. The optimal input was designed with respect to the five short period parameters, indicated by the underlining. For a speci- fied length of 2 sec., the optimal input (energy = 58.0) and the doublet input of equal energy, are shown in Fig. 2.

The stability and control derivatives which were used to generate the simu- lated flight data are illustrated below

$$F = \begin{bmatrix} \text{same} & | & & \\ \hline & | & \underline{-2.238} & \underline{-.28} \\ \text{same} & | & & \\ & | & 1 & \underline{-.368} \end{bmatrix} \quad ; \quad G = \begin{bmatrix} \text{same} \\ \hline -.829 \\ -.0075 \end{bmatrix}$$

Fifty noise sequences, of 100 points each, were generated and added to the deterministic state vector values generated by both the optimal and doublet inputs. (It must be noted that the 100 data sequence length is not really adequate for accurate parameter identification.) Each of these 100 sets of simulated flight test data (50 for optimal input, 50 for doublet input) were processed in exactly the same manner using the maximum likelihood identification algorithm (Reference 3). The parameters to be identified were the five short period derivatives and the measurement noise covariance matrix.

RESULTS OF THE MONTE CARLO SIMULATION

The Monte Carlo simulation, as stated, consisted of 100 runs of the identification procedure. After each, the parameter estimates, information matrix and its eigenvalues, and the parameter covariance matrix and its eigenvalues were saved. The ensemble results are tabulated in Table 1.

The theoretical values of the trace of the information matrix using the actual values for the stability and control derivatives were computed to be 2.12×10^7 and 4.74×10^5 for the optimal and suboptimal inputs, respectively. The average values, from Table 1, were 2.15×10^7 and 4.79×10^5 indicating that 50 runs were sufficient for obtaining accurate parameter estimate and information matrix statistics.

By almost all measures of performance, the optimal input has performed better than the suboptimal input. The determinant of the sample covariance, giving a measure of the overall parameter uncertainty (volume of uncertainty ellipsoid) based on the actual derived parameter estimates, was four orders of magnitude smaller for the optimal input. The eigenvalues of the sample covariance were smaller, on a one-to-one basis, for the optimal input, indicating a smaller dimension for each axis of the uncertainty eillipsoid.

The histograms of some of the parameter estimates are shown in Figs. 3-5. For the M_α, Z_α and Z_δ parameters, the optimal input definitely produced a better ensemble of parameter estimates. The mean estimate was much closer to the actual parameter value and the standard deviations and mean square errors were smaller. The performance for the two inputs was about the same for M_q, while the suboptimal input did perform better for M_δ. However, it should be kept in mind that the accuracy of the parameter estimates themselves, however desirable, was not a direct performance objective. Considerably more work is involved for that performance objective due to its nonlinear nature. For this evaluation, the determinant* and trace of the parameter estimate covariance matrix were the criterion of interest.

II. ANALYSIS OF FLIGHT INSTRUMENTATION ERROR EFFECTS

A principal source of error in identifying stability and control derivatives is the errors in the flight instrumentation. A method for quantitatively assessing the effects of instrumentation errors such as biases and mounting

---

* The weighted trace criterion can be made to very closely approximate the determinant of the information matrix.

246

misalignment on the accuracy in estimates of individual stability and control derivatives is now summarized. More complete details can be found in References 4 and 5. This method is useful for specifying both the accuracy and the type of flight sensors that should be utilized to satisfy the objectives of the flight test program.

ERROR ANALYSIS TECHNIQUE

The maximum likelihood identification technique (Reference 3) chooses parameters $\hat{p}$ (stability and control derivatives) which minimize the cost function

$$J = \sum_{i=1}^{n} (y_i - \hat{y}_i)^T R^{-1} (y_i - \hat{y}_i) \tag{1}$$

where y_i are the measured outputs of aircraft motion as perturbed from steady flight

$\hat{y}_i$ are the estimated outputs based on the estimated parameters, $\hat{p}$

R is the covariance of an assumed white noise in the output measurements.

Minimization of Eq. (1) is done via modified Newton-Raphson optimization with

$$\hat{p}_{k+1} = \hat{p}_k - \left[\frac{\partial^2 J}{\partial p^2}\right]^{-1} \frac{\partial J}{\partial p}^T \tag{2}$$

and

$$\frac{\partial J}{\partial p} = - 2 \sum_{i=1}^{n} (y_i - \hat{y}_i)^T R^{-1} \frac{\partial y_i}{\partial p} \tag{3}$$

$$\frac{\partial^2 J}{\partial p^2} = 2 \sum_{i=1}^{n} \left[\left(\frac{\partial y_i}{\partial p}\right)^T R^{-1} \left(\frac{\partial y_i}{\partial p}\right)\right] \tag{4}$$

In this analysis, it is assumed that the true value of p is known, and the effect of instrument errors is small. Thus, only one application of Eq. (2), with $\hat{p}$ set at the true value p is needed to determine the effect of the measurement error on the estimate of p. From Eq. (2), the error due to instrument errors is then

$$\delta p \approx \hat{p} - p = - \left[\frac{\partial^2 J}{\partial p^2}\right]^{-1} \frac{\partial J}{\partial p}^T \tag{5}$$

$$= \left[\sum_{i=1}^{n} \left\{\frac{\partial y_i}{\partial p}^T R^{-1} \frac{\partial y_i}{\partial p}\right\}\right]^{-1} \sum_{i=1}^{n} \frac{\partial y_i}{\partial p}^T R^{-1} (y_i - \hat{y}_i)$$

247

Output errors only affect the value of y_i in Eq. (5); the desired parameter sensitivity is then

$$\frac{\partial}{\partial e}(\delta p) = \left[\frac{\partial^2 J}{\partial p^2}\right]^{-1} \sum_{i=1}^{n} \frac{\partial y_i}{\partial p} R^{-1} \frac{\partial y_i}{\partial e} \qquad (6)$$

where $\frac{\partial y_i}{\partial e}$ is obtained by differentiating the output with respect to any additive or multiplicative non-noise error source e.

The parameter errors due to known measurement errors are then

$$E\{\delta p\} = \frac{\partial(\delta p)}{\partial e} E\{e\} \qquad (7)$$

For random measurement errors, the expected covariance of the parameter estimates is

$$E\{\delta p\ \delta p^T\} = E\{\delta p\ \delta p^T\}_{noise} + \frac{\partial(\delta p)}{\partial e} E\{e\ e^T\} \frac{\partial(\delta p)}{\partial e}^T \qquad (8)$$

MEASUREMENT SYSTEM

The output measurements can be expressed by the equation:

$$y_i = Ty_T + B + w_i \qquad (9)$$

where w_i is white noise with $E\{w_i\} = 0$; $E\{w_i\ w_j^T\} = R\delta_{ij}$

y_T is the actual output vector

y_i is the recorded measurement vector

B is the instrument bias vector

T is a matrix whose diagonal terms are scale factor errors and off-diagonal terms represent cross-coupling errors.

Some specific off-diagonal terms of the T matrix include

1) α and β boom correction
2) accelerometer location corrections
3) misalignments (accelerometers and gyros)

The resulting identification process including these instrument errors is depicted in Figure 6.

The following seven instruments were assumed to be available for longitudinal measurements:

248

1. pitch altitude gyro
2. pitch rate gyro
3. angle-of-attack vane
4. air speed indicator (pitot tube)
5. longitudinal accelerometer
6. normal accelerometer
7. pitch angular accelerometer

The accuracy range of these instruments is summarized in Table 2. As a means of having a reference set of instruments with which to conduct the study, a "base-line" set of instrument accuracies was chosen within the range of Table 2. This set of accuracies is listed in the "base" columns, and was assumed to represent instrumentation quality of a typical flight test program.

RESULTS FOR LONGITUDINAL MEASUREMENT ERRORS

The time history of pitch angle, pitch rate, angle-of-attack, and longitudinal speed perturbations about the reference flight path used to identify the longitudinal stability and control derivatives of an F4 (and a similar shaped input for a DC-8) is shown in Figure 7. The measurement data sequence consisted of 300 points taken every 0.05 sec. over a 15 sec. time span. The elevator deflection consisted of a doublet of $\pm 2.5°$ followed by step inputs of $-0.5°$ and $0.5°$.

In studying the effect of instrument errors, two different identification setups were used. In the first, only the stability and control derivatives were identified so that all bias errors affected the total estimation uncertainty. In the second, it was assumed that all instrument biases but one were estimated so that their contributions were essentially eliminated. In both cases, initial conditions were not used as error sources.

Table 3 presents the results of using the ensemble analysis technique to compute the standard deviations of the longitudinal parameters for both aircraft using the baseline set of errors. It can be seen that the addition of non-noise error sources has a substantial effect on the standard deviation of the parameter estimate accuracy. The errors in parameters M_u, Z_u, X_w, and X_u are increased by over an order of magnitude by the non-modeled instrument errors.

This growth in the standard deviations is illustrated more distinctly in the bar graphs in Figure 8. It must be pointed out that the largest errors are in the parameters associated with the phugoid mode. This is because the 15 sec data span taken does not have the necessary information content to obtain better accuracies for the parameters which govern the phugoid motion. From Figure 8 it can be also seen that the effects of the instrument errors on the parameter accuracy are quite similar for both aircraft. Thus, these results appear to be mainly dependent upon the instrument error magnitudes, rather than the aircraft tested.

Also, from Table 3, the estimation of biases increases the parameter deviation due to noise only, but generally reduces the total deviation of each parameter. Thus, it might be better to structure the identification scheme so that other errors, such as the accelerometer misalignments, are also estimated, in addition to the biases.

Parameter sensitivities $\frac{\partial(\delta p)}{\partial e}$ are useful in specifying the instrumentation accuracy required. As an example of this application, Figure 9 illustrates the deviation of the parameter Z_{δ_e} due to the uncertainty in the longitudinal position of the F-4 aircraft center-of-gravity. For the Z_{δ_e} uncertainty to be less than 10% of the nominal value, the position of the center-of-gravity must be known to within 0.4 ft.

CONCLUDING REMARKS

The process of extracting aircraft stability and control derivatives has been shown to be a much broader problem than just specifying an identification algorithm. The accuracy of the identified parameters depend very strongly on both the control input applied and the measurement system used. New techniques for the design of optimal control inputs and the specification of an instrumentation system have been presented. It has been shown that, when evaluated in a simulation of the overall identification process, the optimal input greatly reduced the uncertainties about and coupling between the unknown parameters. The evaluation of the instrumentation system design technique has shown that unmodeled instrumentation errors can have a large effect on identification accuracy and the improvements in parameter accuracies can be achieved by identifying the dominant errors.

REFERENCES

1. Mehra, R.K., "Optimal Input Design for Linear System Identification", 1972 JACC, Stanford, California, August 1972.

2. Mehra, R.K. and Stepner, D.E., "Dual Control and Identification Methods for Avionic Systems - Part II, Optimal Input Design for Linear System Identification", Final Report to AFOSR under contract F44620-71-C-0077

3. Stepner, D.E. and Mehra, R.K., " Maximum Likelihood Identification and Optimal Input Design for Identifying Aircraft Stability and Control Derivatives", Final Report to NASA LRC under contract NAS1-10700, October 1972.

4. Sorensen, J.A., "Analysis of Instrumentation Error Effects on the Identification Accuracy of Aircraft Parameters", NASA CR-112121, Washington, D.C., May 1972.

5. Sorensen, J.A., Tyler, J.S., and Powell, J.D., "Evaluation of Flight Instrumentation for the Identification of Stability and Control Derivatives", 2nd AIAA Atmospheric Flight Mechanics Conf., Moffett Field, Calif, September 1972

TABLE 1. MONTE CARLO RESULTS BASED ON IDENTIFICATION
FOR 50 SETS OF SIMULATED DATA

	Optimal Input	Suboptimal Input
Trace of Sample Covariance	.242	.315
Determinant of Sample Covariance	1.62×10^{-19}	1.501×10^{-15}
Eigenvalues of Sample Covariance	.234	.262
	$.725 \times 10^{-2}$	$.514 \times 10^{-1}$
	$.252 \times 10^{-3}$	$.115 \times 10^{-2}$
	$.188 \times 10^{-4}$	$.766 \times 10^{-4}$
	$.202 \times 10^{-7}$	$.126 \times 10^{-5}$
Parameter Standard Deviations	.407	.307
	.295	.491
	.00925	.0235
	.0771	.0537
	.000570	.00168
Average Trace of Information Matrix	2.15×10^{7}	4.79×10^{5}
Eigenvalues of Average Information Matrix	2.14×10^{7}	4.79×10^{5}
	2.95×10^{4}	8.46×10^{3}
	6.56×10^{3}	4.77×10^{2}
	1.39×10^{2}	2.18×10^{1}
	1.12	4.14
Average Determinant of Information Matrix	4.70×10^{18}	1.95×10^{14}
Average Trace of the Covariance Matrix (Cramer-Rao Lower Bound)	.182	.312
Lower bound on parameter standard deviations (from Cramer-Rao Lower Bound)	.351	.303
	.234	.441
	.00876	.0311
	.0665	.0568
	.000247	.00262

TABLE 2.   STANDARD DEVIATIONS OF LONGITUDINAL INSTRUMENT ERRORS

Instrument	Units	Full Scale Deflection	Random Noise			Random Biases			Random Scale Factors		
			Min	Base	Max	Min	Base	Max	Min	Base	Max
Pitch attitude gyro	deg	30—90	.015	.15	.45	.015	.15	.45	.05%	.5%	.5%
Pitch rate gyro	deg/sec	30—60	.015	.10	.30	.015	.10	.30	.05%	.5%	.5%
Angle-of-attack vane	deg	20	.01	.10	.10	.010	.10	.10	.05%	.5%	.5%
Pitot tube	ft/sec	1000	.50	1.00	2.50	.50	1.00	2.50	.05%	.5%	.5%
Forward accelerometer	g's	1	.0005	.005	.005	.0005	.005	.005	.05%	.5%	.5%
Vertical accelerometer	g's	2.5	.0025	.005	.025	.0025	.005	.025	.05%	.5%	.5%
Pitch Accelerometer	deg/sec^2	30—60	.015	.10	.30	.015	.10	.30	.05%	.5%	.5%

TABLE 3.   STANDARD DEVIATIONS OF PARAMETER ESTIMATES DUE TO
INSTRUMENT ERRORS FOR LONGITUDINAL EQUATIONS OF MOTION

Parameter	F-4C					DC-8				
	Nominal[a] Value	Biases Not Estimated		Biases[b] Estimated		Nominal[a] Value	Biases Not Estimated		Biases Estimated	
		Noise Only	Total Errors	Noise Only	Total Errors		Noise Only	Total Errors	Noise Only	Total Errors
Mq	$-.719$ s^{-1}	.182-2[c]	.704-2	.189-2	.683-2	$-.792$ s^{-1}	.224-2	.666-2	.231-2	.653-2
Mw	$-.591$d/f·s	.359-3	.130-2	.455-3	.495-3	$-.498$d/f·s	.641-3	.322-2	.734-3	.165-2
Mu	$-.0295$d/f·s	.257-2	.353-1	.442-2	.588-2	$-.0004$d/f·s	.462-3	.582-2	.823-3	.207-2
Mδe	-16.2 s^{-2}	.106-1	.788-1	.117-1	.778-1	-1.35 s^{-2}	.899-3	.603-2	.932-3	.611-2
Zw	$-.762$ s^{-1}	.914-3	.790-2	.103-2	.717-2	$-.628$ s^{-1}	.103-2	.829-2	.124-2	.709-2
Zu	$-.0617$ s^{-1}	.345-2	.484-1	.617-2	.753-2	$-.251$ s^{-1}	.772-3	.105-1	.127-2	.546-2
Zδe	-1.24f/d·s^2	.167-1	.150	.183-1	.144	$-.178$f/d·s^2	.143-2	.120-1	.147-2	.118-1
Xw	$.0273$ s^{-1}	.494-3	.675-2	.769-3	.852-2	$.0629$ s^{-1}	.534-3	.640-2	.880-3	.608-2
Xu	$.00701$ s^{-1}	.478-3	.596-2	.116-2	.376-2	$-.0291$ s^{-1}	.315-3	.421-2	.661-3	.196-2

a Dimensions are: s—sec; d—deg; f—ft.
b Biases of six instruments estimated. Pitot tube bias is not directly estimated.
c Deviations are in the form .182-2 which means .182 x 10^{-2}

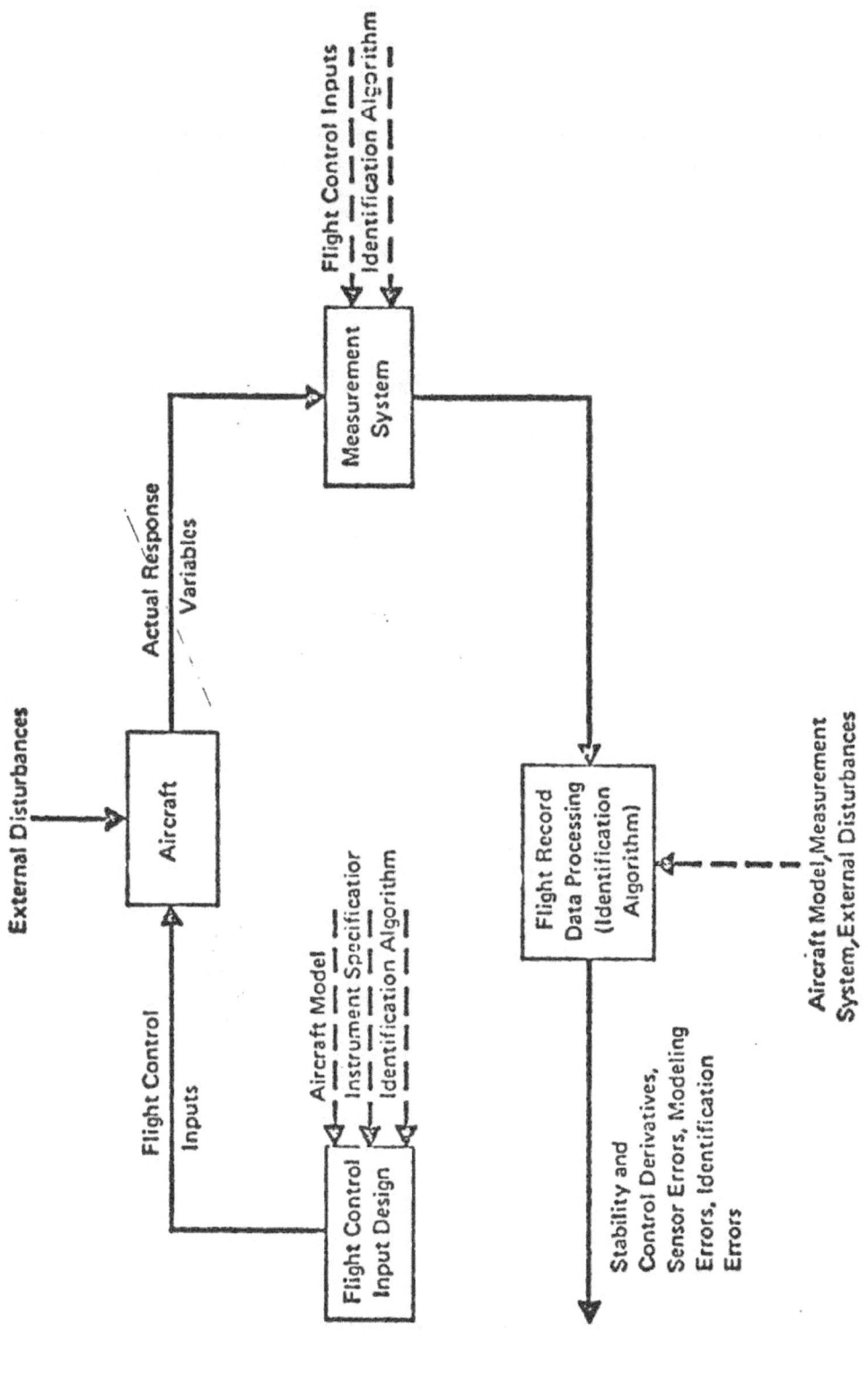

FIGURE 1. THE INTEGRATED AIRCRAFT IDENTIFICATION PROCESS

253

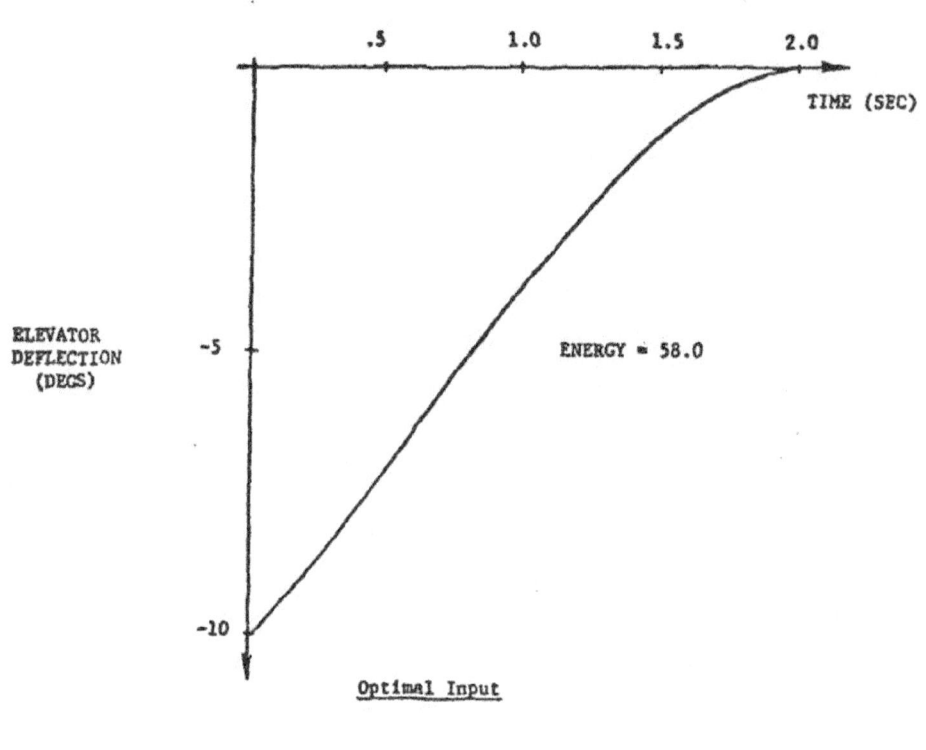

Optimal Input

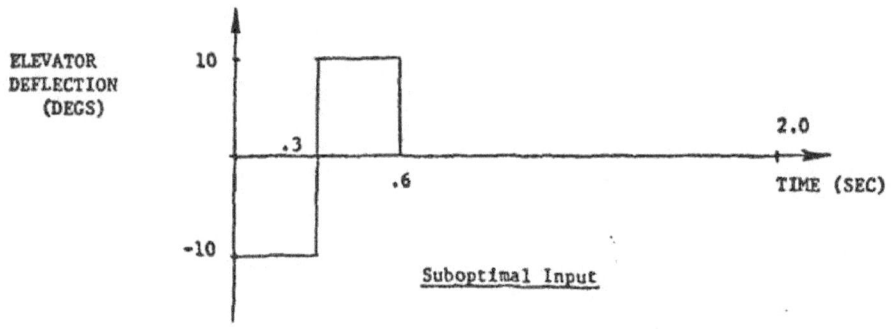

Suboptimal Input

FIGURE 2.   OPTIMAL AND SUBOPTIMAL INPUT FOR
MONTE CARLO SIMULATION

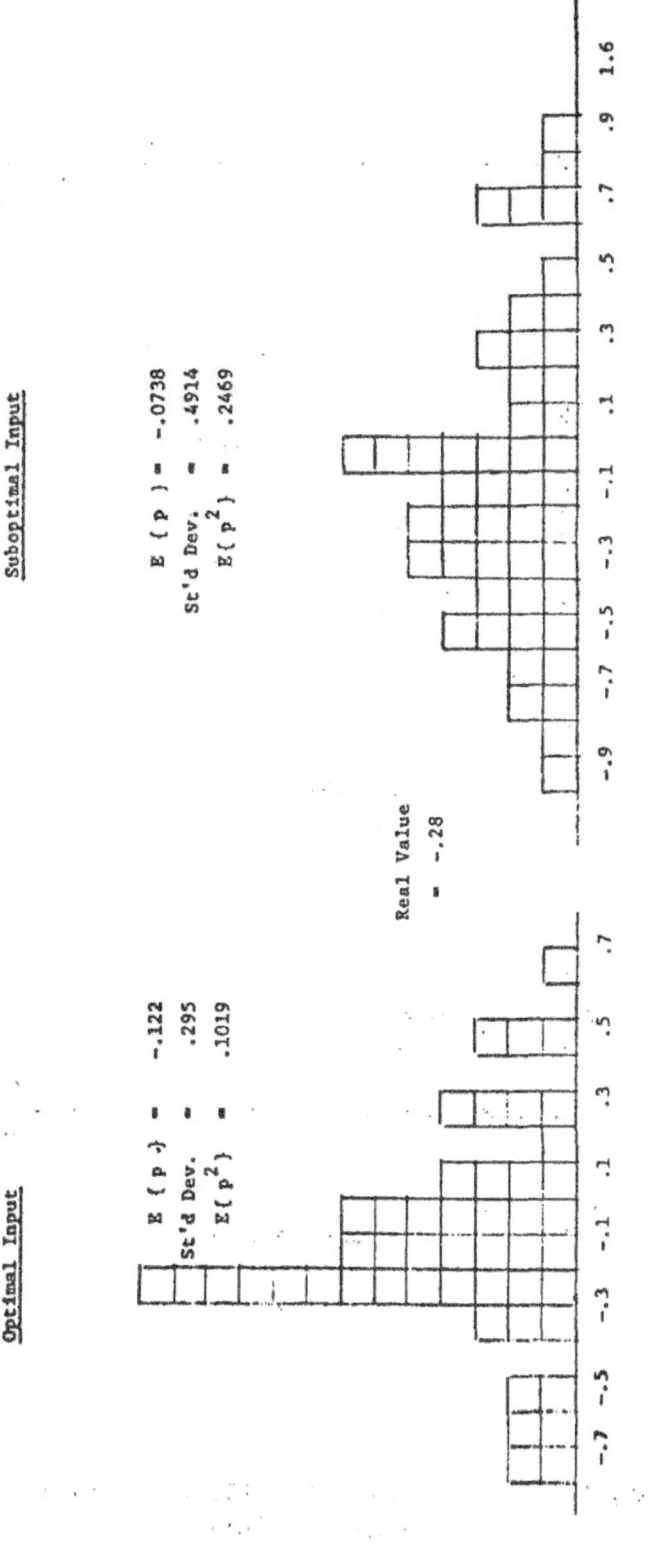

FIGURE 3. PARAMETER ESTIMATE HISTOGRAMS FOR M_α

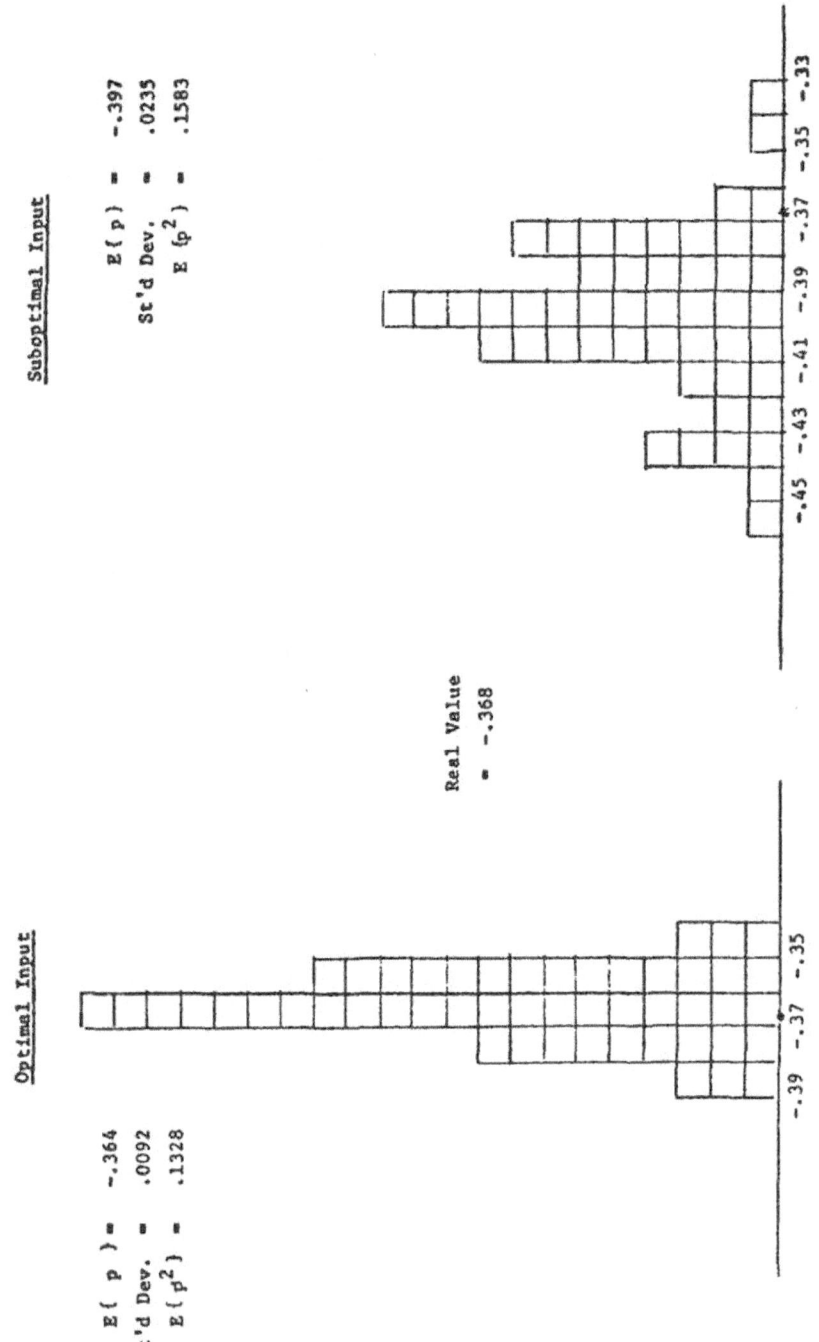

FIGURE 4. PARAMETER ESTIMATE HISTOGRAMS FOR Z_α

256

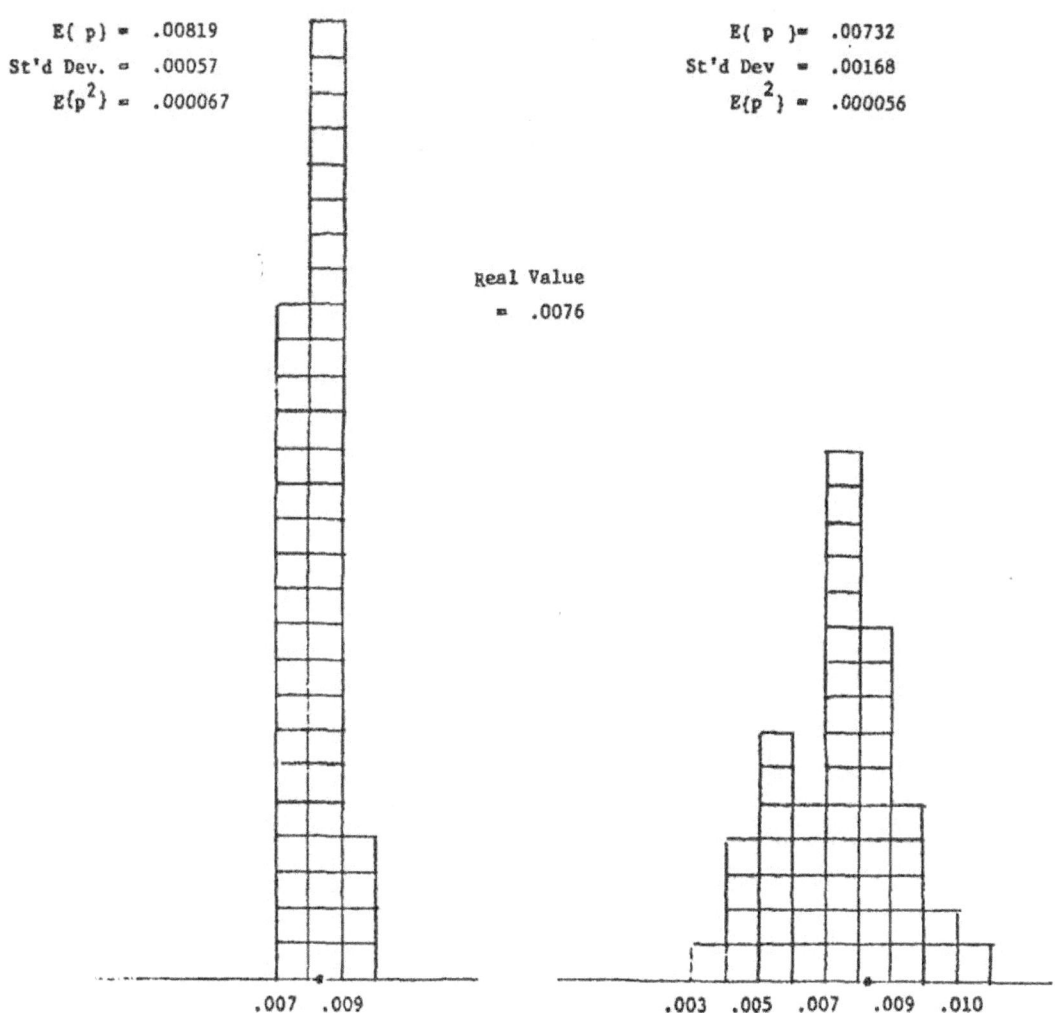

Optimal Input

Suboptimal Input

E{ p} = .00819
St'd Dev. = .00057
E{p^2} = .000067

E{ p } = .00732
St'd Dev = .00168
E{p^2} = .000056

Real Value
= .0076

.007 .009

.003 .005 .007 .009 .010

FIGURE 5. PARAMETER ESTIMATE HISTOGRAMS FOR Z_{δ_e}

257

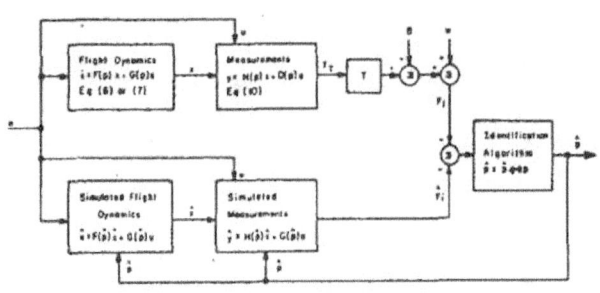

FIGURE 6.   EFFECT OF MEASUREMENT ERRORS ON THE
IDENTIFICATION PROCESS

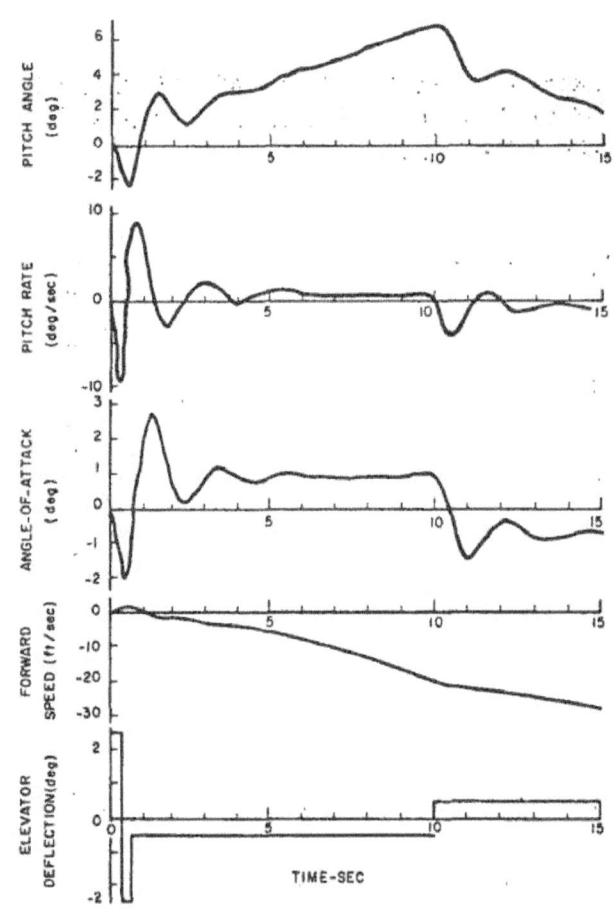

FIGURE 7.   LONGITUDINAL REFERENCE TRAJECTORY AND
ELEVATOR DEFLECTION INPUT

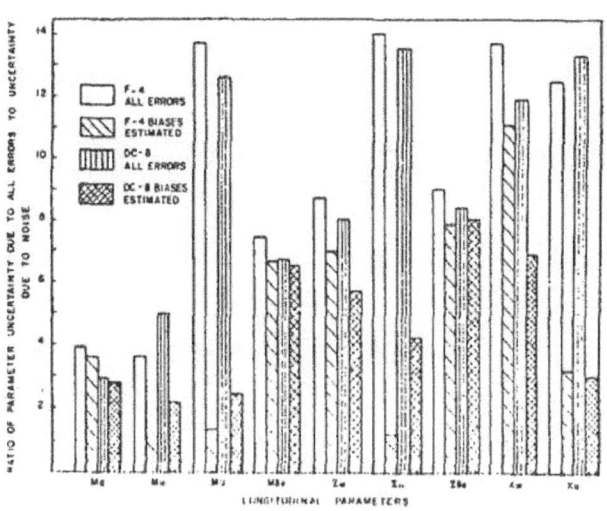

FIGURE 8.   COMPARISON OF STANDARD DEVIATIONS OF LONGITUDINAL
MOTION OF PARAMETER ESTIMATES FOR BASELINE
INSTRUMENT ERRORS

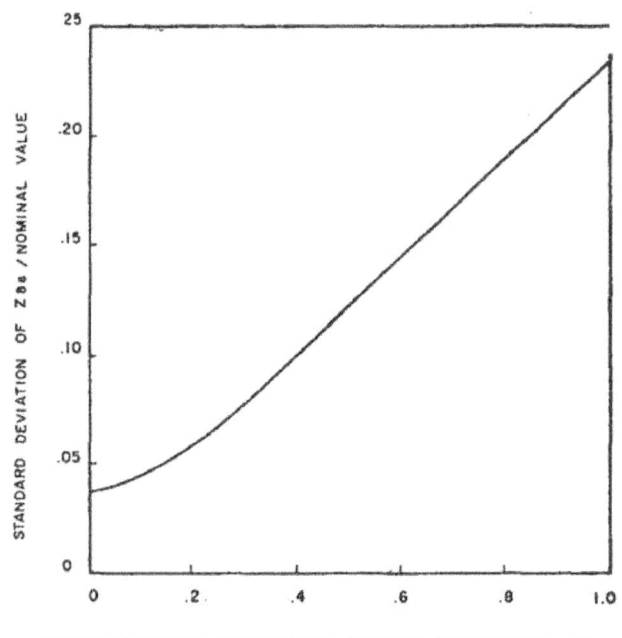

CENTER-OF-GRAVITY LONGITUDINAL POSITION UNCERTAINTY - FT.

FIGURE 9.   EFFECT OF CENTER-OF-GRAVITY POSITION UNCERTAINTY ON
THE ESTIMATION ACCURACY OF THE PARAMETER Z_{δ_e}

EFFECTS OF FLIGHT INSTRUMENTATION ERRORS ON THE ESTIMATION OF
AIRCRAFT STABILITY AND CONTROL DERIVATIVES

Wayne H. Bryant and Ward F. Hodge

NASA Langley Research Center
Hampton, Virginia

SUMMARY

An error analysis program based on an output error estimation method was used to evaluate the effects of sensor and instrumentation errors on the estimation of aircraft stability and control derivatives. A Monte Carlo analysis was performed using simulated flight data for a high-performance military aircraft, a large commercial transport, and a small general aviation aircraft for typical cruise flight conditions. The effects of varying the input sequence and combinations of the sensor and instrumentation errors were investigated. The results indicate that both the parameter accuracy and the corresponding measurement trajectory fit error can be significantly affected. Of the error sources considered, instrumentation lags and control measurement errors were found to be most significant.

INTRODUCTION

One of the important tasks associated with current efforts to improve the estimates of stability and control derivatives obtained from flight data is to evaluate the effects of unmodeled errors in the measurements. The estimated quantities may be used in a variety of contexts each with its own accuracy requirements, and the measurement systems and flight maneuvers used may be specified primarily for other purposes. For these reasons, it is desirable to be able to evaluate the effect of a given instrumentation set on the accuracy of estimated stability and control parameters and, conversely, to synthesize an instrumentation set which will permit the aircraft parameters to be identified to a desired level of accuracy. The purpose of the present study was to investigate the suitability of presently utilized instrumentation by establishing the parameter and trajectory fit error statistics for a nominal instrumentation set and to identify the most significant measurement errors.

The relationship between instrumentation errors and the accuracy of derivatives estimated from flight data has been reported in references 1 and 2 where the measurement system errors evaluated included random white noise and static error sources (such as bias, scale factor errors, misalinement, center-of-gravity uncertainty, and vane correction). It was determined in that effort that the static measurement errors can be a much greater source of error than the white noise. The current effort is based on the approach employed in reference 1, but differs by including dynamic errors in the measurements of the states and both static and dynamic errors in the measurement of the control input.

261

In the following sections, the method used for the analysis is described
and the scope of the study detailed. Results are presented which show the
effect of varying the input sequence and the aircraft type for the three basic
dynamic modes. Finally, the major sources of errors in the estimation of the
derivatives are identified.

METHOD OF ANALYSIS

Error Analysis Algorithm

The process of estimating stability and control derivatives by minimizing
an appropriate quadratic performance function $J(p)$ provides a natural approach
for analyzing the effects of unmodeled errors in the measurement data y_m. The
function $J(p)$ is defined as

$$J(p) = \sum_{i=1}^{NOP} (y_{mi} - y_i)^T R^{-1} (y_{mi} - y_i) \tag{1}$$

where R^{-1} is a weighting matrix, y_i is a function of the estimated derivatives, and the product is summed over the number of data points in the measured
trajectory. Referring to figure 1, the dashed curve represents the ideal case
for error-free measurements where the minimum of $J(p)$ is zero and the true
values of the parameters p_t would be obtained (assuming no algorithm
or computational errors). For other situations where the measurements contain
error sources which are not estimated or otherwise accounted for in the calculation of the estimate $y(p)$, the minimum of $J(p)$ will not be zero and the
algorithm will seek values of p which differ from p_t as indicated by Δp
in figure 1. The minimum value of J is a measure of the trajectory fit error
caused by the unmodeled errors in y_m, and Δp are the corresponding errors
in p. The technique used in reference 1 and continued here is to seek esti-
mates ($\Delta \hat{p}$ and $\hat{J} = J(\hat{p})$) of these quantities for simulated flight measurements
developed for various combinations of instrument error sources. While the
modified Newton-Raphson technique employed in references 1 and 2 was used to
obtain these estimates, the results presented here are applicable to any
algorithm which minimizes the performance function defined in (1). The modified Newton-Raphson algorithm has the form

$$\Delta \hat{p}_j = - \left[\frac{\partial^2 J}{\partial p^2} \right]^{-1} \frac{\partial J}{\partial p} \tag{2}$$

and the total parameter estimation error ($\Delta \hat{p}$) is obtained by summing the contributions from each iteration until the convergence criteria are met. For
this study, convergence is met when the current change in the estimate ($\Delta \hat{p}_j$)
is within 1% of the current value of $\hat{p}$ for each parameter estimated.

A Monte Carlo analysis was performed to obtain the means and standard
deviations of $\Delta \hat{p}$ and $J(\hat{p})$. In the majority of error cases, 50 samples of

262

simulated flight data were used; however, in a few cases statistics were formed on as few as 25 samples. The smaller number of samples was determined to have a negligible effect on the results obtained. Consequently, no distinction is made between data as to the number of samples used.

Aircraft Equations

The equations of motion used to represent the aircraft dynamics in the present study are:

$$
\begin{bmatrix} \Delta\dot{\theta} \\ \Delta\dot{q} \\ \Delta\dot{w} \\ \Delta\dot{u} \end{bmatrix} = \begin{bmatrix} 0 & 1 & 0 & 0 \\ 0 & M_q & M_w & M_u \\ -g\sin\theta_0 & V\cos\alpha_0 & Z_w & Z_u \\ -g\cos\theta_0 & -V\sin\alpha_0 & X_w & X_u \end{bmatrix} \begin{bmatrix} \Delta\theta \\ \Delta q \\ \Delta w \\ \Delta u \end{bmatrix} + \begin{bmatrix} 0 \\ M_{\delta e} \\ Z_{\delta e} \\ 0 \end{bmatrix} \begin{bmatrix} \Delta_{\delta e} \end{bmatrix}
\tag{3}
$$

for the longitudinal mode and

$$
\begin{bmatrix} \Delta\dot{\beta} \\ \Delta\dot{p} \\ \Delta\dot{r} \\ \Delta\dot{\phi} \end{bmatrix} = \begin{bmatrix} Y_\beta \sin\alpha_0 & - \cos\alpha_0 & g\cos\theta_0/V \\ L^*_\beta & L^*_p & L^*_r & 0 \\ N^*_\beta & N^*_p & N^*_r & 0 \\ 0 & 1 & \tan\theta_0 \end{bmatrix} \begin{bmatrix} \Delta\beta \\ \Delta p \\ \Delta r \\ \Delta\phi \end{bmatrix} + \begin{bmatrix} Y_{\delta a} & Y_{\delta r} \\ L^*_{\delta a} & L^*_{\delta r} \\ N^*_{\delta a} & N^*_{\delta r} \\ 0 & 0 \end{bmatrix} \begin{bmatrix} \Delta_{\delta a} \\ \Delta_{\delta r} \end{bmatrix}
\tag{4}
$$

for the lateral directional mode. The short-period equations are obtained from (3) by eliminating the state Δu and all its factors. The unknown parameters estimated in the longitudinal mode are M_q, M_w, M_u, $M_{\delta e}$, $Z_{\delta e}$, Z_w, Z_u, X_w, and X_u. In the short-period mode, M_q, M_w, Z_w, $M_{\delta e}$, and $Z_{\delta e}$ are estimated and in the lateral mode Y_β, $Y_{\delta a}$, $Y_{\delta r}$, L^*_β, L^*_p, $L^*_{\delta a}$, $L^*_{\delta r}$, N^*_β, N^*_p, N^*_r, $N^*_{\delta r}$, and L^*_r are estimated. Aircraft considered in the study were the DC-8, the F4-C, and Cessna 172. This selection permits the evaluation of a high-performance aircraft, a large transport, and a small general-aviation aircraft. Derivative values for the DC-8 and F4-C were obtained from reference 2 and for the C-172 from reference 3.

Measurement Equations

The ideal measurement equations are represented as

$$
y = H(p)\ x + D(p)\ u
\tag{5}
$$

The simulated longitudinal measurements are:

(1) pitch attitude (θ)
(2) pitch rate (q)
(3) angle of attack (α)
(4) longitudinal velocity (u)
(5) longitudinal acceleration (N_x)
(6) normal acceleration (N_z)
(7) pitch acceleration ($\dot{q}$)

In the short-period mode, longitudinal velocity and acceleration are not used. In the lateral mode, the simulated measurements are:

(1) angle of sideslip (β)
(2) roll rate (p)
(3) yaw rate (r)
(4) roll attitude (ϕ)
(5) lateral acceleration (N_y)
(6) roll acceleration ($\dot{p}$)
(7) yaw acceleration ($\dot{r}$)

These measurements are corrupted by errors in the following order:

$$y_I = T\,y + b \tag{6}$$

where T is a matrix of scale factor, cross-coupling, and misalinement errors, and b represents measurement biases. Details concerning the structure of the T matrix are contained in reference 1. State measurement lags are modeled as

$$\dot{y}_L = F_m\,(y_I - y_L) \tag{7}$$

where F_m is a diagonal matrix of measurement time constants. The simulated flight data measurements are then obtained by adding white noise

$$y_m = y_L + w \tag{8}$$

In a similar fashion, control surface position measurement errors are modeled as

$$u_I = T_c\,u + b_c \tag{9}$$

where T_c is a matrix of scale factor errors, and b_c are measurement biases. Control measurement lags are included as

$$\dot{u}_L = F_c\,(u_I - u_L) \tag{10}$$

with F_c a diagonal matrix of measurement system time constants. The simulated control measurement is obtained by adding white noise as

$$u_m = u_L + w_c \qquad (11)$$

and is used in the development of the estimated state measurements.

Table I lists the nominal values used in the measurement error models. All values except measurement time constants were obtained from reference 1. The measurement time constants were obtained from reference 4 and are indicative of the magnitudes when onboard filtering is desired. Those members of the table given as random errors are used for measurement errors which are assumed to be constant for a particular flight test (one Monte Carlo pass in this analysis), but which vary from flight test to flight test, or for different measurement systems. The investigation was carried out for the five measurement error configurations listed in Table II. This grouping of error sources allows the effects of the individual error types to be assessed.

RESULTS AND ANALYSIS

Derivative Estimation Errors for the F4-C

Two control input maneuvers were used in the following investigation. Input sequence 1 was a 15-second trajectory resulting in the motions illustrated in figures 2(a) and 2(b), and was used as a standard for each of the three aircraft so the effects of aircraft type could be observed. Input sequence 2 was a 10-second trajectory producing the motions shown in figures 2(c) and 2(d).

Figure 3 shows the percentage change in each of the stability and control derivatives from the nominal or true value for the F4-C obtained using input sequence 1. This information is plotted for the five error configurations listed in Table II for the full longitudinal (fig. 3(a)), the short-period approximation (fig. 3(b)), and the lateral-directional set of equations (fig. 3(c)). For convenience, a hatched bar is used to denote the mean value of the plotted quantity and a solid bar its standard deviation (1σ) as indicated on the figures. This convention is used throughout the paper.

The comparison of Cases 1 and 2 (standard deviation) indicates that the derivatives are sensitive to the control measurement errors while the comparison of Cases 2 and 4 (means) indicates that the derivatives are also generally sensitive to lags. As an illustration, the error in the derivative M_W for Case 0 (fig. 3(a)) is very small (0.02%) while the standard deviation is somewhat larger (0.08%), which is expected since the white noise is random with zero mean. Case 1 adds state measurement static errors, modeled primarily as random errors, which increase the standard deviation of the error in M_W while the mean value remains nearly the same as Case 0. Addition of static control measurement errors results in a further increase in the standard deviation of the M_W estimation error (2.2%) while the mean error is still very small (0.13%). These results show static errors in the control measurements to be relatively large sources of error for this particular derivative, while center-of-gravity uncertainty, misalinement and misplacement of sensors, scale factors, and measurement biases do not appreciably affect the estimation accuracy.

In Case 3 dynamic errors are added to Case 1 and the mean error in the estimation of M_w approaches the magnitude of the random error, although the magnitude of each is only about 0.24%. Thus state measurement lags cause no particular problem in estimating this derivative. When control measurement lags are also added to obtain the most general measurement error model considered (Case 4), the mean value of the estimation error again approaches the standard deviation (3.8%), indicating the static and dynamic control measurement errors have about the same effect.

While the accuracy required depends on the specific application, the magnitudes of the errors observed indicate that extra care may be required in the design and installation of the measurement system. For example, the mean errors in the estimation of M_q for Cases 2, 3, and 4 are 2.8%, 26.6%, and 23.8%, respectively, while the standard deviations are 9.8%, 0.93%, and 10.2%. Additional cases can be pointed out in the short-period mode (fig. 3(b)) and in the lateral mode (fig. 3(c)). Caution should be exercised in drawing conclusions about the weak derivatives such as M_u, X_u, $Y_{\delta a}$, and $Y_{\delta r}$. The effect of measurement system errors is large in terms of percentage of the true value even for the noise only case (Case 0). However, the actual effect of these errors on the trajectory is very small. The large values are indicative of the difficulty in estimating these derivatives.

Figure 4 shows the ratio of the fit error statistics for each measurement error configuration relative to those for white noises only (Case 0). In general, for a fixed number of data points, the fit error statistics increase as more error sources are added and are largest for Case 4. These results indicate the difficulty in matching flight measurements when control measurements are present.

Effects of Control Input Sequence

Figure 5 gives the ratio of the parameter estimation error with input sequence 2 to the error with input sequence 1 for the F4-C and shows the effect of input sequence on the magnitude of the derivative estimation error as a function of measurement error combination. In the full longitudinal mode (fig. 5(a)), the derivatives associated with the phugoid motion are quite sensitive to the reduced data sequence. This increased sensitivity to measurement errors is a consequence of having insufficient information upon which the algorithm bases its estimates of the phugoid derivatives. Since figure 3(a) also shows large errors in the phugoid derivatives for input sequence 1, only the short-period approximation will be used to represent longitudinal motion for the remainder of this paper.

Figure 5(b) shows the effect of the change in input control sequence for the short-period mode. In general, the ratios cluster around unity indicating no definite advantage for either sequence. In several cases, the ratios differ significantly from unity but the total error is still a small percentage of p_t.

The largest changes are the mean error in M_q for Case 3 which drops from 26.4% for input sequence 1 to 18.9% for input sequence 2 as shown by the ratio of 0.72 on figure 5(b) for this derivative. Furthermore, the variance in the error of M_q decreases from 11.8% for input sequence 1 to 8.2% for input sequence 2 in Case 4, shown by the ratio of 0.70 on figure 5(b). Figure 5(c) shows similar results for the lateral-directional mode.

Comparison of the Effects of Measurement System

Errors for Different Classes of Aircraft

Although this study was based primarily on the F4-C, a comparison was made to determine the effect of instrumentation errors on a general aviation and a large transport aircraft. To represent the light aircraft, the Cessna 172 was used, and for the large transport, the McDonnell Douglas DC-8. Results using input sequence 1 were obtained for each aircraft for both the short-period and lateral-directional modes. The same five measurement system error configurations were also used so that ratios of the corresponding parameter estimation error statistics for the DC-8 and the C-172 to those previously obtained for F4-C could be formed. Figure 6(a) shows the ratio of the parameter estimation error statistics for the DC-8 to those for the F4-C in the short-period mode and figure 6(b) shows this ratio for the C-172 and the F-4C. Figures 6(c) and (d) show similar ratios for the lateral mode. For a few derivatives, the true value for either the DC-8 or C-172 is zero so that the parameter uncertainty ratio was not formed.

In the short-period mode the parameter uncertainty ratios for the DC-8 and the C-172 (figs. 6(a) and 6(b)) indicate that order of magnitude changes can occur in the parameter uncertainties for the different aircraft. These ratios may represent a very small total error for some of the derivatives, however. For Case 0, the largest change observed in the mean was from 0.03% to 0.27% (Z_w for the C-172), and in the standard deviation from 0.07% to 1.36% (M_w for the C-172). The parameter Z_w for Case 2 has a ratio of 218 (mean error) for the C-172, representing an increase in uncertainty for this derivative from 0.02% for the F4-C to 5% for the C-172.

For several parameters significant changes exist in the estimation uncertainty, especially for Cases 3 and 4. For example, the mean for the C-172/F4-C for Case 4 represents a decrease from 21% to 4% for the C-172. In the lateral mode, the mean error in L^*_p for Case 3 increases from 1% for the F4-C to 23.5% for the C-172. Other similar cases can be observed.

Determination of the Dominant Error Sources

A potentially valuable use of this error analysis is the determination of the dominant error source. To achieve this end, the error cases for which the largest parameter estimation errors existed where singled out for examination. For both the short-period and lateral-directional modes, these were Case 2 (which adds static control measurement errors) and Case 3 (which adds state measurement lags). To determine dominant state measurement lags, the lags were considered in groups according to type of instrument and the resultant parameter estimation errors were assessed. Lags were considered first in the gyroscope measurements, then in the angle-of-attack vane measurements, and finally in the acceleration measurements.

In the short-period mode, acceleration measurement lags were responsible for very nearly all the additional parameter uncertainty due to all lags. To determine the relative effects of lags within the acceleration measurement, lags were introduced in the linear acceleration measurement (N_z) and the angular acceleration measurement ($\dot{q}$) independently. The results indicate the angular acceleration measurement lags are responsible for the majority of parameter estimation uncertainty. To check this result, the total fit error statistics for each measurement error configuration was resolved into components corresponding to each measurement. These components are plotted in figure 7, where the values assigned to the individual fit errors are the percentage of total fit error for that error case. For the short-period mode and error Case 3, it is apparent that the pitch acceleration measurement is the primary contributor since the individual fit error is nearly 82% of the total. The linear acceleration measurement accounts for most of the remaining fit error.

A similar analysis was performed on the control measurement errors. Results indicate the white noise alone accounts for nearly 60% of the total parameter uncertainty, while control measurement biases are responsible for approximately 40%. Addition of the scale factor errors resulted in no significant change in the parameter estimation uncertainty. For this measurement error configuration, the white noise and biases on the control measurements each are capable of producing significant errors in the estimation of the derivatives.

Figure 7(a) also shows the effect of static control measurement errors on the individual measurement fit errors. From the figure it can be seen that the majority of the total error comes from the acceleration measurements. The specific mechanism through which the control measurement errors affect the acceleration measurements is given by the measurement equations.

Results similar to the short-period mode have been obtained for the lateral mode. For Case 3, lags in the acceleration measurements cause the majority of parameter estimation uncertainty. However, lags in the linear acceleration measurement N_y are the primary source of estimation uncertainty. Figure 7(b) for Case 3 corroborates this result with the fit error attributed to N_y being 55% of the total and for $\dot{p}$, 27% of the total fit error.

CONCLUSIONS

The results from a Monte Carlo analysis of the effects of unmodeled measurement system errors on the estimation of stability and control derivatives indicate the following conclusions:

1. Static and dynamic control input measurement errors and dynamic state measurement errors can cause parameter estimation and trajectory fit errors which may require extra care in the design and calibration of the measurement system.

2. The dominant source of error in the state measurements is acceleration lags. Of these, the pitch acceleration measurement for the short-period mode causes the majority of parameter uncertainty. For the lateral mode, lags in the N_y measurement result in most of the uncertainty with lags in the roll acceleration measurement being responsible for about half as much error as N_y.

3. For comparable conditions, the measurement error sources appear to affect the three classes of aircraft considered generally in the same manner. However, for some parameters significant changes in the estimation accuracy exist.

REFERENCES

1. Sorenson, John A.: Analysis of Instrumentation Error Effects on the Identification Accuracy of Aircraft Parameters. NASA CR-112121, 1972.

2. Sorenson, J. A., Tyler, J. S., Jr., and Powell, J. David: Evaluation of Flight Instrumentation for the Identification of Stability and Control Derivatives. Preprint 72-963, Am. Inst. Aeron. and Astronaut. September 1972.

3. Leisher, L. L., and Walter, H. L.: Stability Derivatives of Cessna Aircraft. Rep. 1356, Research Dept., Cessna Aircraft Co., May 1957.

4. Hill, R. W., Clinkenbeard, I. L., and Bolling, N. F.: V/STOL Flight Test Instrumentation Requirements for Extraction of Aerodynamic Coefficients. Tech. Report AFFDL-TR-68-154, Vought Aeronautics Division, LTV Aerospace Corp., December 1968.

TABLE I. SUMMARY OF MEASUREMENT SYSTEM ERRORS

| Instrument | Random (2σ) | | | Mean | |
	Bias	Noise	Misalinement	Lag $\left(\frac{1}{\tau}\right)$	Location error
Attitude gyro	0.15° (1)	0.15° (1)	0.6°	0.33 sec^{-1}	----
Rate gyro	.10°/sec	.10°/sec	.6°	.33 sec^{-1}	----
Linear accelerometer	.005 g's (2)	.005 g's (2)	.6°	.10 sec^{-1}	1 ft
Angular accelerometer	.10°/sec^2	.10°/sec^2	.6°	.33 sec^{-1}	1 ft
Vane	.05° (3)	.05° (3)	----	.33 sec^{-1}	1 ft
Pitot tube	1 f/s	1 f/s	----	1.00 sec^{-1}	----
Control measurement	.10°	.10°	----	.50 sec^{-1}	----

Center-of-gravity uncertainty (random) 0.5 ft

Scale factor error (random) 0.005

(1) Roll attitude bias and noise is 0.50°.
(2) N_y bias and noise is 0.0005 g units.
(3) Angle-of-attack vane bias and noise is 0.10°.

TABLE II.   MEASUREMENT ERROR CONFIGURATIONS USED IN THE ANALYSIS

Error configuration	State measurement			Control measurement		
	White noise	Static errors	Dynamic errors	White noise	Static errors	Dynamic errors
Case 0	✓					
Case 1	✓	✓				
Case 2	✓	✓		✓	✓	
Case 3	✓	✓	✓			
Case 4	✓	✓	✓	✓	✓	✓

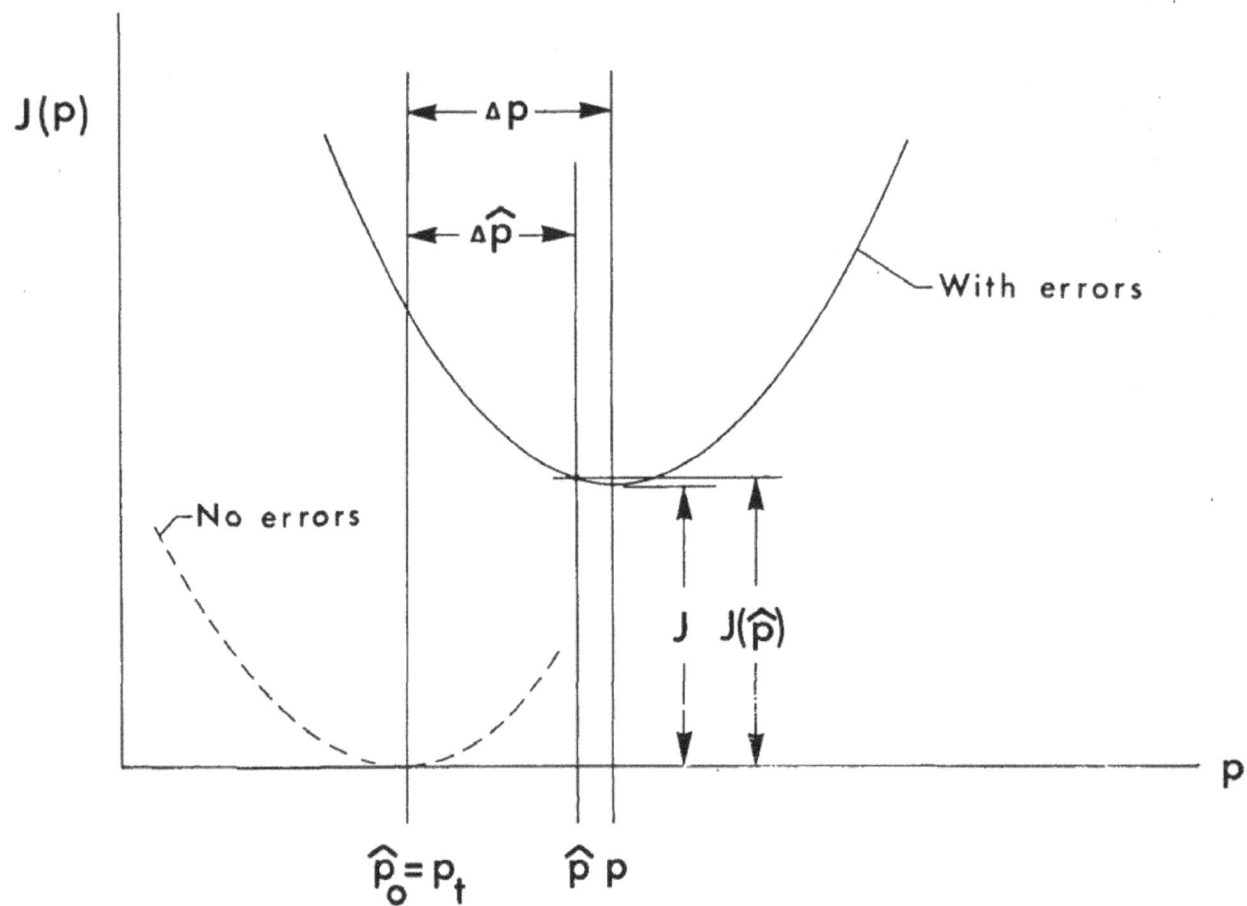

Figure 1. - Parameter and trajectory fit errors.

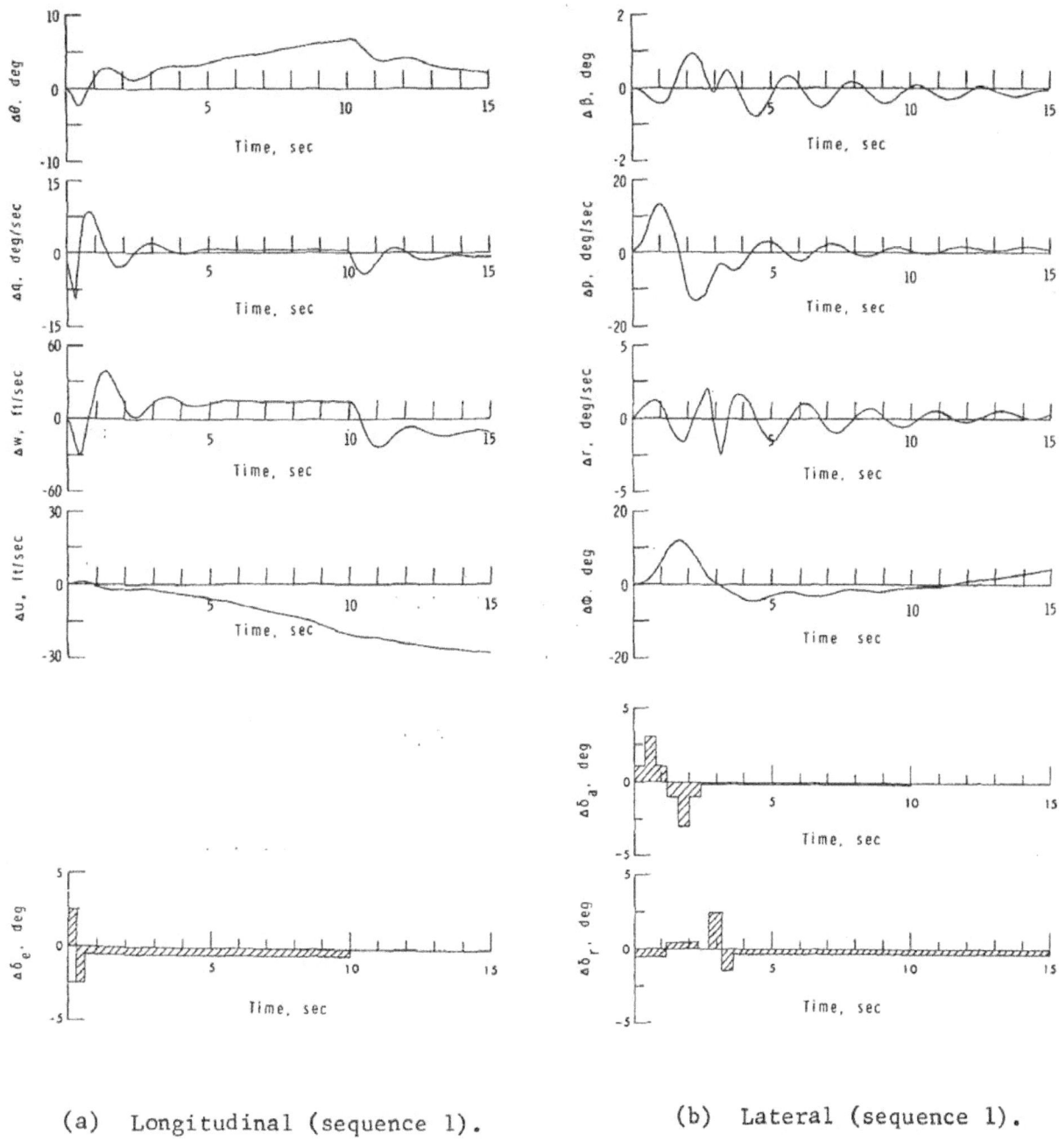

(a) Longitudinal (sequence 1).

(b) Lateral (sequence 1).

Figure 2. - Aircraft state time histories.

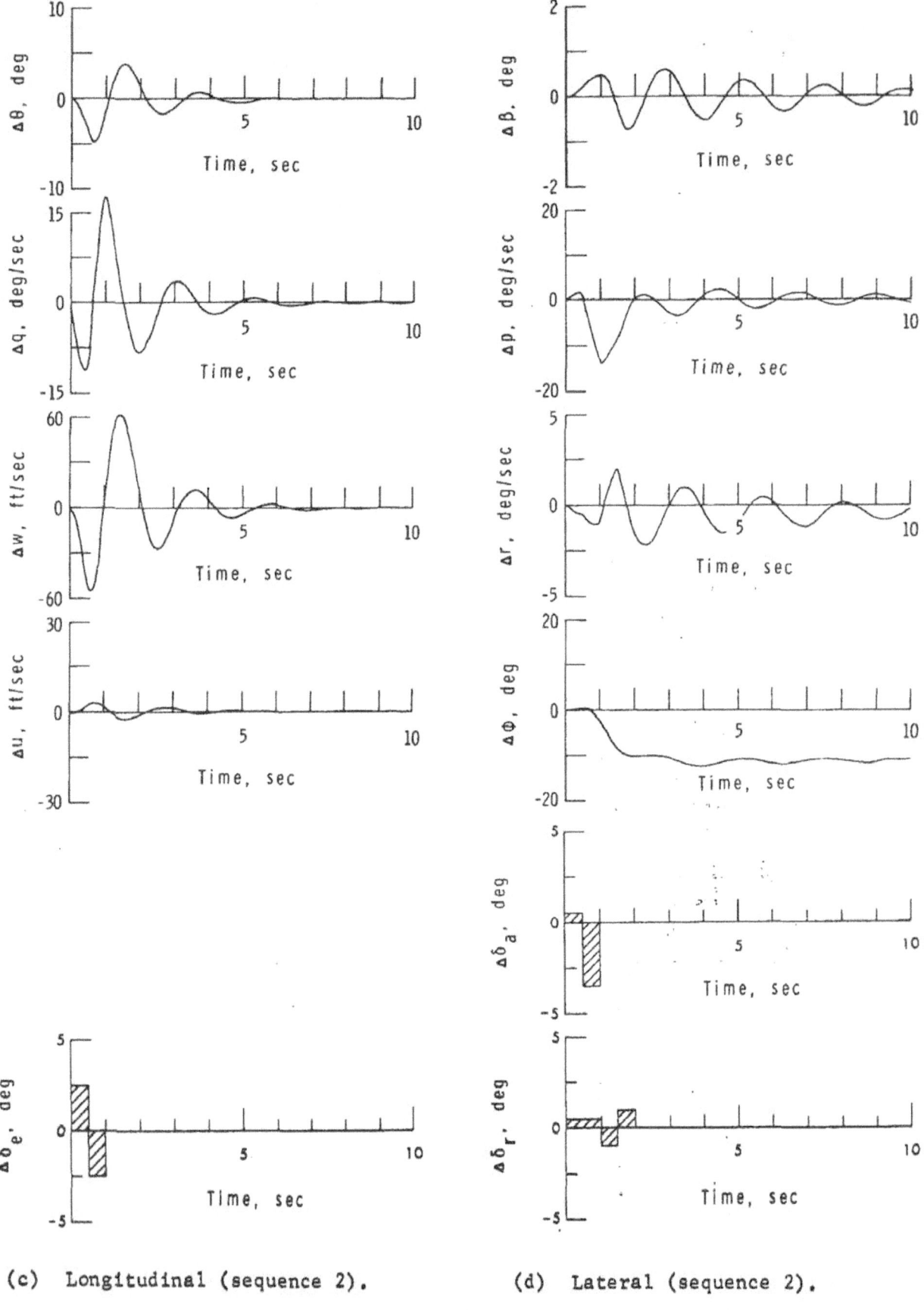

(c) Longitudinal (sequence 2).   (d) Lateral (sequence 2).

Figure 2. - Concluded.

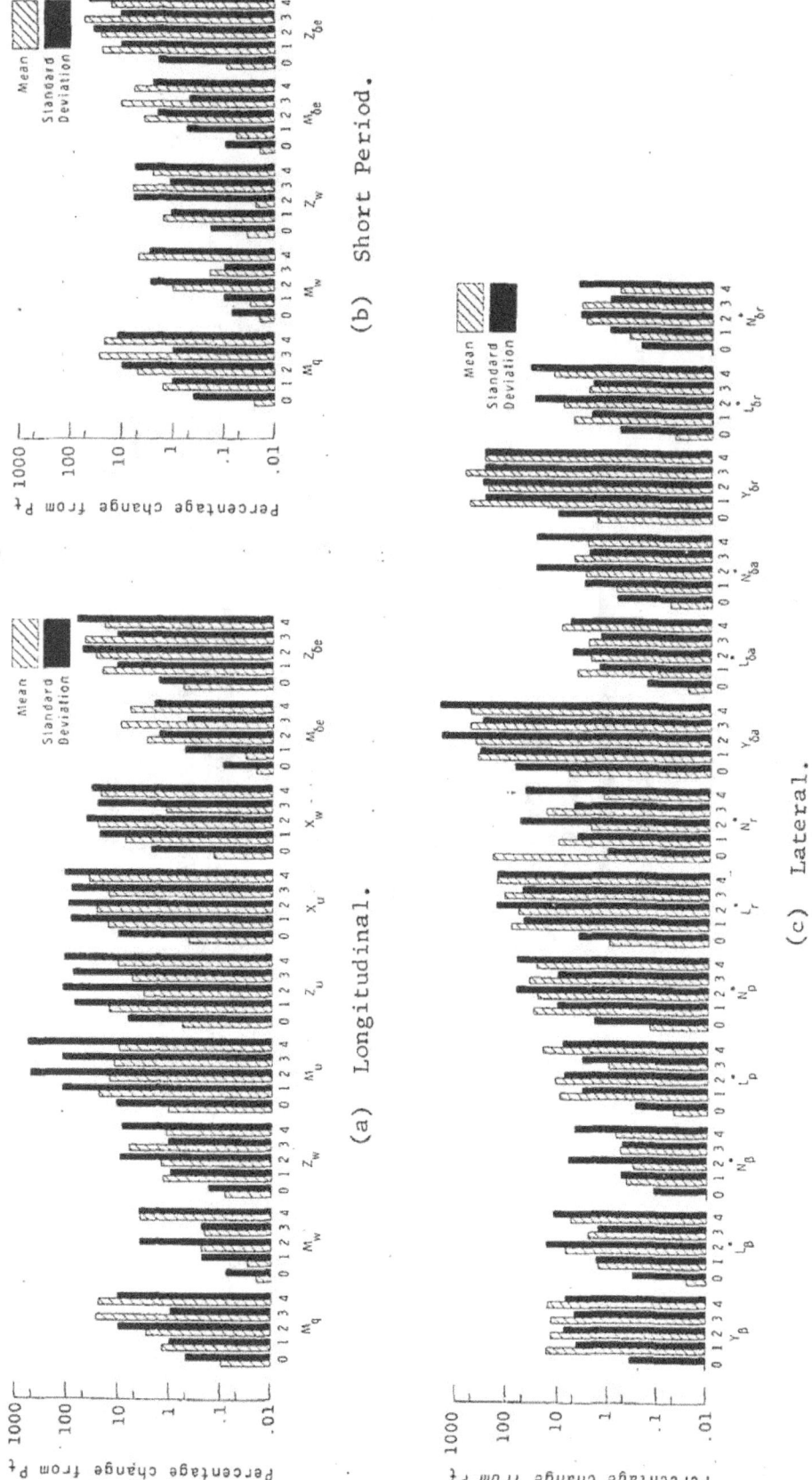

Figure 3. - Percentage parameter estimation uncertainty caused by measurement error for the F4-C.

275

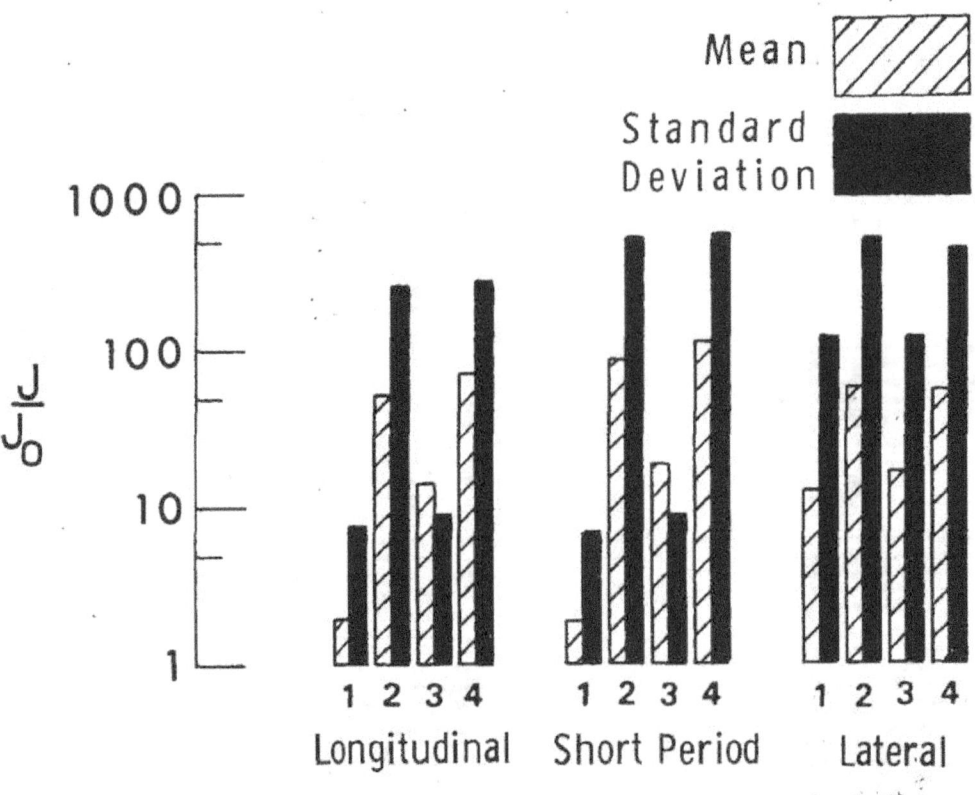

Figure 4. - Fit error magnitude relative to that for white noise only.

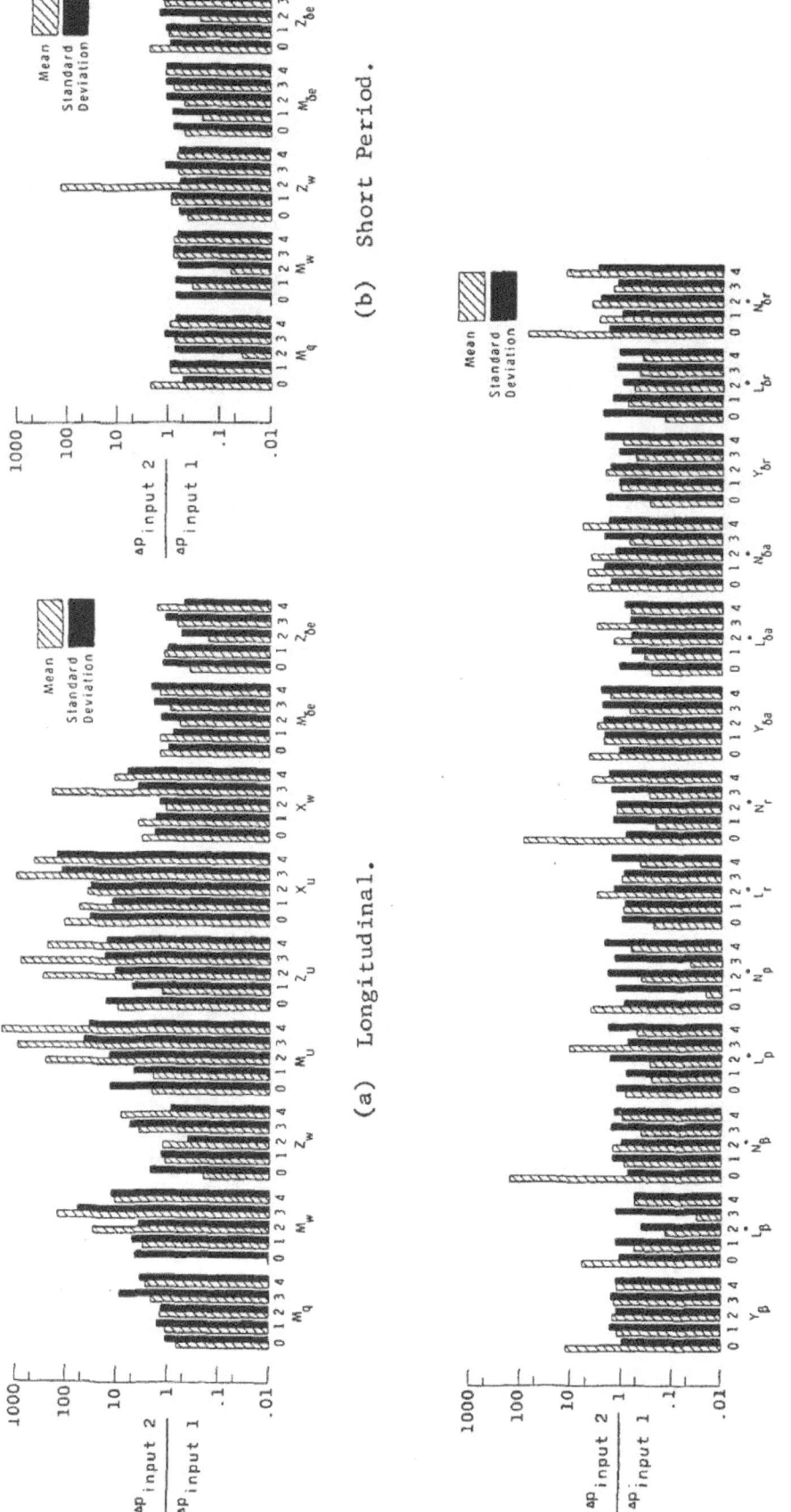

(a) Longitudinal.

(b) Short Period.

(c) Lateral.

Figure 5. - Effect of input sequence on parameter estimation uncertainty.

277

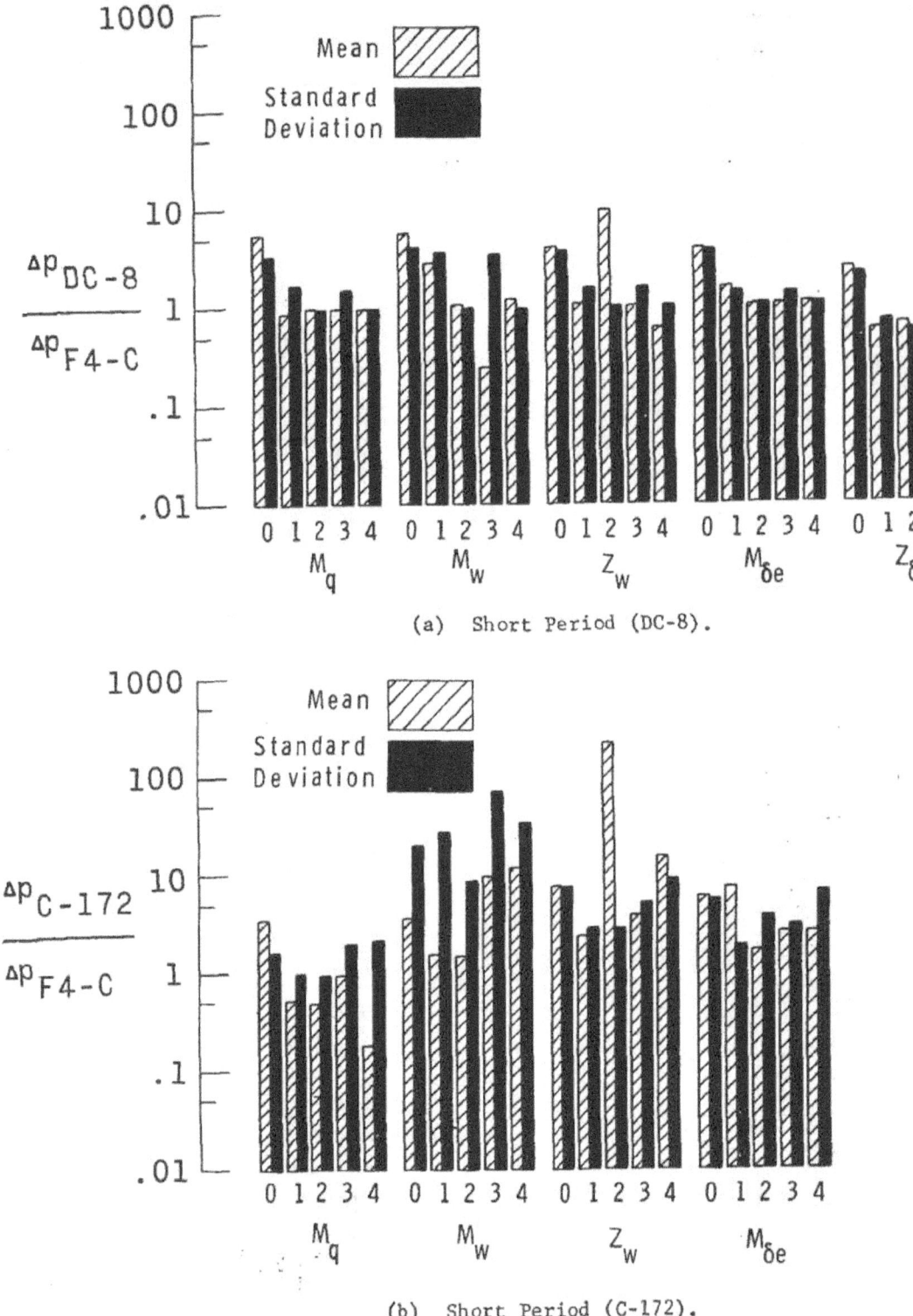

(a) Short Period (DC-8).

(b) Short Period (C-172).

Figure 6. - Comparison of aircraft parameter estimation
uncertainty with that for the F4-C.

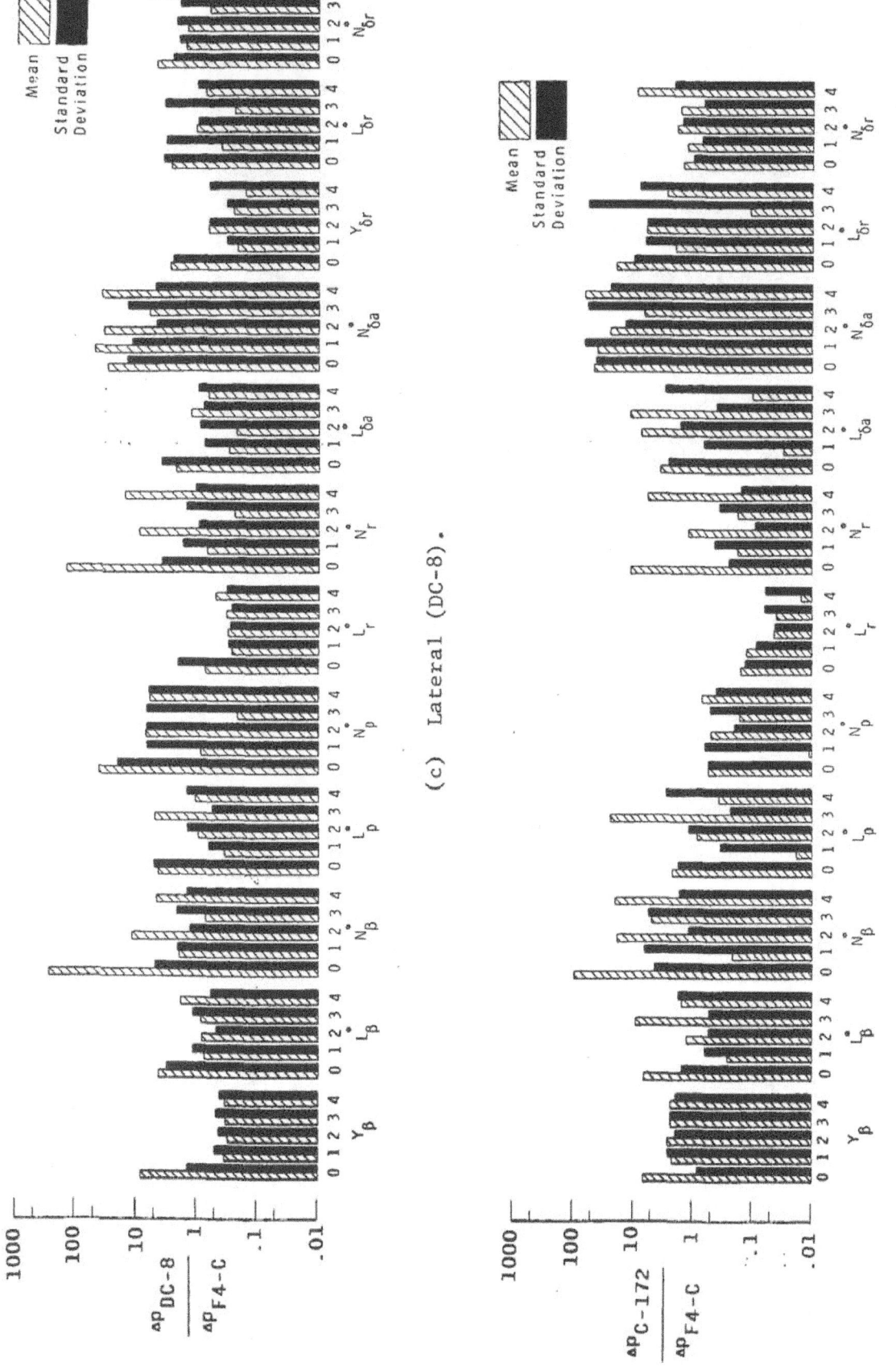

(c) Lateral (DC-8).

(d) Lateral (C-172).

Figure 6. - Concluded.

279

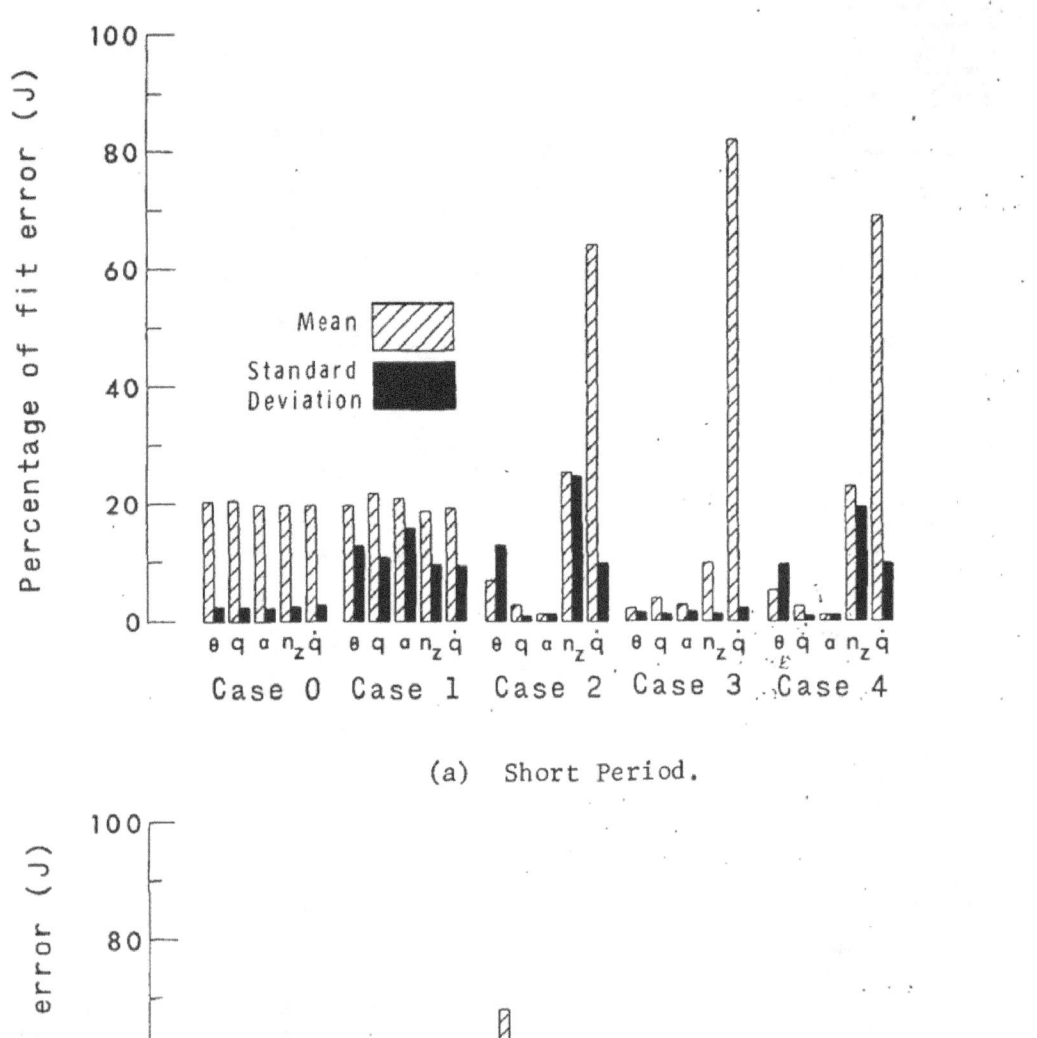

(a) Short Period.

(b) Lateral.

Figure 7. - Resolution of total fit error for the F4-C into
components attributed to each measurement.

PARAMETER IDENTIFIABILITY OF LINEAR

DYNAMICAL SYSTEMS

Keith Glover
Decision and Control Sciences Group
Department of Electrical Engineering
Massachusetts Institute of Technology
Cambridge, Massachusetts 02139

Jan C. Willems
Mathematisch Instituut
Rijksuniversiteit Groningen
Groningen
The Netherlands

SUMMARY

It is assumed that the system matrices of a stationary linear dynamical system have been parametrized by a set of unknown parameters. The question considered here is, when can such a set of unknown parameters be identified from the observed data? Conditions for the local identifiability of a parametrization are derived in three situations: (i) when input/output observations are made, (ii) when there exists an unknown feedback matrix in the system and (iii) when the system is assumed to be driven by white noise and only output observations are made. Also a sufficient condition for global identifiability is derived for case (i).

1. INTRODUCTION

In this paper we will consider the identification of systems described by the linear differential or difference equations:

$$\frac{dx}{dt}(t) = Ax(t) + Bu(t) \qquad \text{or} \qquad x(k+1) = Ax(k) + Bu(k)$$

$$y(t) = Cx(t) + Du(t) \qquad\qquad y(k) = Cx(k) + Du(k)$$

where $x \epsilon R^n$; $u \epsilon R^m$, $y \epsilon R^p$, $A \epsilon R^{n \times n}$, $B \epsilon R^{n \times m}$, $C \epsilon R^{p \times n}$.

In practical identification problems such equations may often be postulated on the basis of a priori knowledge on the structure and physics of the system, with the elements of the matrices A, B, C and D, either zero, known physical constants, or certain known functions of unknown parameters. Thus, if the unknown parameters are denoted $\alpha \epsilon \Omega \subset R^q$, then the system matrices may be written as $A(\alpha)$, $B(\alpha)$, $C(\alpha)$, and $D(\alpha)$, where $A: R^q \rightarrow R^{n \times n}$, $B: R^q \rightarrow R^{n \times m}$, $C: R^q \rightarrow R^{p \times n}$ and $D: R^q \rightarrow R^{p \times m}$

Research supported by NASA under Contract No. NGL-22-009-124 with the Electronics Systems Laboratory of the Massachusetts Institute of Technology, Cambridge, Massachusetts 02139 U.S.A.

281

A natural question which arises in the context of such identification problems is whether or not these unknown parameters (A, B, C, D)(α), can be identified from observations of the system?

When such a model can be formulated it has two main advantages over using canonical forms (as given for example in references 1, 2, and 3). Firstly the parameters being identified have a physical interpretation and secondly for multiple input/multiple output systems, the canonical forms have the disadvantage that a set of integers (e.g. the Kronecker indices) must be determined before the real valued parameters can be identified.

Three situations will be considered here. In section 2 both input and output observations are assumed to be available and the effect of a feedback matrix is also considered. In section 3, the system is assumed to be driven by white noise and only the outputs are observed.

2. IDENTIFIABILITY FROM INPUT/OUTPUT OBSERVATION

In this section it will be assumed that *both* the input and the output of a linear time invariant system are observed and that the input is persistently exciting (as defined for example in reference 4). That is, there are sufficient assumptions so that the transfer function of the system can be identified from input and output observations. Consider now the following definition:

Definition 1:

Let (A, B, C, D) (α) $:\Omega \subset R^q \rightarrow R^{n(n+m+p) + mp}$ be a parametrization of the system matrices (A, B, C, D). This parametrization is said to be *locally identifiable* at $\hat{\alpha}\epsilon\Omega$ if there exists an $\epsilon > 0$ such that

(i) $||\alpha-\hat{\alpha}|| < \epsilon, ||\beta-\hat{\alpha}|| < \epsilon; \alpha,\beta \epsilon\Omega$ and

(ii) $C(\alpha) (Is-A(\beta))^{-1}B(\alpha) + D(\alpha) = C(\beta) (Is-A(\beta))^{-1}B(\beta) + D(\beta)$

for all $s \epsilon \mathbb{C}$ (= complex plane). imply $\alpha = \beta$

In other words, in the neighborhood of $\hat{\alpha}$, there are no two systems with distinct parameters which will have the same transfer function. This definition is similar to the definition of *"non-degeneracy"* as given in reference 5.

The following theorem gives a sufficient condition for local identifiability as defined above:

Theorem 1

Let (A, B, C, D) $(\alpha): \Omega \subset R^q \to R^{n(n+m+p)+mp}$ (with Ω an open set in R^q) be a C' (i.e., continuously differentiable on Ω) parametrization of the system matrices (A, B, C, D) and let $(A, B, C, D)(\hat{\alpha})$ be minimal. Then (A, B, C, D) (α) is locally identifiable at $\hat{\alpha}$ if and only if there exists an $\epsilon > 0$ such that $(TA(\alpha)T^{-1}, TB(\alpha), C(\alpha)T^{-1}, D(\alpha)) \neq (A, B, C, D)(\beta)$ for all $T \in GL(n) = \{T \mid \det T \neq 0\}$ and $\alpha, \beta \in N_\epsilon(\hat{\alpha}) = \{\alpha \mid \|\alpha - \hat{\alpha}\| < \epsilon\}$.

A sufficient condition for this is that $\det(X'X) \neq 0$ where

$$
X = \begin{bmatrix}
I \otimes A'(\hat{\alpha}) - A(\hat{\alpha}) \otimes I & \\
I \otimes B'(\hat{\alpha}) & \\
-C(\hat{\alpha}) \otimes I & M \\
0 &
\end{bmatrix}
, \quad
M = \begin{bmatrix}
\dfrac{\partial \alpha_{ij}(\hat{\alpha})}{\partial \alpha} \\
\vdots \\
\dfrac{\partial b_{ij}(\hat{\alpha})}{\partial \alpha} \\
\vdots \\
\dfrac{\partial c_{ij}(\hat{\alpha})}{\partial \alpha} \\
\vdots \\
\dfrac{\partial d_{ij}(\hat{\alpha})}{\partial \alpha}
\end{bmatrix}
$$

The indices of M are such that the elements of A, B, C and D are listed by rows. (see Proof).

$\times$ denotes Kronecker product.

Proof:

The necessary and sufficient condition follows immediately from the facts that minimal systems form an open set in parameter space, and that any two equivalent minimal systems are related by a similarity transformation, $(A, B, C, D) \to (TAT^{-1}, TB, CT^{-1}, D)$, (see reference 6).

A complete proof of the sufficient condition will appear in reference 7. The method of proof is as follows:

Let $F: GL(n) \times \Omega \times \Omega \to R^{n(n+m+p)+mp}$ be defined as

$$F(T, \alpha, \beta) = (TA(\alpha)T^{-1} - A(\beta), TB(\alpha) - B(\beta), C(\alpha)T^{-1} - C(\beta), D(\alpha) - D(\beta))$$

It is then shown that X is the Jacobian of F evaluated at $(I, \hat{\alpha}, \hat{\alpha})$ which, by the implicit function theorem, implies that $F(T, \alpha, \beta) = 0$ has a unique solution for (T, α) as a function of β in a neighborhood of $(I, \hat{\alpha}, \hat{\alpha})$. The result then follows by extending the neighborhood of I to all of $GL(n)$.

283

Theorem 1 gives a straightforward test for local identifiability which is significantly simpler to apply than methods based on the so-called *information matrix*.

Identification in the presence of unknown feedback can cause practical difficulties and Corollary 1 gives conditions for identifiability with feedback.

Corollary 1

Consider the linear feedback system,

$$\frac{dx}{dt}(t) = A(\alpha)\ x\ (t) + B(\alpha)\ u(t)$$

$$y(t) = C(\alpha)\ x(t)\ ,\ u(t) = -Kx(t) + v(t)\ ,\ K\ \varepsilon\ R^{mxn}$$

where $(A, B, C)\ (\alpha)$ *is a C' parametrization. Then the parameter α and the feedback matrix K are locally identifiable at $\alpha = \hat{\alpha}$ and $K = \hat{K}$ from observations of $u(t)$ and $y(t)$ if* $\det [\ \bar{X}'\ \bar{X}\ ] \neq 0$.

where

$$\bar{X} = \begin{bmatrix} I \otimes \tilde{A}' - \tilde{A} \otimes I & & \vdots & & \vdots & \hat{B} \otimes I \\ I \otimes \hat{B}' & & \vdots & M & \vdots & 0 \\ -\hat{C} \otimes I & & \vdots & & \vdots & 0 \end{bmatrix}$$

where

$$\tilde{A} = A(\hat{\alpha}) - B(\hat{\alpha})\ \hat{K},\quad \hat{B} = B(\hat{\alpha}),\quad \hat{C} = C(\hat{\alpha})\ and\ M\ is\ defined\ in\ Theorem\ 1.$$

Proof

The above theorem is a consequence of Theorem 1 since in this situation the feedback matrix simply induces a new parametrization of the system matrix defined by $\tilde{A}(\alpha, K) = A(\alpha) - B(\alpha)K$.

Generalization to situations where the feedback matrix is partly known or where it is not required to be identified can be obtained. Disadvantages of the concept of local identifiability are that the nominal values, $\hat{\alpha}$, must be known and the size of the neighborhood of $\hat{\alpha}$ is in general not easily found. However, if a parametrization is locally identifiable for all $\alpha \varepsilon \Omega$, then one may conclude that identification algorithms will be locally well-behaved but may converge to one of several solutions depending on the initial estimates and on the actual data received. It is thus desirable to attempt to generalize the result of the theorem to global identifiability.

Definition 2

Let $(A, B, C, D) (\alpha): \Omega \subset R^q \to R^{n(n+m+p)+mp}$ be a parametrization of the system matrices (A, B, C, D). This parametrization is said to be *globally identifiable* if

(i) $C(\alpha) (Is-A(\alpha))^{-1} B(\alpha) + D(\alpha) = C(\beta) (Is-A(\beta))^{-1} B(\beta) + D(\beta)$
    for all $s \epsilon \mathbb{C}$.

and

(ii) $(A, B, C, D) (\alpha)$ is minimal.

imply that $\alpha = \beta$.

Condition (ii) could be deleted in the above definition but then the definition would be very restrictive since most useful parametrizations admit multiple representations of non-minimal systems.

The following proposition gives a sufficient condition for global identifiability, when the parametrization is affine, (i.e., a linear map plus an offset).

Proposition 1:

An *affine parametrization* $(A, B, C, D) (\alpha): \Omega \subset R^q \to R^{n(n+m+p)+mp}$, *is globally identifiable if* $\det [Y'(\alpha,\beta) Y(\alpha,\beta)] \neq 0$ *for all* $\alpha, \beta \epsilon \Omega$, *where*

$$Y(\alpha,\beta) = \begin{bmatrix} Z(\alpha,\beta) & 0 & M \\ 0 & Z(\beta,\alpha) & -M \end{bmatrix}$$

$$Z(\alpha,\beta) = \begin{bmatrix} I \otimes A'(\alpha) - A(\beta) \otimes I \\ I \otimes B'(\alpha) \\ -C(\beta) \otimes I \\ 0 \end{bmatrix}$$

and M is given in Theorem 1.

Proof:

Since we are only concerned with minimal systems global identifiability is implied if the equations $TA(\alpha) = A(\beta)T$, $TB(\alpha) = B(\beta)$, $C(\alpha) = C(\beta)T$, $D(\alpha) = D(\beta)$, have a unique solution for all $\alpha, \beta \epsilon \Omega$ and $T \epsilon GL(n)$. Let q_1 and q_2 be the vectors formed by listing respectively the elements of $(T-I)$ and $T^{-1}-I)$ by rows. Then it is easily verified that the above equations are equivalent to $[q_1, q_2, \alpha-\beta]Y'(\alpha,\beta) = 0$, since $(A, B, C, D) (\alpha)$ is affine. Therefore since $\det (Y'(\alpha,\beta) Y(\alpha,\beta)) \neq 0$ the nullspace of $Y(\alpha,\beta) = N(Y(\alpha,\beta)) = \{0\}$ and the result is thus verified.

Remarks:

1. The condition is not necessary since $(\bar{q}_1, \bar{q}_2, \bar{\alpha} - \bar{\beta}) \epsilon N (Y(\hat{\alpha}, \beta))$ does not imply that $(I+Q_1)^{-1} = I + Q_2$, $\bar{\alpha} = \hat{\alpha}$ and $\bar{\beta} = \beta$, which is required for a system not to be globally identifiable.

2. A somewhat more restrictive sufficient condition for global identifiability is that $N(Z(\alpha, \beta), M) = \{0\}$ for all $\alpha, \beta \epsilon \Omega$. We remark that this condition is in fact satisfied by the canonical forms given in references 1 and 3.

3. Note the similarity between the condition in Remark 2 and the condition or Theorem 1. However local identifiability for all $\alpha \epsilon \Omega$ does not in general imply global identifiability. An open conjecture is that local identifiability for all $\alpha \epsilon R^{n(m+p)}$ implies global identifiability when the parametrization is affine, and $D = 0$.

3. IDENTIFIABILITY FROM OUTPUT OBSERVATION

In this section we will consider the identifiability of a continuous time linear stationary system under the following assumptions.

A1. The input $u(t)$ is not observed directly, but is assumed to be a white noise process with $E(u(t)u'(T)) = I\delta(t-T)$.

A2. The matrix A is asymptotically stable, (i.e., the eigenvalues of A are strictly in the left half plane).

A3. The system has reached steady state when the observations begin (i.e., the output process $y(t)$ is a stationary process).

A4. The system to be identified is globally minimal, i.e., the dimension of the state is less than or equal to that of any other system having the same output spectral density when driven by white noise (see reference 8).

Under these assumptions the most information that may be obtained from the output observations is the output spectral density, $\phi(s) = G(s)G'(-s)$, where $G(s) = C(Is-A)^{-1}B + D$. This motivates the following definition.

Definition 3

Let (A, B, C, D) $(\alpha): \Omega \subset R^q \to R^{n(n+m+p)+mp}$ be a parametrization of the system matrices (A, B, C, D). This parametrization is said to be *locally identifiable from its output spectral density* at $\hat{\alpha} \epsilon \Omega$ if there exists an $\epsilon > 0$ such that

(i) $||\alpha - \hat{\alpha}|| < \varepsilon, ||\beta - \hat{\alpha}|| < \varepsilon; \alpha, \beta \varepsilon \Omega.$

and

(ii) $G(s, \alpha) \ G'(-s, \alpha) = G(s, \beta) G'(-s, \beta)$ for all $s \varepsilon \mathbb{C}$.

imply $\alpha = \beta$.

where $G(s, \alpha) = C(\alpha) \ (Is - A(\alpha))^{-1} B(\alpha) + D(\alpha)$

A condition for local identifiability in this sense can be obtained via the characterization of all globally minimal solutions to the spectral factorization problem given in the following lemma.

Lemma 1

Let (A, B, C, D) and (F, G, H, J) be the system matrices of two globally minimal continuous time systems satisfying conditions (A1) and (A2). Then these systems have the same output spectral density function if and only if there exist matrices $T \varepsilon GL(n)$ and $P = P'$ such that

$$TAT^{-1} = F, \ CT = H, \ AP + PA' = BB' - TGG'T'$$
$$PC' = BD' - TGJ', \ DD' = JJ'.$$

The proof of Lemma 1 is a straightforward consequence of Lemma 2 in reference 8.

The local identifiability problem considered in this section thus reduces to verifying whether or not the following equations have the unique solution $\alpha = \beta$, $T = I$, $P = 0$, for all $\alpha, \beta \varepsilon N_\varepsilon(\hat{\alpha})$.

$$P = P', \ TA(\alpha)T^{-1} = A(\beta), \ C(\alpha)T = C(\beta),$$
$$A(\alpha)P + PA'(\alpha) = B(\alpha)B'(\alpha) - TB(\beta)B'(\beta)T',$$
$$PC'(\alpha) = B(\alpha)D'(\alpha) - TB(\beta)D'(\beta),$$
$$D(\alpha)D'(\alpha) = D(\beta)D'(\beta).$$

The following theorem can be proved in an analogous manner to the proof of Theorem 1, using the implicit function theorem to show that (β, T, P) is a unique function of α.

Theorem 2

Let $(A, B, C, D) \ (\alpha) : \Omega \subset R^q \to R^{n(n+m+p)+mp}$ (with Ω an open set in R^q) be a C' parametrization of the system matrices (A, B, C, D) of a continuous time system satisfying (A1)-(A4). Then this parametrization is locally identifiable from its output spectral density at $\hat{\alpha} \varepsilon \Omega$, if the following linear equations in $(\delta B, \delta D, \delta T, \delta P, \delta \beta)$, have a unique solution (i.e. zero).

(i) $\delta P = \delta P'$

(ii) $(\hat{A}\delta P + \delta T \hat{B} \hat{B}' + \delta B \hat{B}') + (\hat{A}\delta P + \delta T \hat{B} \hat{B}' + \delta B \hat{B}')' = 0$

(iii) $\delta P \hat{C}' = (\delta T \hat{B} \hat{D}' + \delta B D' + \hat{B} \delta D')$

(iv) $\hat{D} \delta D' + \delta D \hat{D}' = 0$

(v) $(\hat{A}\delta T - \delta T \hat{A}, \delta B, \hat{C}\delta T, \delta D) = \left(\left(\dfrac{\partial \alpha_{ij}}{\partial \alpha}(\hat{\alpha})\delta \beta \right), \left(\dfrac{\partial b_{ij}}{\partial \alpha}(\hat{\alpha})\delta \beta \right), \right.$

$$\left. \left(\dfrac{\partial c_{ij}}{\partial \alpha}(\hat{\alpha})\delta \beta \right), \left(\dfrac{\partial d_{ij}}{\partial \alpha}(\hat{\alpha})\delta \beta \right) \right)$$

where $(\hat{A}, \hat{B}, \hat{C}, \hat{D}) = (A, B, C, D)(\hat{\alpha})$.

We remark that the condition of Theorem 2 may be restated as a non-zero determinant condition for a matrix of dimension less than or equal to

$\dfrac{n}{2}(3n + 2m + 1) + mp$. Analagous results can be derived for the discrete time situation.

4. CONCLUSIONS

In this paper we have presented some tests for the identifiability of parametrizations of stationary linear dynamical systems. These conditions should be of great value in situations where sufficient a priori knowledge is available so that state space equations can be written down with relatively few unknown parameters (i.e., $\leq n(m+p)+mp$).

An open problem, presently under investigation, is that of finding weaker sufficient conditions for global identifiability.

5. REFERENCES

1. V.M. Popov, "Invariant Description of Linear, Time-Invariant Controllable Systems", SIAM J. Control, Vol. 10, No. 2, 1972, pp. 252-264.

2. D. Q. Mayne, "A Canonical Model for Identification of Multivariable Linear Systems", IEEE Trans. Automatic Control, Vol. AC-17, 1972, pp. 728-729.

3. D. G. Luenberger, "Canonical Forms for Linear Multivariable Systems", IEEE Trans. on Automatic Control, Vol. AC-12, 1967, pp. 290-293.

4. K.J. Astrom and T. Bohlin, "Numerical Identification of Linear Dynamic Systems from Normal Operating Records", Paper, IFAC Symposium Theory of Self-adaptive Control Systems, Teddington, England. Also, in Theory of Self-Adaptive Control Systems (Ed. P.H. Hammond), Plenum Press, New York, 1966.

5. R.E. Kalman, "On Structural Properties of Linear, Constant, Multi-variable Systems", Paper 6.A, Proceedings of Third IFAC Congress, London, 1966.

6. R.W. Brockett, Finite Dimensional Linear Systems, John Wiley, 1970.

7. K. Glover and J.C. Willems, "On the Identifiability of Linear Dynamical Systems", Third IFAC Symposium on Identification and System Parameter Estimation, The Hague, The Netherlands, June 12-15, 1973.

8. B.D.O. Anderson, "The Inverse Problem of Stationary Covariance Generation", J. Of Statistical Physics, Vol. 1, 1967.

A NEW CRITERION FOR MODELING SYSTEMS

By Lawrence W. Taylor, Jr.

NASA Langley Research Center
Hampton, Virginia

SUMMARY

It has long been recognized that models of systems based on a set of
data do not always adequately predict the system response. It is known that
the measured response can be matched more closely with more adjustable model
parameters, but also that there is a limit to the number of parameters which
if exceeded produce a model which is poorer at predicting the system response
than if a simpler model was used. The analyst has not had at his disposal
a definitive means of knowing how complex the model should be, short of
dividing the data in two parts - one for parameter selection and the other
for testing.

The criterion that is proposed is an expected value of the mean-square
response error as an alternative to testing a model against new data. Modeling with respect to this new criterion does not change the estimate for a
given model format from a maximum likelihood estimate or mean-square response
error estimate. The new criterion does, however, provide a means of comparing
models with different formats and varying complexity.

A numerical example is used to illustrate the application of the proposed
criteria and the problem of searching for the "best" model. For all but the
most trivial system identification problems, it is shown that a prohibitive
number of combinations of terms of the model must be investigated to ensure
the final model is best. Although the computations can be greatly reduced,
the problem of efficiently searching for the best candidate model remains to
be worthy of attention.

INTRODUCTION

The problem of modeling systems continues to receive considerable attention because of its importance and because of the difficulties involved. A
wealth of information has accumulated in the technical literature on the
subject of systems identification and parameter estimation; references 1
through 5 are offered as summaries. The problem that has received most attention is one in which the form of the system dynamics is known, input and noisy
output data are available, and only the values of the unknown model parameters
are sought which optimize a likelihood function or the mean-square response
error. It would seem with the variety of estimates, the algorithms, the
error bounds, and the convergence proofs, and the numerical examples that
exist for these systems identification problems, that any difficulties in

obtaining accurate estimates of the unknown parameter values should be a thing
of the past. Unfortunately, difficulties continue to confront the analyst.
Perhaps the most important reason for difficulties in modeling systems is
that we do not know the form of the system dynamics. Consequently, the
analyst must try a number of candidate forms and obtain estimates for each.
He is then confronted with the problem of choosing one of them with little
or no basis on which to base his selection. It is tempting to use a model
with many parameters since it will fit the measured response error best.
Unfortunately, it is often the case that a simpler model would be better for
predicting the system's response because the fewer number of unknown parameters could be estimated with greater accuracy. It is this problem of the
analyst that this paper addresses and offers a criterion which can be used to
select the best candidate model. Specifically, a numerical example will be
used to illustrate the notions expressed in references 4, 6, and 7.

SYMBOLS

Program variables are given in a separate section.

A	systems matrix (MX by MX)
B	control matrix (MX by MU)
c	parameter vector (MC by 1)
c_0	a priori parameter vector (MC by 1)
$\hat{c}$	estimate of parameter vector (MC by 1)
D_1	weighting matrix for measured response
D_2	weighting matrix for a priori estimate vector
$E\{\}$	expected value
F	state transformation matrix (MZ by MX)
G	control transformation matrix (MZ by MU)
h	time increment
I	identity matrix
i,k	time and iteration indices, respectively
J	cost function to be minimized
N	number of time points
$P()$	probability density function

p   roll rate

r   yaw rate

T   record length, sec; as superscript, transpose

t   time, sec

u   control vector

x   state vector

$\dot{x}$ time derivative of state vector

y   calculated response vector

z   measured response vector used to determine the model parameters

z^1 measured response vector independent of z

β sideslip angle

δ_a aileron deflection

δ_r rudder deflection

ϕ bank angle

$\nabla_c()$ first variation (gradient) with respect to parameter vector

$\nabla_c^2()$ second variation (gradient) with respect to parameter vector

Dots over symbols denote derivatives with respect to time.

The program variables used in the definitions of the symbols are as follows:

MC   number of unknown parameters

MU   number of control variables

MX   number of state variables

MZ   number of response variables

SYSTEMS IDENTIFICATION WITH MODEL FORMAT KNOWN

The systems identification problem of determining the parameters of a linear, constant-coefficient model will be considered for several types of estimates. It will be shown that maximum-likelihood estimates can be identical to those which minimize the mean-square response error. The subject algorithm, therefore, can be used to obtain a variety of similar estimates. Attention is also given to the calculation of the gradient that is involved in the algorithm and to the Cramer-Rao bound which indicates the variance of the estimates.

Problem Statement

The problem considered is that of determining the values of certain model parameters which are best with regard to a particular criterion, if the input and noisy measurements of the response of a linear, constant-coefficient system are given. The system to be identified is defined by the following equations:

$$\dot{x} = Ax + Bu \tag{1}$$

$$y = Fx + Fu + b \tag{2}$$

$$z = \dot{y} + n \tag{3}$$

where

x   state vector

u   control vector

y   calculated response variable

b   constant-bias vector

n   noise vector

z   measured response variable

The unknown parameters will form a vector c. The matrices A, B, F, and G and the vectors b and $x(0)$ are functions of c.

Minimum Response Error Estimate

One criterion that is often used in systems identification is the mean-square difference between the measured response and that given by the model. A cost function which is proportional to the mean-square error can be written as

$$J = \sum_{i=1}^{N} (z_i - y_i)^T D_1 (z_i - y_i) \tag{4}$$

294

where D_1 is a weighting matrix and i is a time index. The summation approximates a time integral. The estimate of c is then

$$\hat{c} = \underset{c}{ARG\ MIN}(J) \tag{5}$$

where ARG MIN means that vector c which minimizes the cost function J.

Linearize the calculated response y with respect to the unknown parameter vector c

$$y_1 = y_{1_0} + \nabla_c y_1 (c - c_0) \tag{6}$$

where

y_{1_0} nominal response calculated by using c_0

$\nabla_c y_1$ gradient of y with respect to c

c_0 nominal c vector

Substituting y_1 into the expression for J and solving for the value of c which minimizes J yields

$$\hat{c} = c_0 + \left[\sum_{i=1}^{N} (\nabla_c y_1)^T D_1 \nabla_c y_1 \right]^{-1} \left[\sum_{i=1}^{N} (\nabla_c y_1)^T D_1 (z_1 - y_{1_0}) \right] \tag{7}$$

If this relationship is applied iteratively to update the calculated nominal response and its gradient with respect to the unknown parameter vector, the minimum-response error estimate $\hat{c}$ will result. The method has been called quasi-linearization and modified Newton-Raphson. The latter seems more appropriate since the Newton-Raphson (ref. 1) method would give

$$c_{k+1} = c_k + \left[\nabla_c^2 J_k \right]^{-1} \left[\nabla_c J_k \right] \tag{8}$$

where

$$\nabla_c J_k = -2 \sum_{i=1}^{N} (\nabla_c y_{1_k})^T D_1 (z_1 - y_{1_k}) \tag{9}$$

$$\nabla_c^2 J_k = 2 \sum_{i=1}^{N} (\nabla_c y_{1_k})^T D_1 \nabla_c y_{1_k} - 2 \sum_{i=1}^{N} (\nabla_c^2 y_{1_k})^T D_1 (z_1 - y_{1_k}) \tag{10}$$

295

The second term of $\nabla_c^2 J_k$ diminishes as the response error $(z_i - y_{i_k})$ diminishes. The modified Newton-Raphson method is identical to quasi-linearization if the term is neglected.

Maximum Conditional Likelihood Estimate

Another criterion that is often used is to select c to maximize the likelihood of the measured response when c is given

$$\hat{c} = \underset{c}{\text{ARG MAX}}\{P(z|c)\} \qquad (11)$$

If the noise n has a Guassian distribution with zero mean, the probability distribution of the measured response at any single time point can be written as

$$P(z_i|c) = \frac{1}{(2\pi)^{(MZ)/2}|M_1|^{1/2}} \exp\left[-\frac{1}{2}(z_i - y_i)^T M_1^{-1}(z_i - y_i)\right] \qquad (12)$$

where $M_1 = E\{nn^T\}$ and MZ is the number of elements in the response vector.

If the values of n are uncorrelated at different time samples, the conditional probability (given the value of c) of the entire set of measured response z can be written as

$$P(z|c) = \frac{1}{(2\pi)^{N(MZ)/2}|M_1|^{N/2}} \exp\left[-\frac{1}{2}\sum_{i=1}^{N}(z_i - y_i)^T M_1^{-1}(z_i - y_i)\right] \qquad (13)$$

The maximum conditional likelihood estimate of the unknown parameters will be the set of values of c which maximize $P(z|c)$, if it is recognized that y is a function of c. If it is noted that the maximization of $P(z|c)$ occurs for the value of c which minimizes the exponent, the maximum conditional likelihood estimate is the same as that which minimizes the mean-square response error provided the weighting matrix D_1 equals M_1^{-1}.

Maximum Unconditional Likelihood (Bayesian) Estimate

The unconditional probability density function of z can be expressed as

$$P(z) = P(z|c)P(c) \qquad (14)$$

The probability of c relates to the a priori information available for c before use is made of the measurements z. If $P(c)$ is Gaussian

$$P(c) = \frac{1}{(2\pi)^{(MC)/2}|M_2|^{1/2}} \exp\left[-\frac{1}{2}(c - c_0)^T M_2^{-1}(c - c_0)\right] \qquad (15)$$

where

$$M_2 = E\left\{(c - c_0)(c - c_0)^T\right\}$$

$$c_0 = E\{c\}$$

The unconditional probability of z is then

$$P(z) = \frac{1}{(2\pi)^{[(MC)+N(MZ)]/2}|M_1|^{N/2}|M_2|^{1/2}} \exp\left[-\frac{1}{2}\sum_{i=1}^{N}(z_i - y_i)^T M_1^{-1}(z_i - y_i)\right.$$

$$\left.-\frac{1}{2}(c - c_0)^T M_2^{-1}(c - c_0)\right] \qquad (16)$$

Again, the expression is maximized by minimizing the exponent. Setting the gradient with respect to c equal to zero and solving yields

$$\hat{c}_{k+1} = \hat{c}_k + \left[\sum_{i=1}^{N}(\nabla_c y_{i_k})^T M_1^{-1} \nabla_c y_{i_k} + M_2^{-1}\right]^{-1}\left[\sum_{i=1}^{N}(\nabla_c^T y_{i_k})^T M_1^{-1}(z_i - y_{i_k})\right.$$

$$\left. - M_2^{-1}(\hat{c}_k - c_0)\right] \qquad (17)$$

An identical estimate would have resulted if a weighted sum of mean-square response error and the mean-square difference of c and its a priori value are minimized, provided the weighting matrices used equaled the appropriate inverse error covariance matrices. Consequently, the same algorithm can be used for both estimates

$$\hat{c} = \underset{c}{\text{ARG MIN}}\left[\sum_{i=1}^{N}(z_i - y_i)^T M_1^{-1}(z_i - y_i) + (c - c_0)^T M_2^{-1}(c - c_0)\right] \qquad (18)$$

Variance of the Estimates

A very important aspect of systems identification is the quality of the estimates. Since the estimates themselves can be considered to be random numbers, it is useful to consider their variance, or specifically their error covariance matrix. The Cramer-Rao bound (ref. 1) provides an estimate of the error covariance matrix and is identical to that developed in the following discussion.

Note that when the maximum unconditional likelihood estimate of c converges, it is necessary for the following equation to be satisfied:

$$\sum_{i=1}^{N} \nabla_c y_i^T M_1^{-1}(z_i - y_i) - M_2^{-1}(\hat{c} - c_0) = 0$$

If c_{true} is introduced, and y_i is linearized with respect to c, it follows that

$$\sum_{i=1}^{N} \nabla_c y_i^T M_1^{-1}\left(z_i - y_{i_{true}} - \nabla_c y_i(\hat{c} - c_{true})\right)$$

$$- M_2^{-1}(\hat{c} - c_{true}) - M_2^{-1}(c_{true} - c_0) = 0$$

After solving for $\hat{c} - c_{true}$

$$\hat{c} - c_{true} = Q^{-1}\left[\sum_{i=1}^{N} \nabla_c y_i^T M_1^{-1}(n_i) - M_2^{-1}(c_{true} - c_0)\right]$$

where

$$Q \triangleq \left[\sum_{i=1}^{N} \nabla_c y_i^T M_1^{-1} \nabla_c y_i + M_2^{-1}\right]$$

$$n_i = z_i - y_{i_{true}}$$

Squaring and taking the expected value yields

$$E\left\{(\hat{c} - c_{true})(\hat{c} - c_{true})^T\right\} = -2Q^{-1}\left[\sum_{i=1}^{N} (\nabla_c y_i)^T M_1^{-1} E\left\{n_i(c_{true} - c_0)^T\right\} M_2^{-1} Q^{-1}\right]$$

$$+ Q^{-1}\left[\sum_{i=1}^{N}\sum_{j=1}^{N} (\nabla_c y_i)^T M_1^{-1} E\left\{n_i n_j^T\right\} M_1^{-1} \nabla_c y_j\right] Q^{-1}$$

$$+ Q^{-1} M_2^{-1} E\left\{(c_{true} - c_0)(c_{true} - c_0)^T\right\} M_2^{-1} Q^{-1}$$

298

Since

$$E\left\{n_i(c_{true} - c_0)^T\right\} = 0$$

$$E\left\{n_i n_j^T\right\} = 0 \qquad (i \neq j)$$

$$= M_1 \qquad (i = j)$$

and

$$E\left\{(c_{true} - c_0)(c_{true} - c_0)^T\right\} = M_2$$

$$E\left\{(\hat{c} - c_{true})(\hat{c} - c_{true})^T\right\} = Q^{-1}\left[\sum_{i=1}^{N} (\nabla_c y_i)^T M_1^{-1} \nabla_c y_i + M_2^{-1}\right] Q^{-1} = Q^{-1}$$

which is the inverse that is required in obtaining the maximum unconditional likelihood estimate of c. If the a priori information is not used, M_2^{-1} will be the null matrix so that

$$E\left\{(c - c_{true})(c - c_{true})^T\right\} = \left[\sum_{i=1}^{N} (\nabla_c y_i)^T M_1^{-1} \nabla_c y_i\right]^{-1}$$

The resulting estimate of the parameter error covariance matrix is useful in assessing the quality of the estimated parameter values.

Importance of Testing Results

The importance of testing a model once estimates of the unknown parameters have been made, cannot be over emphasized. The test can be to predict response measurements not included as part of the data used in obtaining the model parameters. If the fit error, J, is greater for this test than for a similar test using a model with fewer unknown parameters, it would indicate that the data base is insufficient for the more complicated model. The testing should be repeated and the number of model parameters reduced until the fit error for the test is greater for a simpler model. The testing procedure should be used even if the model format is "known" since the data base can easily lack the quantity and the quality for estimating all of the model parameters with sufficient accuracy.

Reference 4 provides an example of a model which failed a test of predicting the system's response. Figure 1 shows how the fit error varied for both the data used to determine the model parameters and that of the test, versus the number of model parameters. The lower curve shows that as additional unknown parameters are included, the fit error decreases. When the models were tested by predicting independent response measurements, however, the upper curve shows that models having more unknown parameters performed more poorly than simpler models. Fortunately, more data were available but had not been used in the interest of reducing the computational effort. When four times the data were used to determine the parameters of the same models, the results shown in figure 2 were more successful. The fit error for the predicted response tests continued to decrease as the number of unknown model parameters increased, thereby validating the more complicated models.

Unless such tests are performed, the analyst is not assured of the validity of his model. Unfortunately, reserving part of the data for testing is costly. A means of making such tests without requiring independent data is discussed in a later section of the paper.

SYSTEM IDENTIFICATION WITH MODEL FORMAT UNKNOWN

One never knows in an absolute sense what the model format is of any physical system because it is impossible to obtain complete isolation. Even if the model format was known with certainty, it is necessary to test one's results against a simpler model, a point made in an earlier section. Regardless of the cause, the requirement is the same: a number of candidate models must be compared on a basis which reflects the performance of each in achieving the model's intended use. For the purpose of this discussion, it will be assumed that the model's intended use is to predict the response of a system and that a meaningful measure of the model's performance is a weighted mean-square error.

A Criterion for Comparing Candidate Models

Let us continue the discussion of the problem stated earlier using the same notation. The weighted mean-square response error which was minimized by the minimum response error estimate was

$$J = \sum_{i=1}^{N} (z_i - y_i)^T D_i (z_i - y_i)$$

Let us denote the weighted mean-square response error which corresponds to testing the model's performance in predicting the system's response as

$$J^1 = \sum_{i=1}^{N} (z^1_i - y_i)^T D_i (z^1_i - y_i)$$

where z^1 is measured response data that is not part of z which is used to determine the model parameters. For convenience, we consider the input, u, to be identical in both cases.

The criterion suggested for comparing candidate models is the expected value of J^1. If it is possible to express the expected value, $E\{J^1\}$, in terms not involving actual data for z^1, then a considerable saving in data can be made and improved estimates will result from being able to use all available data for estimating.

Let us examine first the expected value of the fit error with respect to the data used to determine estimates of the unknown parameters.

We can express the response error as

$$z - y = y_{true} + n - y \approx y_{true} + n - y_{true} - \nabla_c y(\hat{c} - c_{true})$$

$$= n - \nabla_c y(\hat{c} - c_{true})$$

assuming the response can be linearized with respect to the model parameters over the range in which they are in error.

The expected value of the fit error, $E\{J_1\}$, becomes

$$E\{J\} = E\left\{\sum_{i=1}^{N}(z_i - y_i)^T D_1 (z_i - y_i)\right\}$$

$$= E\left\{\sum_{i=1}^{N}\left[n - (\nabla_c y_i)(\hat{c} - c_{true})\right]^T D_1 \left[n - (\nabla_c y_i)(\hat{c} - c_{true})\right]\right\}$$

Expanding we get

$$E\{J\} = E\left\{\sum_{i=1}^{N} n_i^T D_1 n_i\right\} - 2E\left\{\sum_{i=1}^{N} n_i^T D_1 (\nabla_c y_i)(\hat{c} - c_{true})\right\}$$

$$+ E\left\{\sum_{i=1}^{N}(\hat{c}^T - c_{true}^T)(\nabla_c y_i)^T D_1 (\nabla_c y_i)(\hat{c} - c_{true})\right\}$$

If a maximum likelihood estimate is used, or if the minimum mean-square response error estimate is used with a weighting equal to the measurement error covariance matrix, then we can write

$$D_1 = M^{-1}$$

$$\hat{c} = c_{true} + \left[\sum_{i=1}^{N} (\nabla_c y_i)^T M^{-1} (\nabla_c y_i)\right]^{-1} \left[\sum_{i=1}^{N} (\nabla_c y_i)^T M^{-1} n_i\right]$$

again linearization is assumed and it is noted that

$$z_i - y_{true_i} = n_i$$

After substituting we get

$$E\{J\} = E\left\{\Sigma\ n_i{}^T M^{-1} n_i\right\} - 2E\left\{P^T Q^{-1} P\right\} + E\left\{P^T Q^{-1} Q Q^{-1} P\right\}$$

$$= E\left\{\Sigma\ n_i{}^T M^{-1} n_i\right\} - E\left\{P^T Q^{-1} P\right\}$$

where

$$P = \sum_{i=1}^{N} (\nabla_c y_i)^T M^{-1} n_i$$

$$Q = \sum_{i=1}^{N} \nabla_c y_i\ M^{-1} \nabla_c y_i$$

Next, let us examine expected fit error $E\{J\}$ of a model used to predict response measurements, z^1, which are independent of the data z, used to determine the estimates of the model.

We can again express the expected fit error as

$$E\left\{J^1\right\} = E\left\{\sum_{i=1}^{N} n_i^1{}^T D_1 n_i^1\right\} - 2E\left\{\sum_{i=1}^{N} n_i^1{}^T D_1 \nabla_c y_i (\hat{c} - c_{true})\right\}$$

$$+ E\left\{\sum_{i=1}^{N} \left(\hat{c}^T - c_{true}^T\right)(\nabla_c y_i)^T D_1 \nabla_c y_i (\hat{c} - c_{true})\right\}$$

Note that the only difference between the above expression and that obtained earlier for $E\{J\}$ is that the noise vector is n^1 instead of n. The same expression can be used for $\hat{c}$ as before since it is the estimate of c based on the data, z, that is desired.

$$\hat{c} = c_{true} + \left[\sum_{i=1}^{N}(\nabla_c y_i)^T M^{-1}\nabla_c y_i\right]^{-1}\left[\sum_{i=1}^{N}(\nabla_c y_i)^T M^{-1}n_i\right]$$

Substituting the above expression for $\hat{c}$, and M^{-1} for D_1 we get

$$E\left\{J^1\right\} = E\left\{\sum_{i=1}^{N}n_i^{1\,T}M^{-1}n_i^1\right\} - 2E\left\{\sum_{i=1}^{N}n_i^{1\,T}M^{-1}\nabla_c y_i Q^{-1}\sum_{i=1}^{N}(\nabla_c y_i)^T M^{-1}n_i\right\}$$

$$+ E\left\{P^T Q^{-1}P\right\}$$

where P and Q are defined as before. Since the noise vector, n^1 and n, are uncorrelated, that is

$$E\left\{n_i n_i^{1\,T}\right\} = 0 \quad \text{for all } i \neq j$$

then the second term is zero.

Since the noise vectors n and n^1 have the same covariance matrix, M, we can write

$$E\left\{\sum_{i=1}^{N}n_i^{1\,T}M^{-1}n_i^1\right\} = \text{TRACE}\left[E\left\{\sum_{i=1}^{N}n_i^{1\,T}M^{-1}n_i^1\right\}\right]$$

$$= \text{TRACE}\left[E\left\{\sum_{i=1}^{N}n_i^1 n_i^{1\,T}\right\}M^{-1}\right]$$

$$= \text{TRACE}\left[NI\right] = N \cdot MZ$$

where N is the number of time samples and MZ is the number of measurement quantities. Also

$$E\left\{\sum_{i=1}^{N}n_i^{T}M^{-1}n_i\right\} = N \cdot MZ$$

We can now express $E\{J^1\}$ in terms of $E\{J\}$ as

$$E\{J^1\} = E\{J\} + 2E\{P^T Q^{-1} P\}$$

Examining the second term

$$E\{P^T Q^{-1} P\} = E\left\{\sum_{i=1}^{N} n_i^T M^{-1} \nabla_c y_i Q^{-1} \sum_{i=1}^{N} (\nabla_c y_i)^T M^{-1} n_i\right\}$$

After taking the trace of the scalar and reordering the vectors we get

$$E\{P^T Q^{-1} P\} = \text{TRACE}\left[E\left\{\sum_{i=1}^{N} \sum_{j=1}^{N} n_j n_i^T M^{-1} \nabla_c y_i Q^{-1} (\nabla_c y_j)^T M^{-1}\right\}\right]$$

Because the noise is uncorrelated at unlike times, the term simplifies to

$$E\{P^T Q^{-1} P\} = \text{TRACE}\left[\sum_{i=1}^{N} \nabla_c y_i Q^{-1} \sum_{j=1}^{N} (\nabla_c y_j)^T M^{-1}\right]$$

Finally, we have that the expected fit error for the case of testing the model's prediction of the system's response as

$$E\{J^1\} = J + 2\,\text{TRACE}\left[\sum_{i=1}^{N} \nabla_c y_i Q^{-1} \sum_{j=1}^{N} (\nabla_c y_j)^T M^{-1}\right]$$

Since it is available, the actual fit error, J, is used instead of its expected value. The intent of the new criterion, $E\{J^1\}$, is that it be used instead of J in determining the level of model complexity that is best.

An Application of the New Modeling Criterion

The control input and response time histories of the numerical example used to demonstrate the use of the new modeling criterion are shown in figure 3. The dynamics resemble that of the lateral-directional modes of an airplane and are described in detail in reference 5. The level of noise that has been added to the calculated time histories is also indicated in figure 3. The sampling rate was 10 samples per second. The unknown parameters in the A, B, and b matrices can number as many as 10, 9, and 4, respectively. Only 16 parameters have values other than zero.

304

Five sets of calculated responses to which noise was added, were analyzed using the algorithm of reference 5. The analysis was repeated, allowing a different number of parameters to be determined by the algorithm. The fit error, J, in each case was averaged for the five sets of responses, and plotted as the lower curve of figure 4 as a function of the number of unknown parameters allowed to vary. As the number of unknown parameters was increased, terms were always added and never substituted. Because of this it can be argued that the fit error, J, should be monotonically decreasing as the number of unknown parameters increases. It can be seen in figure 4, however, that a point is reached beyond which J increases. The reason for the increase is because there is a point reached where convergence is a problem because of the essentially redundant unknown parameters. Although some of the five cases for 23 unknown parameters showed a decrease in J, others did not. The analyst might be tempted to settle for the model having 19 unknown parameters since the calculated response best fits the "measured response." It would be a mistake to do so; one that is often made.

In addressing the question "Which of the candidate models best predict system response?" a test was applied. The results of the test were then compared to the modeling criterion in the following way. For each set of unknown parameters determined, the corresponding calculated response was compared to the true response to which independent noise was added. The resulting fit errors were again averaged and plotted in figure 4 as the upper curve. The test performed simulates the practice of reserving actual response data for model testing purposes only. As a result of these tests it can be seen that a model with only 10 unknown parameters is best in terms of predicting system response. This is six parameters fewer than the "true model." That is to say there was less error in setting six parameters to zero than the error resulting from trying to determine more than 10 unknown parameters. The number of parameters that best predict system response will increase if additional data are used or if the noise is decreased.

The purpose of the new modeling criterion is to eliminate the need of independent response measurements comparing the model candidates, thereby allowing all of the response data to be used in determining the unknown parameters. The new criterion was calculated as part of the systems identification algorithm and the results are shown in figure 5. The results of testing the candidate models given in the previous figure are included in figure 5 for comparison. The minimum in the curve for the new criterion occurs at 10 unknown parameters as was the case for the previous test results. This indicates that the new criterion can be used instead of testing the model candidates against independent response data.

The Problem of Too Many Candidate Models

Although it is a great help to have a criterion for comparing candidate models, there remains a problem of an excessive number of candidate models. Figure 6 illustrates the enormous number of candidate models that result from the combinations of unknown parameters that can be used for the simple example used to illustrate the modeling criterion. The total number of possible candidate models exceeds 8 million. Even though it is possible to greatly reduce the amount of calculation effort by neglecting changes in the gradient of the response with respect to the unknown parameters, $\nabla_c y$, solving a set of simultaneous, algebraic equations would still be required for each model. Consequently, the calculations involved for 8 million candidate models becomes an economic consideration. It is estimated that testing all of the 8 million candidate models would require about one hour on a CDC-6600 computer. As the maximum number of unknown parameters increases, the number of possible candidate models rapidly becomes astronomical.

In practice, the analyst has enough understanding of the dynamics of the system being modeled to know to some degree which terms are primary and which are less important. It would be valuable, however, if one did not have to rely on the analyst's judgment. The problem of searching for the best candidate model, therefore, remains a formidable problem worthy of attention.

CONCLUDING REMARKS

The analyst often faces the problem of selecting a model's level of complexity in addition to determining the model's unknown parameters. If a model is selected solely on the basis of fit error or a likelihood function, the model will probably be less accurate in predicting system response than a simpler one.

Several models of varying complexity should always be examined and at least tested by predicting system response using measurements not used in determining the unknown model parameters. Unfortunately, this form of test requires reserving part of the total data for testing only.

A new criterion was developed by expressing the expected fit error that would result from testing a model. The proposed criterion was shown to give the same results as testing against independent data. By using the modeling criterion instead of testing a better model will result because all of the data can be used to determine its unknown parameters.

A problem exists because of the large number of possible candidate models caused by the numerous combination of terms. The example which involved up to 23 unknown parameters, corresponds to over 8 million candidate models or combinations of parameters. Although it is possible to reduce the computation effort involved, the problem of efficiently searching for the best candidate model remains an area worthy of attention.

REFERENCES

1. Balakrishnan, A. V.: Communication Theory. McGraw-Hill Book Co., C. 1968. IFAC Symposium 1967 on the Problems of Identification in Automatic Control Systems (Prague, Czechoslovakia), June 12-17, 1967.

2. Cuenod, M., and Sage, A. P.: Comparison of Some Methods Used for Process Identification. IFAC Symposium 1967 on the Problems of Identification in Automatic Control Systems (Prague, Czechoslovakia), June 12-17, 1967.

3. Eykhoff, P.: Process Parameter and State Estimation. IFAC Symposium 1967 on the Problems of Identification in Automatic Control Systems (Prague, Czechoslovakia), June 12-17, 1967.

4. Taylor, Lawrence W., Jr.: Nonlinear Time - Domain Models of Human Controllers. Hawaii International Conference on System Sciences (Honolulu, Hawaii), January 29-31, 1968.

5. Taylor, Lawrence W., Jr., and Iliff, Kenneth W.: Systems Identification Using a Modified Newton-Raphson Method - A Fortran Program. NASA TN D-6734, May 1972.

6. Taylor, Lawrence W., Jr.: How Complex Should a Model Be? 1970 Joint Automatic Control Conference (Atlanta, Georgia), June 24-26, 1970.

7. Akaike, H.: Statistical Predictor Identification. Ann. Inst. Statist. Math., Vol. 22, 1970.

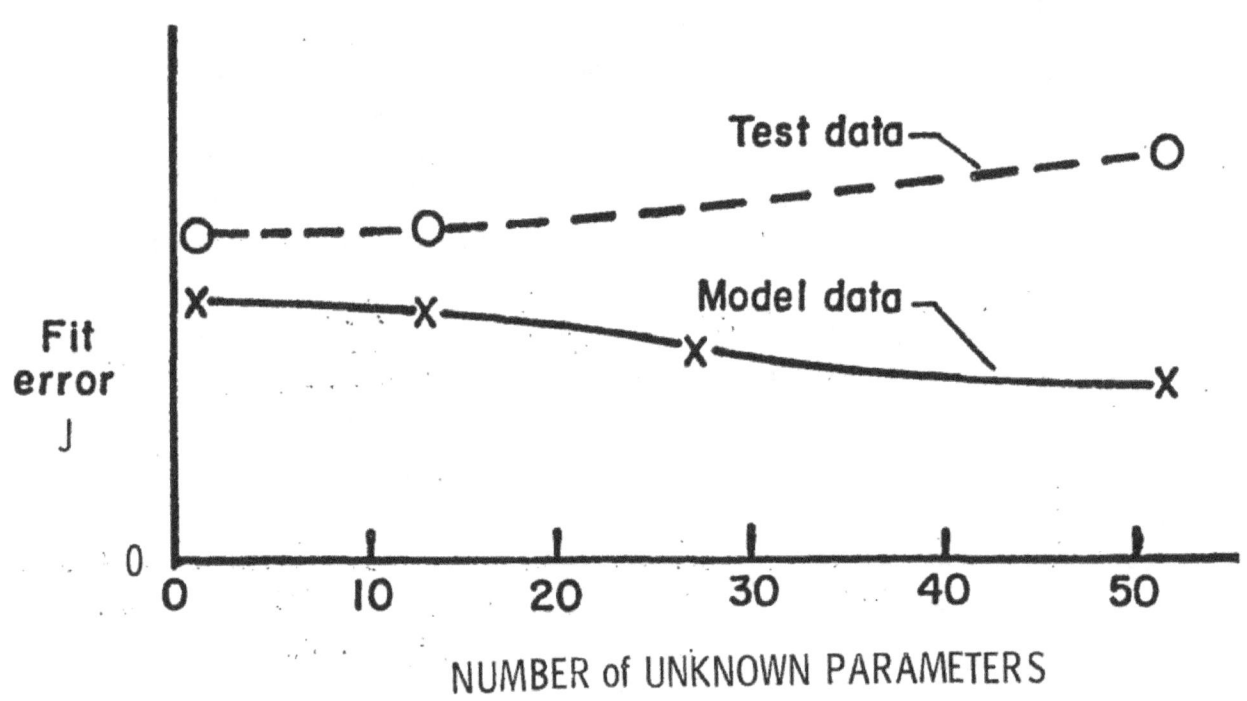

Figure 1.   An example of insufficient data/parameters.   Pilot tracking
case.   T = 60 seconds.

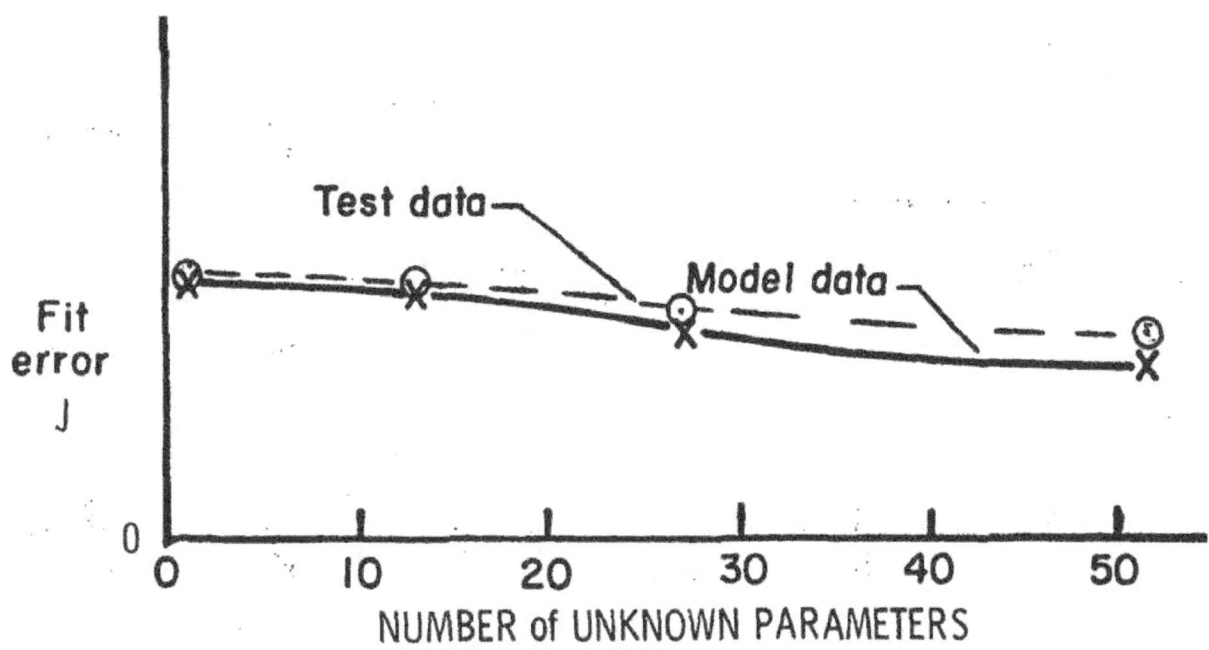

Figure 2. An example of sufficient data/parameters. Pilot tracking
case. T = 240 seconds.

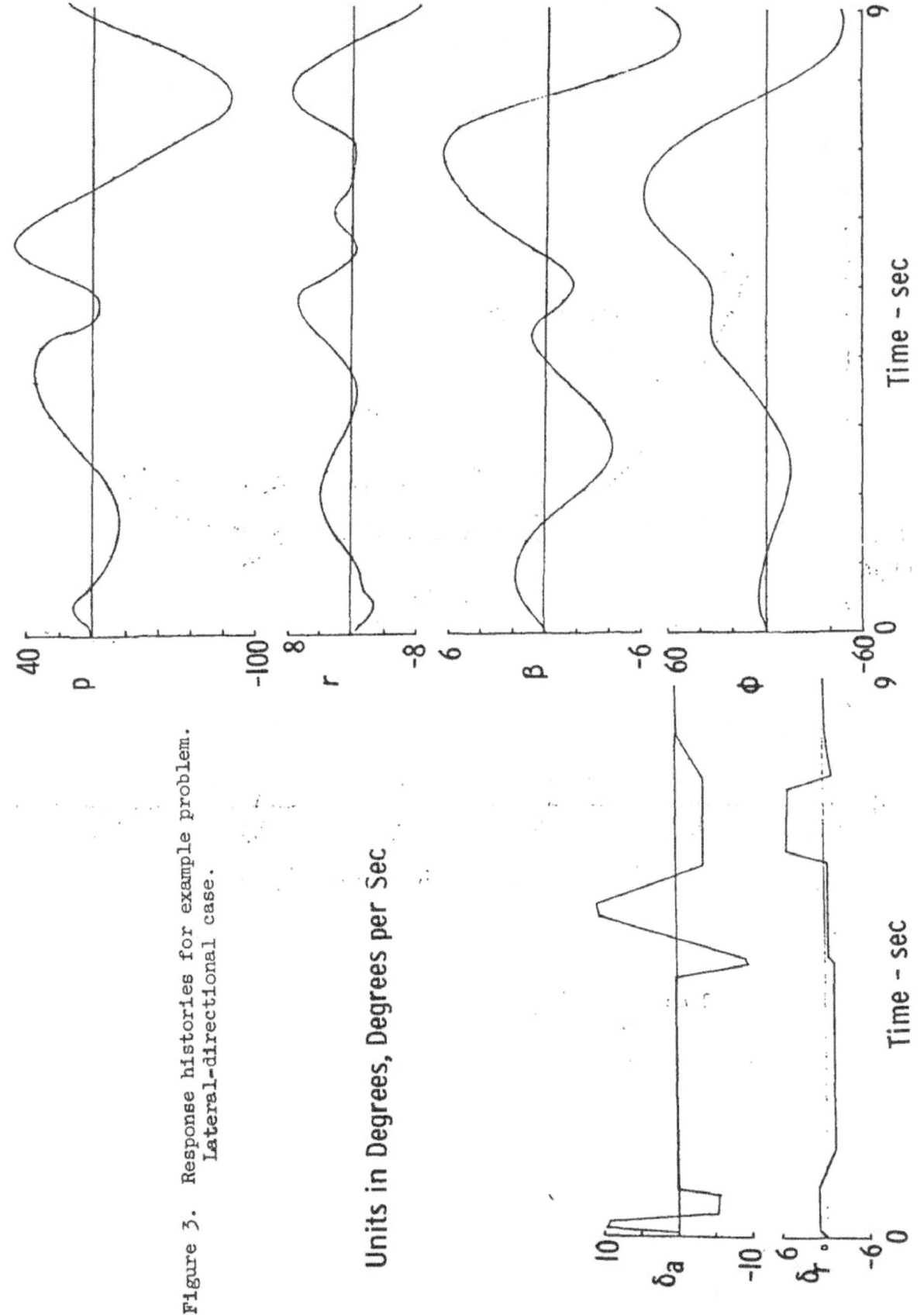

Figure 3. Response histories for example problem. Lateral-directional case.

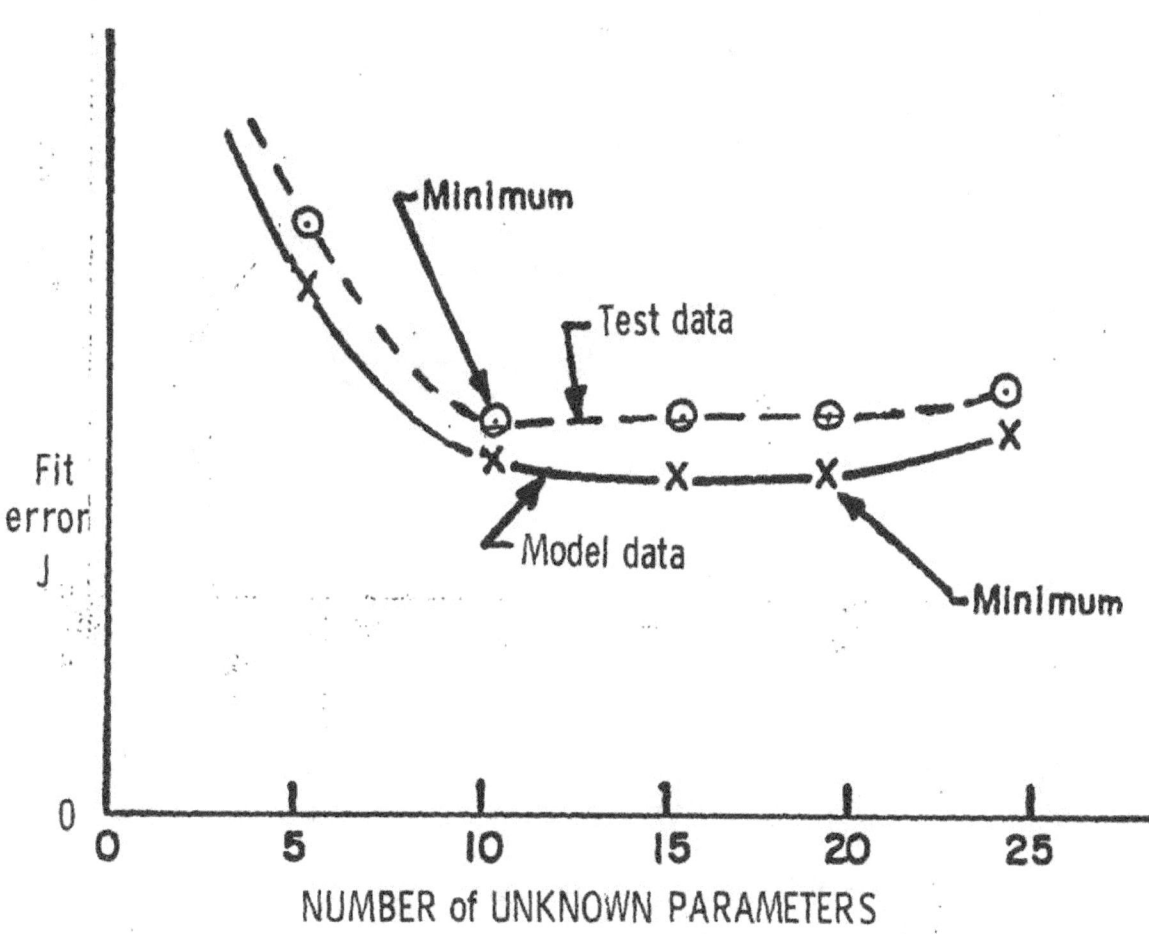

Figure 4. Comparison of fit error using model and test data.
Lateral-directional case.

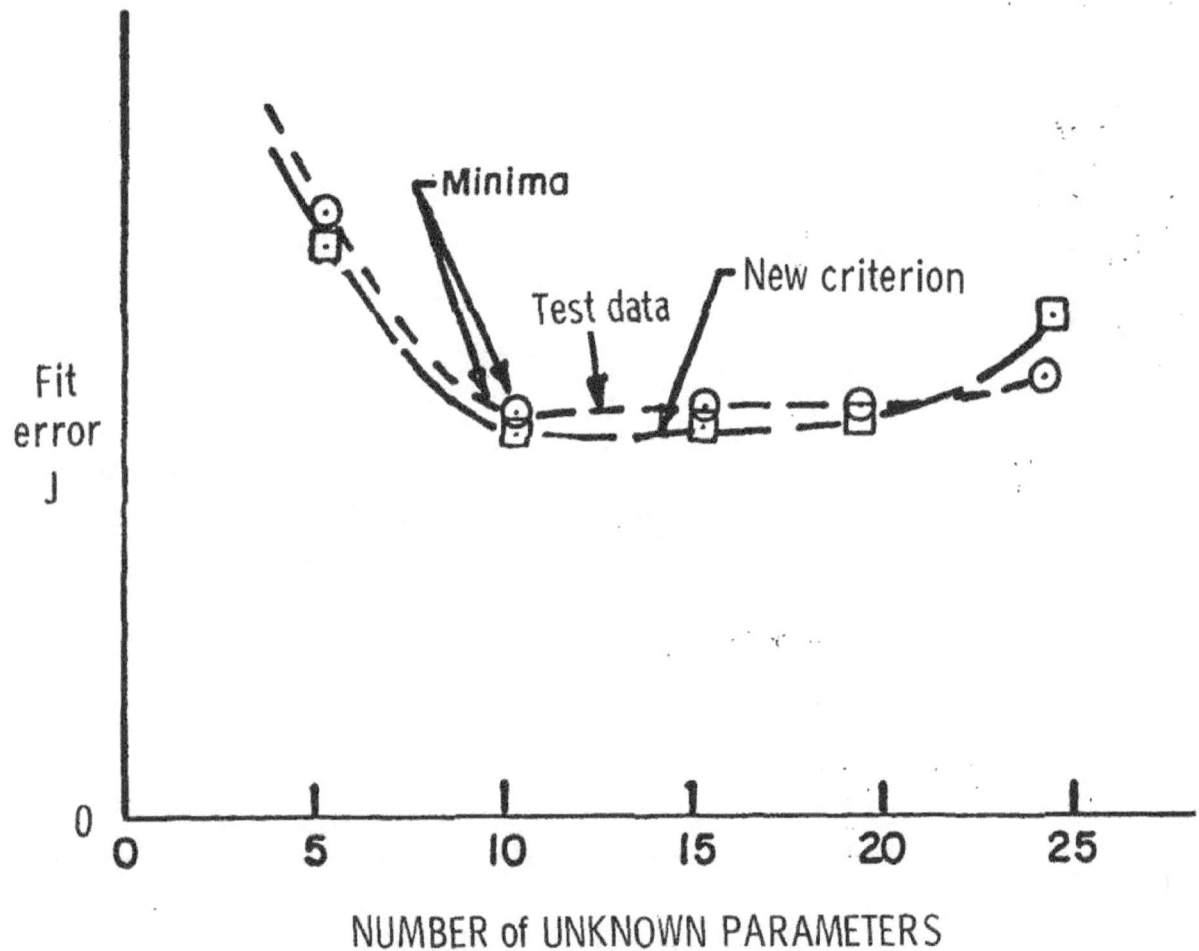

Figure 5.   Comparison of fit error using test data and new criterion.
Lateral-directional case.

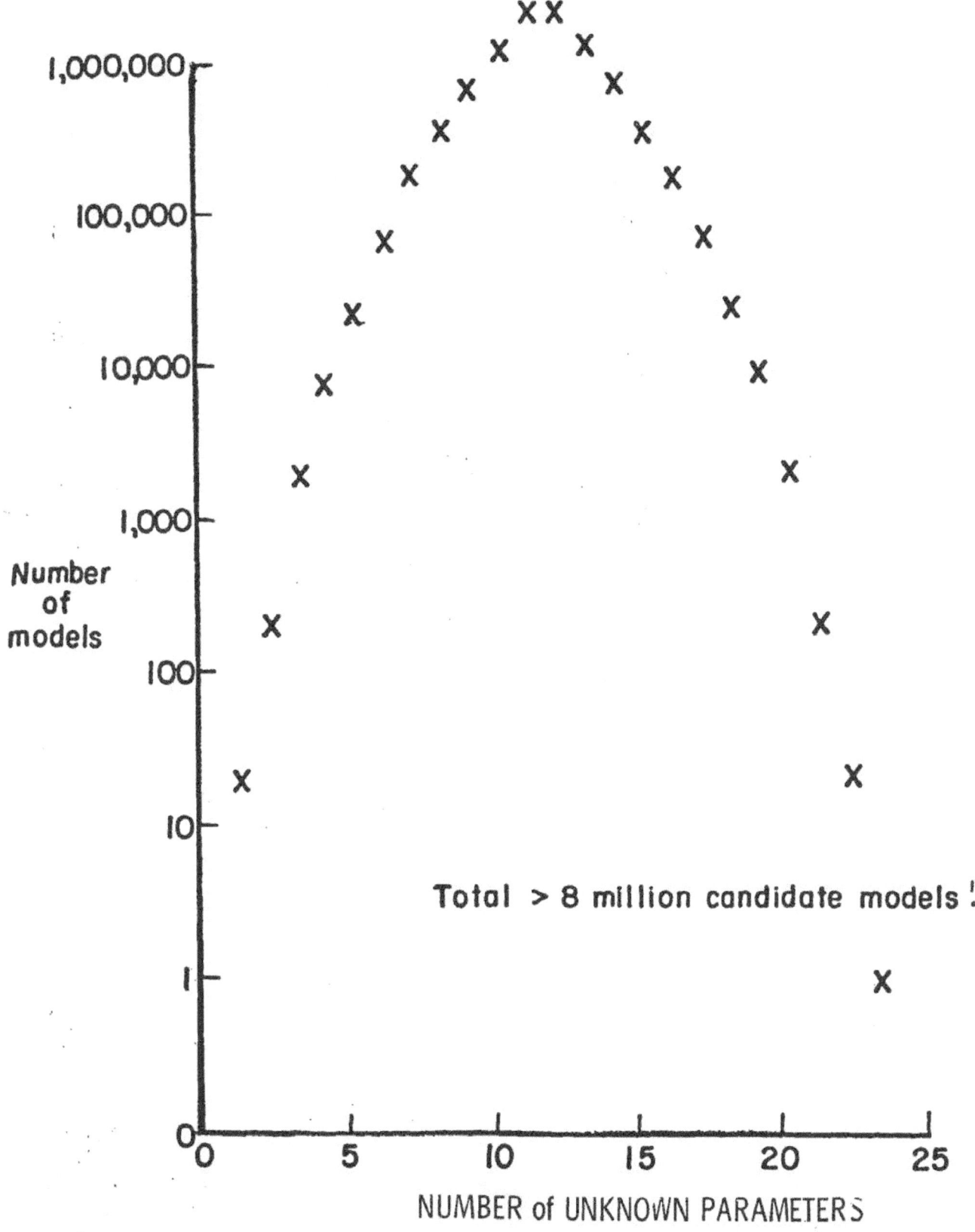

Figure 6. Possible candidate models. Lateral-directional case.

313

A PRACTICAL SCHEME FOR ADAPTIVE AIRCRAFT
FLIGHT CONTROL SYSTEMS*

by

Michael Athans and Dieter Willner
Massachusetts Institute of Technology
Department of Electrical Engineering
Cambridge, Massachusetts 02139

SUMMARY

The purpose of this paper is to present a flight control system design, that
can be implemented by analog hardware, to be used to control an aircraft
with uncertain parameters. The design is based upon the use of modern
control theory. Tne ideas are illustrated by considering control of STOL
longitudinal dynamics.

1. Introduction

This paper is motivated by practical considerations in the design of
stability augmentation systems for high performance and STOL aircraft
whose dynamic characteristics are changing. Such changes in the aerodynamic derivatives have been the primary motivation behind numerous
techniques for parameter identification of aerodynamic parameters.
Although such procedures are useful for off-line parameter estimation,
it is not quite clear how they should be used for on-line closed loop
automatic control.

There seems to be general agreement that on-line computational requirements of least square, maximum likelihood, and extended Kalman filter
methods are far too severe for on-board implementation. If one also
considers the additional computational requirements for the determination
of appropriate autopilot gains (say using the linear quadratic-gaussian
methods of modern control theory), then there is little doubt that
simultaneous identification and control is not a feasible practical
method for aircraft control system design, at least in the near future.

This paper discusses a method that appears practical for on-board implementation. The method presupposes that off-line calculation of aerodynamic derivatives has been made at N distinctly different flight
conditions. For each flight condition one has a linear time-invariant
state variable model which represents approximately the aircraft dynamics.
The flight path of the aircraft, and its changing dynamic characteristics,
may not however, coincide always with one of the N dynamical models.
Hence, a switching type autopilot may not be sufficiently good, especially

*This research was carried out at the Decision and Control Sciences Group
of the MIT Electronic Systems Laboratory with support extended by the NASA
Ames Research Center under grant NGL-22-009-124 and the Air Force Office
of Scientific Research under grant AF-AFOSR-72-2273. The work of D.Willner
was supported by the MIT Lincoln Laboratory, Lexington Mass. This paper was
presented at the Symposium on Parameter Estimation Techniques and Applications in Aircraft Flight Testing, NASA Flight Research Center, Edwards
AFB, California, April 24, 25, 1973.

if the total number N of models available is small. The proposed control system is a type of smooth transitional autopilot, in contradistinction to a switching one. It can be realized using analog hardware. The practicality of the above "analog" scheme, hinges upon the fact that integrated circuits can be used to construct high reliability, low volume, and low weight constant coefficient dynamical systems. Thus, no actual digital computer hardware is needed.

2. Technical Discussion

Suppose that off-line parameter estimation techniques have resulted in N distinct dynamical models for the aircraft under different flight conditions. This corresponds to having a priori available the following state space representations

(1) $\quad \dot{\underline{x}}_i(t) = \underline{A}_i \underline{x}_i(t) + \underline{B}_i \underline{u}(t) + \underline{\xi}_i(t)$

(2) $\quad \underline{z}(t) = \underline{C}_i \underline{x}(t) + \underline{\theta}(t)$

where $\{\underline{A}_i, \underline{B}_i, \underline{C}_i\}$ are constant matrices associated with the i-th model, $\underline{u}(t)$ in the actual control vector, $\underline{z}(t)$ is the actual measurement vector, $\underline{x}_i(t)$ is the state vector of the i-th model, $\underline{\xi}_i(t)$ is plant white noise, and $\underline{\theta}(t)$ is measurement white noise.

Imagine that over some time interval one of the i models is the true one; however, we do not know which one it is. The problem is to generate a control input $\underline{u}(t)$, based on the actual measurements $\underline{z}(t)$, so that the performance of the system is satisfactory. The problem is complicated by the existence of the stationary white noise processes $\underline{\xi}_i(t)$, $\underline{\theta}(t)$;

we assume that they have zero mean and a priori known covariances

(3) $\quad E\{\underline{\xi}_i(t)\underline{\xi}_i'(\tau)\} = \underline{\Xi}_i \delta(t-\tau); \quad E\{\underline{\theta}(t)\underline{\theta}'(\tau)\} = \underline{\Theta}\delta(t-\tau)$

If we knew with absolute certainty which model was the true one, then we could generate the control $\underline{u}(t)$ using the (by now standard) techniques of the Linear-Quadratic-Gaussian (LQG) design that utilizes the steady-state Kalman-Bucy filter and control over the infinite time interval (Ref. [1]). So if it is known that the i-th model is the correct one, one would use a performance index

(4) $\quad J_i = \lim_{T \to \infty} \frac{1}{2T} \int_{-T}^{T} \underline{x}_i'(t)\underline{Q}_i\underline{x}_i(t) + \underline{u}_i'(t)\underline{R}_i\underline{u}_i(t)dt$

and obtain the following solution for the optimal control

(5) $\quad \underline{u}(t) = \underline{u}_i(t) = -\underline{G}_i \hat{\underline{x}}_i(t)$

where $\underline{G}_i$ is a constant control gain matrix and $\hat{\underline{x}}_i(t)$ is generated by the steady state Kalman Bucy filter

(6) $\quad \frac{d}{dt} \hat{\underline{x}}_i(t) = \underline{A}_i\hat{\underline{x}}_i(t) + \underline{B}_i\underline{u}(t) + \underline{H}_i\underline{r}_i(t)$

316

where $\underline{r}_1(t)$ is the residual (innovations) process of the Kalman filter

(7) $\qquad \underline{r}_1(t) = \underline{z}(t) - \underline{C}_1 \hat{\underline{x}}_1(t)$

The control gain matrix $\underline{G}_1$ is found by

(8) $\qquad \underline{G}_1 = \underline{R}_1^{-1} \underline{B}_1' \underline{K}_1$

where $\underline{K}_1$ is the positive definite solution of the algebraic matrix Riccati equation

(9) $\qquad \underline{0} = -\underline{K}_1 \underline{A}_1 - \underline{A}_1' \underline{K}_1 - \underline{Q}_1 + \underline{K}_1 \underline{B}_1 \underline{R}_1^{-1} \underline{B}_1' \underline{K}_1$

The filter gain matrix $\underline{H}_1$ is found by

(10) $\qquad \underline{H}_1 = \underline{\Sigma}_1 \underline{C}_1' \underline{\Theta}^{-1}$

where $\underline{\Sigma}_1$ is the positive definite solution of the algebraic matrix Riccati equation

(11) $\qquad \underline{0} = \underline{A}_1 \underline{\Sigma}_1 + \underline{\Sigma}_1 \underline{A}_1' + \underline{\Xi}_1 - \underline{\Sigma}_1 \underline{C}_1' \underline{\Theta}^{-1} \underline{C}_1 \underline{\Sigma}_1$

The important aspect of this design from a practical point of view is that the control $\underline{u}(t)$ is generated by operating upon the measurement $\underline{z}(t)$ by a linear time invariant dynamical system; this can be implemented with very high reliability and low weight through the use of integrated circuits.

However, as we have remarked before, we do not know which of the i-th models is the correct one. Based upon the operating conditions at some time t_o we may have an idea of which models are most likely and which are not. This is then modelled by assuming initial probabilities

(12) $\qquad p_1(t_o), \ p_2(t_o), \dots, p_N(t_o)$

where $p_i(t_o)$ is the a priori probability that the i-th model is indeed the correct system.

It turns out that for any input $\underline{u}(t)$ one can construct the a posteriori probability $p_i(t)$ that the i-th model is the correct one, given the actual measurement vector $\underline{z}(\tau)$, $t_o \leq \tau \leq t$, and utilizing the residuals $\underline{r}_1(t)$, $\underline{r}_2(t), \dots, \underline{r}_N(t)$ generated by a bank of N Kalman-Bucy filters (as defined by eqs. (6), (7), (10), and (11)). This has been demonstrated by many authors (see for example reference [2]).

The results of Refs. [2] - [4] can be modified to yield the following equations for the computation of the $p_i(t)$. From each Kalman-Bucy filter we generate a scalar time function $q_i(t)$ using the residual vector $\underline{r}_i(t)$ as follows:

(13) $\quad \underline{q}_i(t) \triangleq \frac{1}{2} \underline{r}_i'(t)\underline{\theta}^{-1}\underline{r}_i(t) \qquad\qquad ; \; i=1,2,\ldots,N$

These time functions then can be used to compute the time functions (likelihood ratios) $\lambda_{ji}(t)$, $i \neq j$, $i, u=1,2,\ldots,N$ by solving the scalar differential equations

(14) $\quad \dfrac{d}{dt} \lambda_{ji}(t) = \lambda_{ji}(t)[\underline{q}_i(t) - \underline{q}_j(t)]$

The likelihood ratios $\lambda_{ji}(t)$ are related to the a posteriori probabilities $p_i(t)$ as follows:

(15) $\quad p_i(t) = \dfrac{1}{\displaystyle\sum_{j=1}^{N} \lambda_{ji}(t)} \quad ; \quad \lambda_{ji}(t) = \dfrac{p_j(t)}{p_i(t)}$

From (15) we obtain

(16) $\quad \dot{p}_i(t) = -\dfrac{\displaystyle\sum_{j=1}^{N} \dot{\lambda}_{ji}(t)}{\left(\displaystyle\sum_{j=1}^{N} \lambda_{ji}(t)\right)^2} = -p_i^2(t) \displaystyle\sum_{j=1}^{N} \dot{\lambda}_{ji}(t)$

$\qquad\qquad = -p_i^2(t) \displaystyle\sum_{j=1}^{N} \lambda_{ji}(t)[\underline{q}_i(t)-\underline{q}_j(t)]$

$\qquad\qquad = -p_i^2(t) \displaystyle\sum_{j=1}^{N} \dfrac{p_j(t)}{p_i(t)} [\underline{q}_i(t)-\underline{q}_j(t)]$

$\qquad\qquad = -p_i(t) \displaystyle\sum_{j=1}^{N} p_j(t)[\underline{q}_i(t)-\underline{q}_j(t)]$

But since $\displaystyle\sum_{j=1}^{N} p_j(t) = 1$, eq. (16) reduces to

(17) $\quad \boxed{\dot{p}_i(t) = p_i(t) \left[\left(\displaystyle\sum_{j=1}^{N} p_j(t)\underline{q}_j(t)\right) - \underline{q}_i(t)\right]} \qquad\qquad i = 1,2,\ldots,N$

Eq. (17) shows that the <u>a posteriori</u> probabilities can be constructed by solving a system of N nonlinear differential equations driven by the actual residual signals (via the functions $\underline{q}_i(t)$ -- see eq. (13)).

The set of differential equations (17) since they are time invariant can be solved using purely analog means (integrators, multipliers, and summers) since $0 \leq p_i(t) \leq 1$, there are no scaling problems that will arise. Indeed, asymptotically, one of the $p_i(t)$ will converge to 1 (the one associated with the correct model),and the rest of the $p_i(t)$ will go to zero. Also, note that the term $\sum_{j=1}^{N} p_j(t)\underline{q}_j(t)$ is common to all equations defined by (17) further contributing to the ease of solution.

The control scheme that is then proposed is to generate the actual input $\underline{u}(t)$ to the unknown systems by weighting the subcontrols $\underline{u}_i(t) = -\underline{G}_i\underline{x}_i(t)$ by the <u>a posteriori</u> probabilities, i.e.,

$$(18) \quad \underline{u}(t) = \sum_{i=1}^{N} p_i(t)\underline{u}_i(t) = - \sum_{i=1}^{N} p_i(t)\underline{G}_i\underline{x}_i(t)$$

Note that it is then this control $\underline{u}(t)$ that is used to drive the Kalman Bucy filters (see eq. (6)) rather than the subcontrol $\underline{u}_i(t)$.

The use of the control scheme (18) is appealing because of its inherent simplicity. References [3] and [4] have analyzed its general performance characteristics. From the viewpoint of <u>on-line</u> control of aircraft potential value of this scheme is that it can be realized using time invariant hardware (integrators, multipliers, constant gain amplifiers and adders). It does not require <u>on-line</u> parameter identification and <u>on-line</u> computation of control gains, which in general require digital computers.

As the aircraft moves to a different operating condition, the control system can be reinitialized simply by changing the <u>a priori</u> probabilities $p_i(t_o)$ corresponding to that condition.

The structure of the adaptive system is shown in Figures 1 to 3.

3. <u>Numerical Considerations and Simulation Results</u>

In the first part of this section we discuss possible modifications of the results obtained earlier. These modifications may be necessary to design stable autopilots. In the second part simulation results for a STOL aircraft (longitudinal dynamics) are presented.

3.1 Adjustment to the Filter Gain

The filter gains are computed from Eq. (10). To prevent divergence of the individual filters it may be necessary to increase the state noise covariance $\underline{\Xi}(t)$, $i=1,\ldots,N$. Filter divergence is sometimes a problem in matched systems, i.e., the system model corresponds well to the true system. For mismatched systems, i.e., the system dynamics is $\underline{A}$, $\underline{B}$, $\underline{C}$ and the model is $\underline{A}_i$, $\underline{B}_i$, $\underline{C}_i$, divergence can become a difficult problem. It can be avoided by either decreasing the observation noise or increasing the state noise artificially. Of course in that case the optimality of the individual filters is forfeited.

3.2 Adjustment of the Probability Estimator

The probability estimator is shown in Fig. (3). Its analog network represents Eq. (17) and contains multipliers, adders and integrators. To compensate for integrator drift and inaccuracies of the analog elements it will be necessary to adjust the $p_i(t)$, $i=1,\ldots,N$

periodically. It is also essential to prevent the individual $p_i(t)$ from becoming zero; once a $p_i(t) = 0$ it will stay zero for all future times no matter what happens to the original system. This follows from Eq. (9) as for $p_i(t) = 0$ also $\dot{p}_i(\tau) = 0 \; \forall \; \tau \geq t$.

These difficulties can be circumvented if the $p_i(t)$ $i=1,\ldots,N$ are modified periodically. One such possible modification is described here (the $p_i(t_+)$ are the values of $p_i(t)$ just after the modification expressed in terms of the values $p_i(t_-)$ just before the modification):

$$(19) \quad p_i(t_+) = \frac{\alpha_i(t)}{\displaystyle\sum_{j=1}^{N} \alpha_j(t)}$$

where $\alpha_i(t) = p_i(t_-) + \varepsilon$

and the ε term ($\varepsilon \ll 1$) compensates for integrator drift and avoids $p_i(t)=0$. At the same time this correction assures that $\displaystyle\sum_{i=1}^{N} p_i(t) = 1$.

If ε is chosen small enough (e.g., $\varepsilon \approx .001$) the effect on the resulting probabilities and controller will be negligible. This modification

  a) prevents $p_i(t) = 0$, i.e., all controllers will be considered for all future times

  b) reduces errors due to integrator drift (otherwise the $p_i(t)$ could become larger than one or smaller than zero due to integrator drift)

3.3 Simulation

The simulation was done on a CDC-6600 digital computer. The discretization stepsize was chosen to $\Delta t = .1$ sec. and the simulation length to $\Delta T = 20$ sec.

The model (H_1) is seven dimensional as given in Table (1) and represents the longitudinal dynamics for a STOL-aircraft on a glidepath. Two different models H_2, H_3 (for different points on the flight envelope) were generated by disturbing some of the H_1-parameters. H_2 is intended to be a model for higher altitude, H_3 for lower altitude at approximately the same glidepath and speed (it is not claimed that H_2, H_3 represent exact systems dynamics for these flight conditions). The steady state Kalman-Bucy filter gains $\underline{H}_i$, and the controller gains $\underline{G}_i$ $i=1,\ldots,N$ where approximated by the solutions of the corresponding equations Eqs. (8,9,10,11) after 20 sec (or 200 steps). After 200 steps the individual

320

gain parameters did not change more than .2% per step. The noise covariance matrices Ξ, Θ were chosen to represent reasonable disturbances and measurement accuracies.

The penalty matrices Q, R for the cost criteria were picked after a few trials, no extra effort was expended to find the "best" pair Q, R. These were the same for all three models.

The linearized models of Table (1) were used to design the individual optimal controllers and filters. The resulting control should have been applied to the true nonlinear STOL dynamics. Since these nonlinear dynamics were not available, the linearized model H_1 ($H_1 = H_{true}$) was used.

The first state x_1 of the model was used as a noise shaping stage to generate the vertical wind disturbances. Its input was white noise with the intensity Ξ, the output is $x_1(t)$ as shown in Fig. (4). This wind distrubance is common to all experiments described in this paper.

Figs. 5(a,b,c) show the deviations of the forward velocity $u' = u/v_o$, the height $\delta h' = \delta h/v_o$ and the variation in pitch $\theta(t)$ due to the disturbances and initial conditions $\underline{x}(0)$, for the optimal controller; i.e., the a priori probabilities are $p_1(0)=1$, $p_2(0)=p_3(0)=.0$; i.e., the controller is told that H_1 is the correct system [Case I].

Figs. 6(a,b,c) show the same changes $\delta v'$, $\delta h'$, θ to the same disturbances and initial conditions when the adaptive controller is initialized $p_i(0) = .33$, $i = 1,\ldots,N$; i.e., the controller is not given any information which system generated the data [Case II].

Figs. 6(d,e,f) show the $p_i(t)$, $i = 1,\ldots,N$ (the progress of identification).

Figs. 7(a-f) display a similar experiment as Case II except that the a priori probabilities are set to $p_1(0) = .01$, $p_2(0) = .79$; i.e., the controller is given the bad information that system #3 is the most likely system [Case III].

Comparison of the Results

Figs. 6(d,e,f) and 7(d,e,f) show the speed of identification of this algorithm. Although the system dynamics seem relatively similar, the controller always converged to the correct system (usually within 2-6 seconds). For slightly more dissimilar systems identification was completed usually within 1 second.

From Eq. (17) for $\dot{p}_i(t)$ it is apparent that the speed of identification ($\dot{p}_i(t)$) increases if the systems are more dissimilar, since then the

$q_1(t) = \frac{1}{2} \underline{r}_1'(t)\underline{\theta}^{-1}\underline{r}_1(t)$ also differ more. The u', δh', θ from Figs. 5 (a,b,c), 6(a,b,c), 7(a,b,c) are fairly similar for the different cases; this is mainly due to the rapid identification -- because the controller gains are not very similar at all. From the Figs, it can be seen that the peaks of the states u', δh', θ increase as the controller deviates more (bad $p_1(0)$ initially) from the optimal controller. This deterioration is demonstrated by the increasing cost for increasingly bad initial information in Table (II).

4. Conclusions

An adaptive controller is proposed, that is easy to implement on analog networks. It uses only precomputable steady state filter and controller gains. The controller facilitates a smooth transition from one operating point of the flight envelop to the other. The overall controller converges to the optimal steady state controller if the STOL dynamics correspond to one of the hypothesized system models. Simulation results support the theoretical results and claims that were the basic premises for the design of this adaptive controller.

ACKNOWLEDGEMENT

The authors gratefully acknowledge the support of Dr. J.A. Tabaczynski of the M.I.T. Lincoln Laboratory. We also wish to thank Drs. D.L. Kleinman and W.R. Killingsworth of Systems Control, Inc., Cambridge, Mass., for providing us the nominal linearized STOL dynamics.

REFERENCES

1. M. Athans, "The Role and Use of the Stochastic Linear-Quadratic-Gaussian Problem in Control System Design" IEEE Trans. Auto. Control, Vol. AC-16, 1971, pp. 529-552.

2. A.H. Haddad and J.B. Cruz, Jr., "Nonlinear Filtering for Systems with Random Models" Proc. Second Nonl. Estimation Conf., Sept. 1971, San Diego, Cal.

3. J.G. Deshpande et al., "Adaptive Control of Linear Stochastic Systems" Automatica, Vol. 9, Jan. 1973, pp. 107-116.

4. D. Willner, "Observation and Control of Partially Unknown Stochastic Systems" Ph.D. Thesis, M.I.T. Dept. of Electrical Engineering, May, 1973.

TABLE I

$$\dot{\underline{x}}(t) = \underline{A}\underline{x}(t) + \underline{B}\underline{u}(t) + \underline{\xi}(t); \quad \underline{z}(t) = \underline{C}\underline{x}(t) + \underline{\theta}(t)$$

A. **Physical Significance of Variables**

$x_1(t)$: vertical wind disturbance

$x_2(t) = \dfrac{u}{v_o}$ = normalized perturbation in forward velocity (%)

$x_3(t) = \dfrac{w}{v_o} = \alpha$ = normalized perturbation in angle of attack

$x_4(t)$ = pitch rate (rad/sec)

$x_5(t)$ = pitch perturbation (rad)

$x_6(t) = \dfrac{\delta h}{v_o}$ = normalized height perturbation

$x_7(t)$ = thrust coefficient

$u_1(t)$ = throttle input

$u_2(t)$ = elevator angle (rad)

v_o = nominal forward velocity 126 ft/sec.

B. **System Parameters**

Model #1 (H_1) true model

$$\underline{A} = \begin{bmatrix}
-.25 & .0 & .0 & .0 & .0 & .0 & .0 \\
-.0855 & -.0373 & .0855 & .0 & -.255 & .0 & .0 \\
.556 & -.522 & -.556 & 1.0 & .0268 & .0 & -.1343 \\
1.2 & .2087 & -1.2 & -1.47 & -.0107 & .0 & -.0212 \\
.0 & .0 & .0 & 1.0 & -.001 & .0 & .0 \\
.0 & -.105 & -1.0 & .0 & 1.0 & .0 & .0 \\
.0 & .0 & .0 & .0 & .0 & .0 & -.5
\end{bmatrix}$$

$$\underline{B} = \begin{bmatrix}
.0 & .0 \\
.0 & -.0043 \\
.0 & -.096 \\
.0 & -3.6 \\
.0 & .0 \\
.0 & .0 \\
.5 & .0
\end{bmatrix}$$

$$\underline{C} = \begin{bmatrix}
0.0 & 0.000 & 0.0 & 0.0 & 1. & 0.0 & 0 \\
0.0 & 0.000 & 0.0 & 1.0 & 0. & 0.0 & 0 \\
0.0 & 0.000 & 0.0 & 0.0 & 0. & 1.0 & 0 \\
0.0 & -.105 & -1.0 & 0.0 & 1. & 0.0 & 0
\end{bmatrix}$$

TABLE I (Cont'd)

Model #2 (H_2):   A(2,5) = −.19   Model #33(H_3):   A(2,5) = −.38
                     A(3,7) = −.1   A(3,7) = −.19
                     A(4,3) = −.9   A(4,3) = −1.8
                     B(4,2) = −2.8   B(4,2) = −4.5
                     All other   All other
                     parameters   parameters
                     as above   as above

C. Cost function penalty matrices

   Q = diag [.0, 100.0, 3.0, 3.0, 3.0, 10.0, 200.0, 20.0]

   R = diag [10.0, 5.0]

D. Noise intensity matrices

   Ξ = diag [.001, .0, .0, .0, .0, .0, .0]

   Θ = diag [.0001, .0001, .0068, .0002]

E. Initial state

   $\underline{x}(0)$ = [.0, .0, .0, .0, .0, .0, .0]

   Initial State Estimate

   $\underline{\hat{x}}(0|0)$ = [.0, .0, .0, .0, .0, .0, .0]

TABLE II
Comparison of Expected and Measured Cost

	Expected Cost (precomputed)	Measured Cost
Case I $p_1(0)$ = 1.0 $\quad p_2(0)$ = 0., $p_3(0)$ = 0.	57.376	80.612
Case II $p_1(0)$ = .33... $\quad$ i = 1,2,3	----	80.674
Case III $p_1(0)$ = .01 $\quad p_2(0)$ = .2, $p_3(0)$ = .79	----	83.188

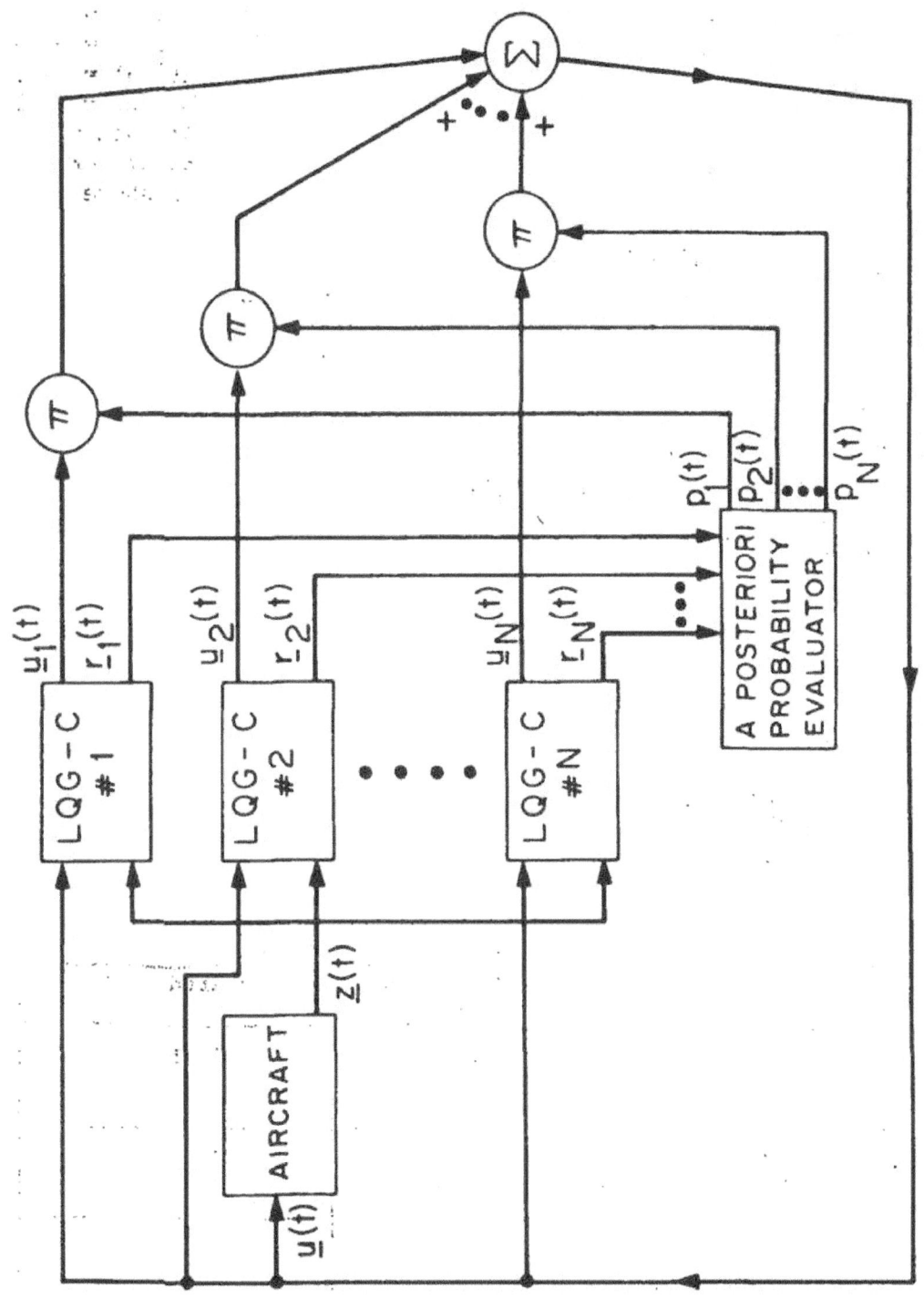

Fig. 1: Structure of Adaptive Autopilot

325

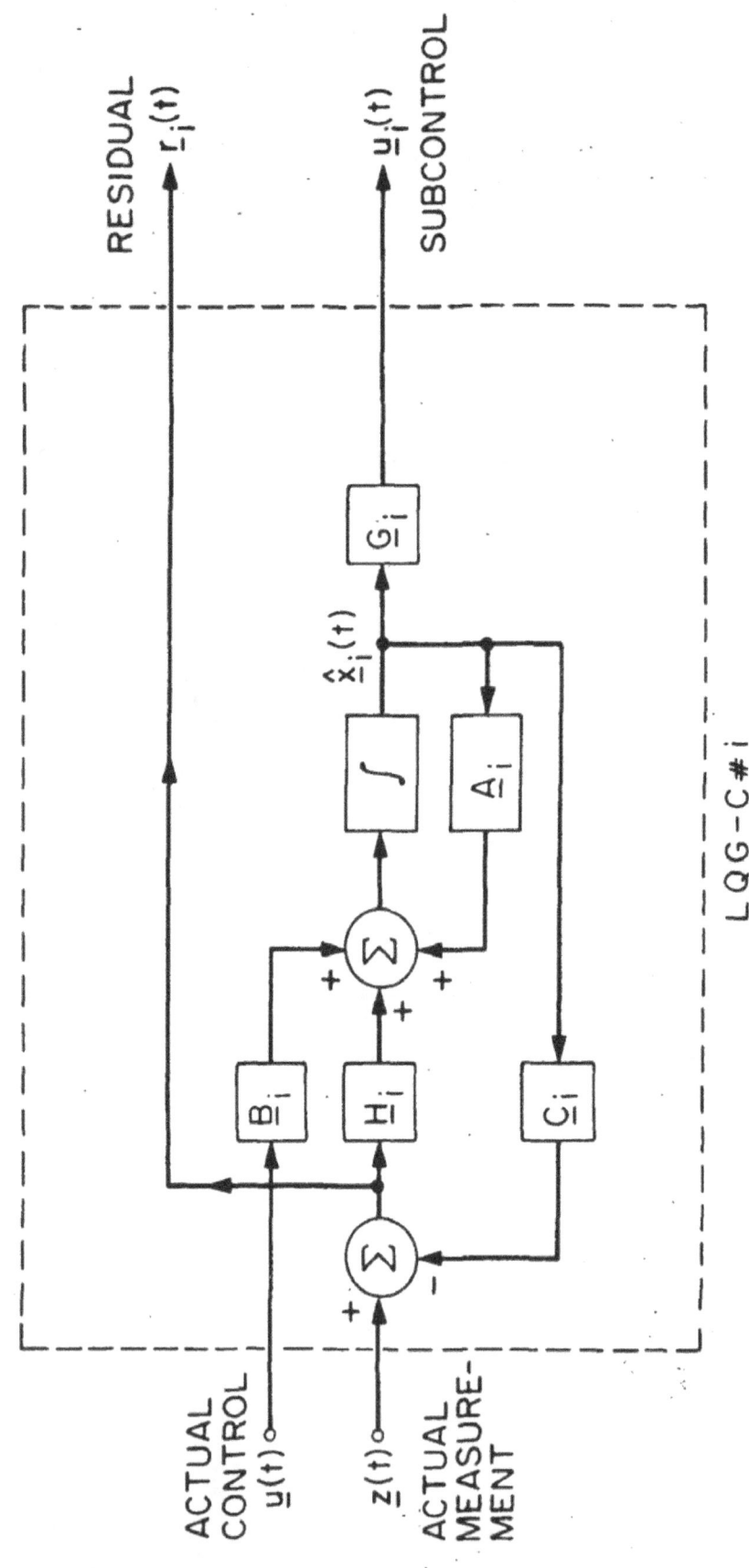

Fig. 2: Structure of LQG Controller #1

326

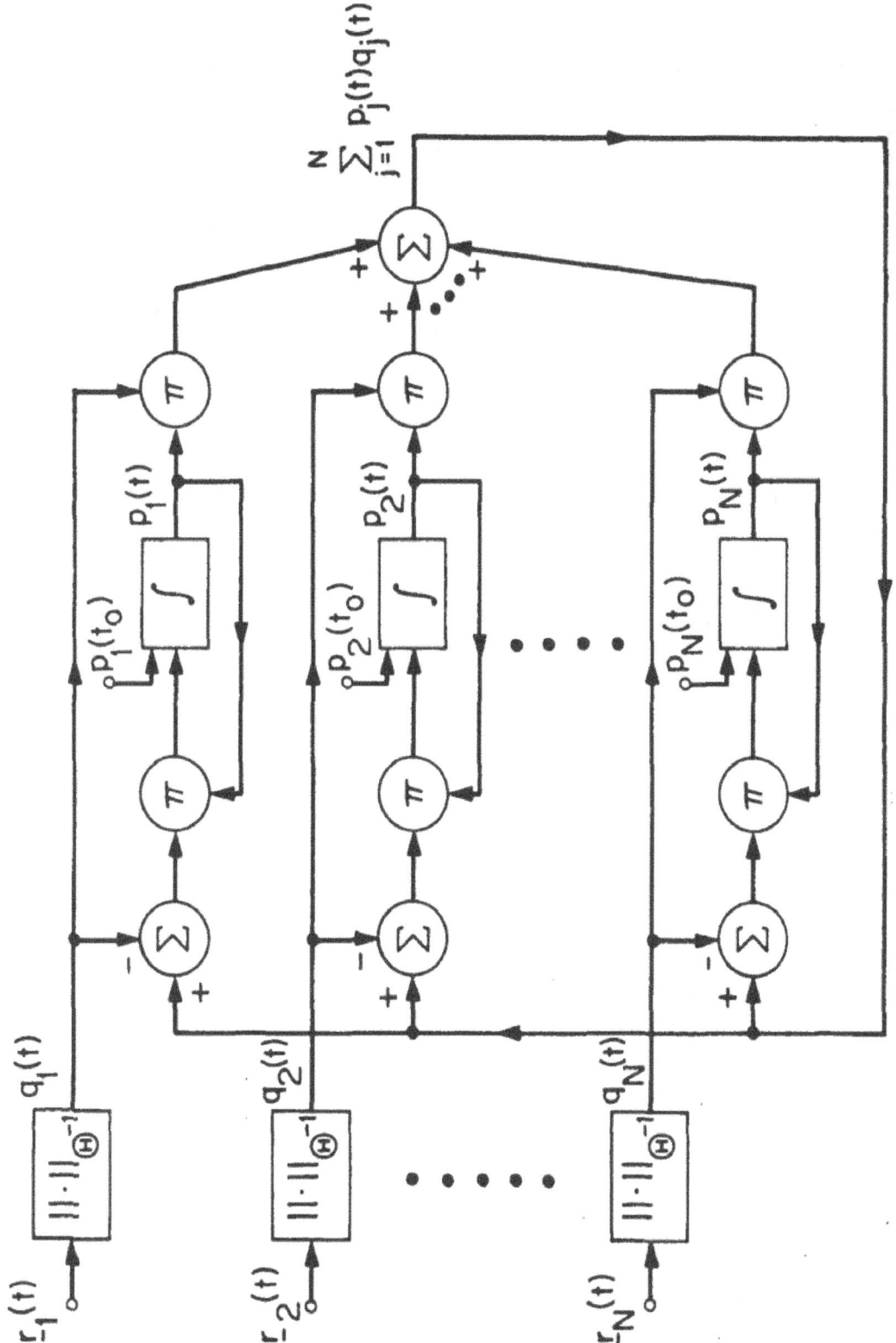

Fig. 3: Structure of A Posteriori Probability Evaluator

327

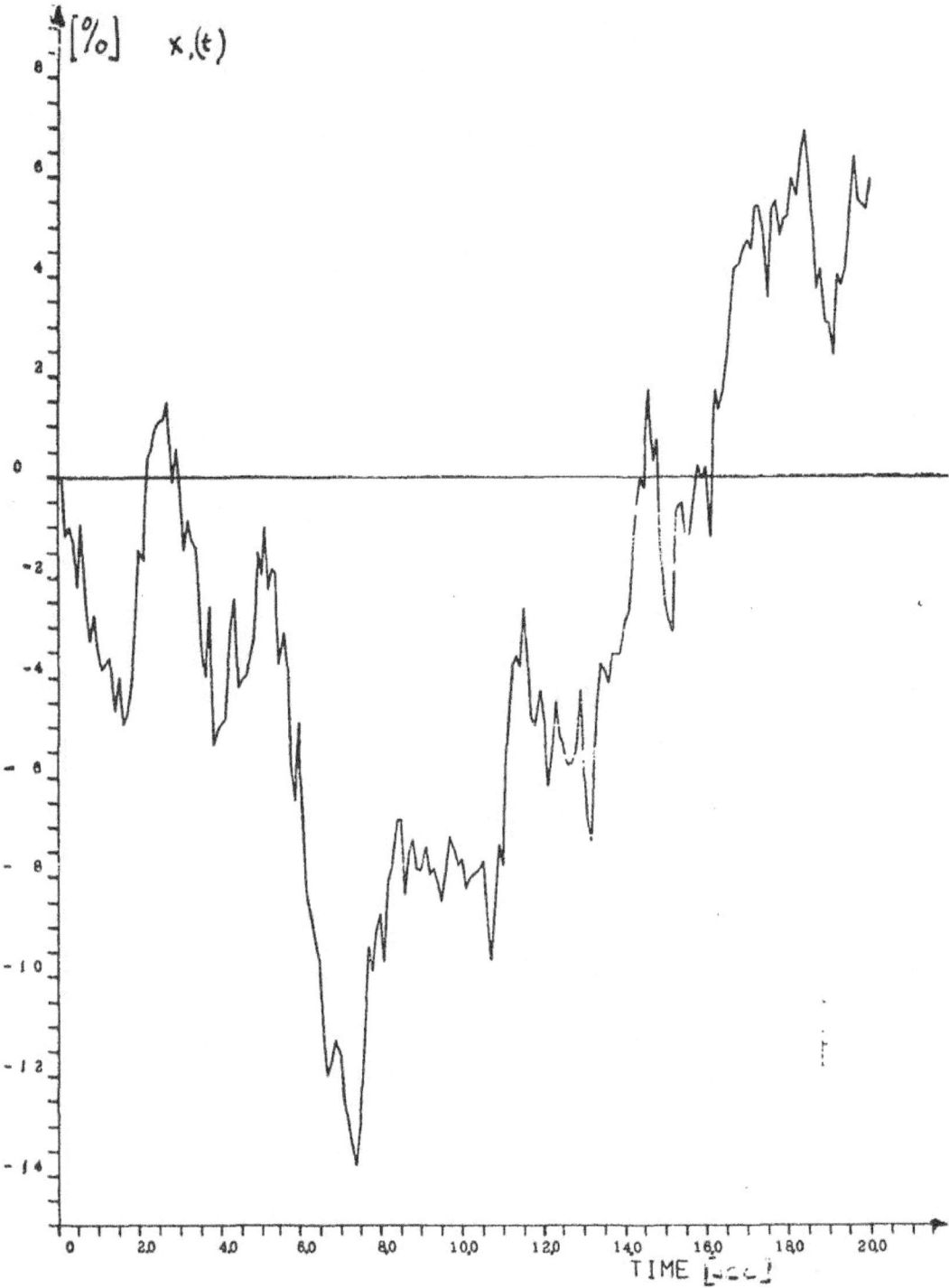

Fig. 4 Vertical Wind Disturbances (x_1) in [%] of the Forward Velocity. This Disturbance is Common to All Three Cases Shown in Figs.(5, 6,7)

328

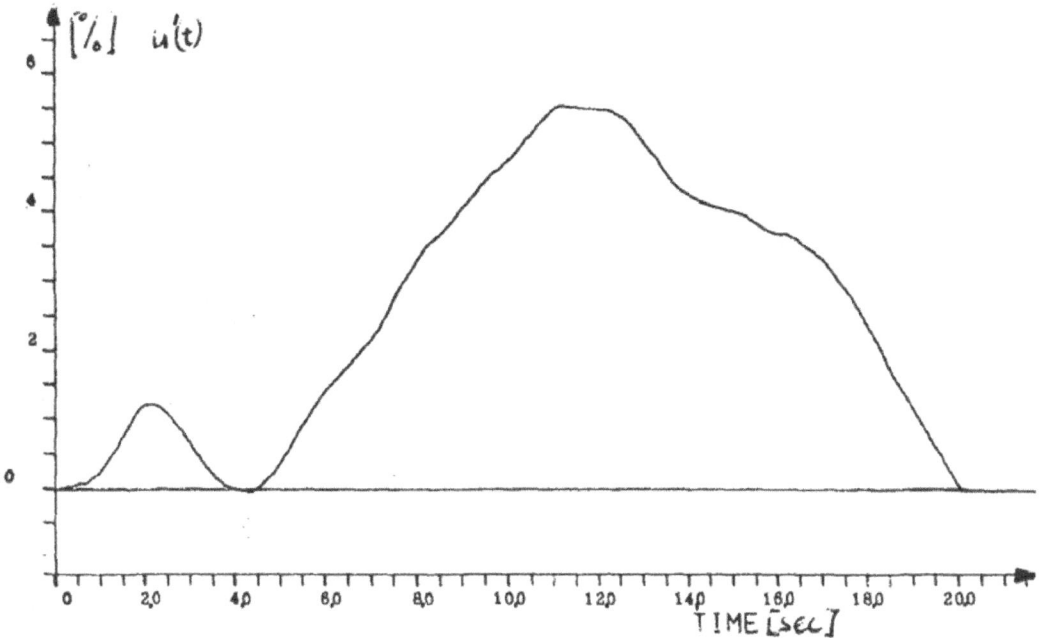

Fig. 5 a: Deviation of the Forward Velocity u' Due to the Disturbances
[Case I]

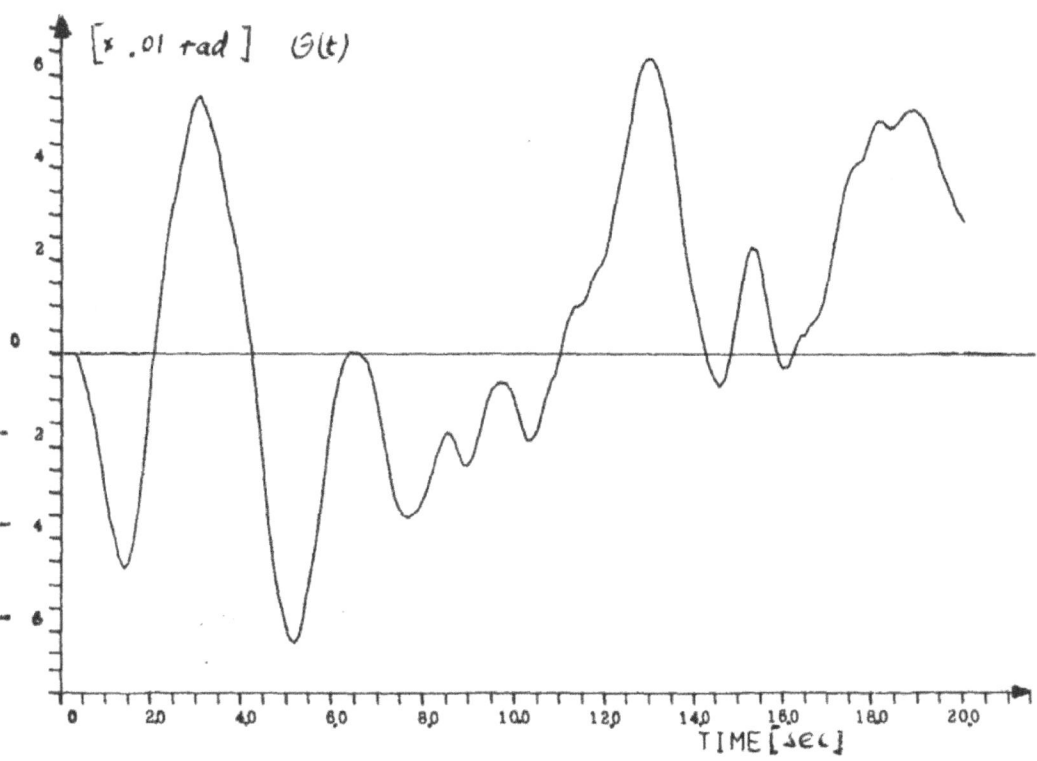

Fig. 5 c: Deviation of the Pitch θ [Case I]

329

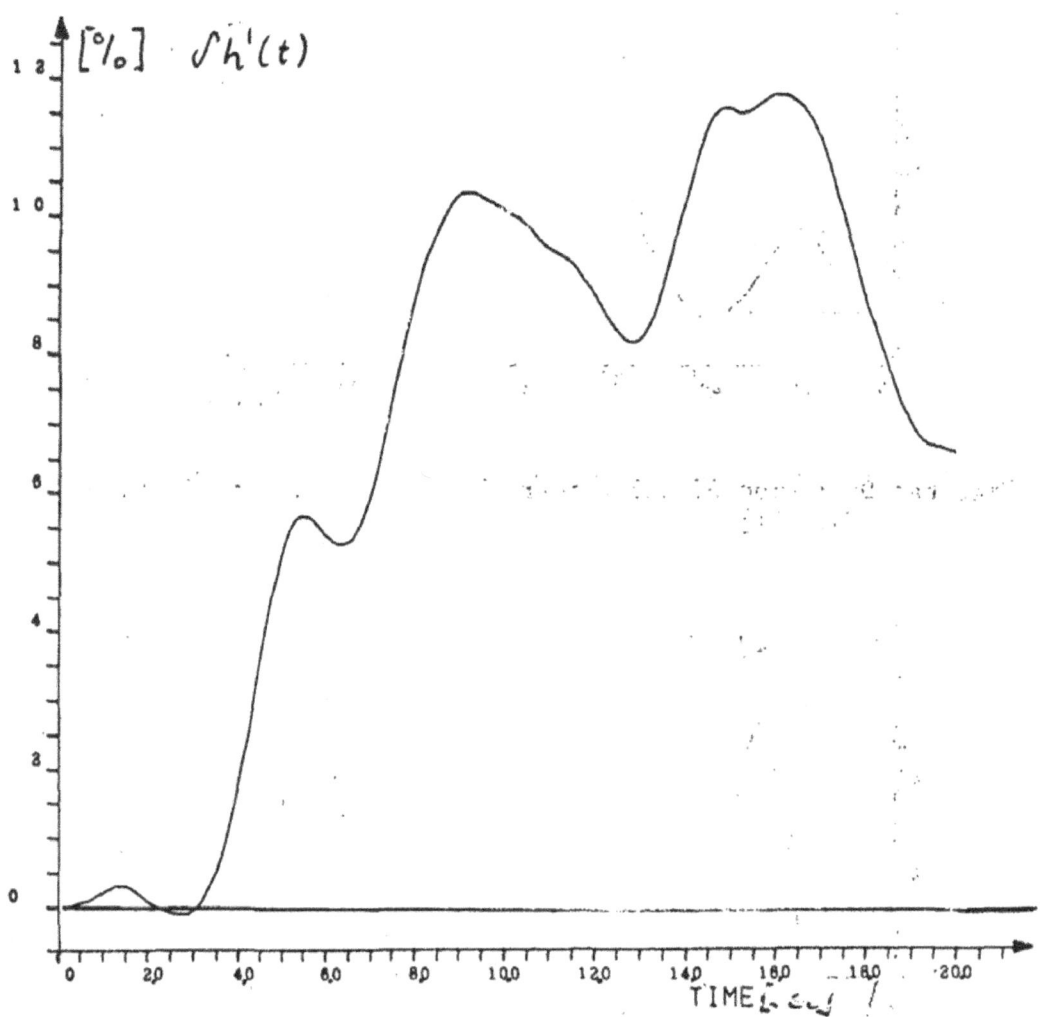

Fig. 5 b: Deviation of the Height δh' Above the Glide Path [Case I]

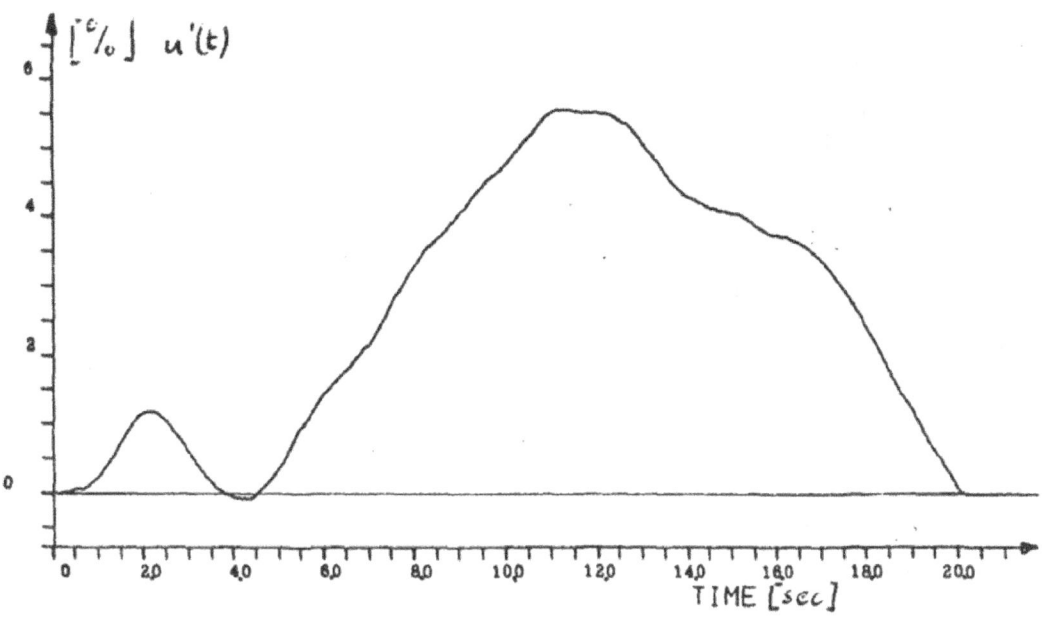

Fig. 6a: Deviation of the Forward Velocity u' Due to Disturbances
[Case II]

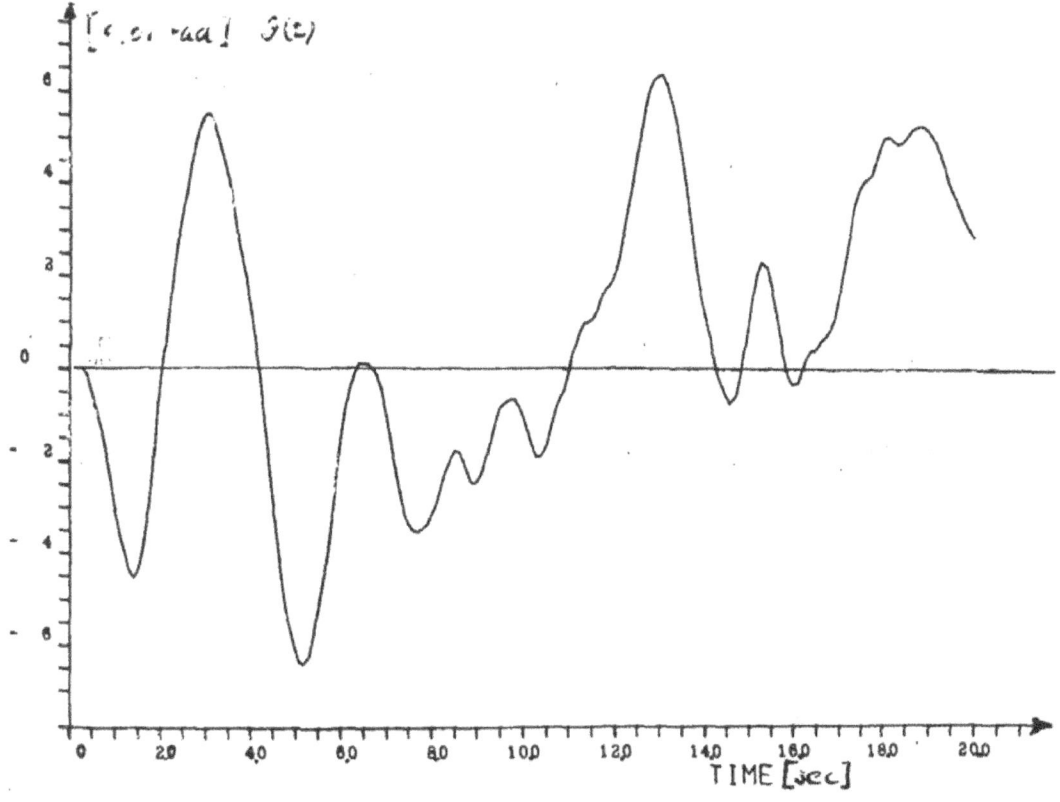

Fig. 6c: Deviation of the Pitch θ [Case II]

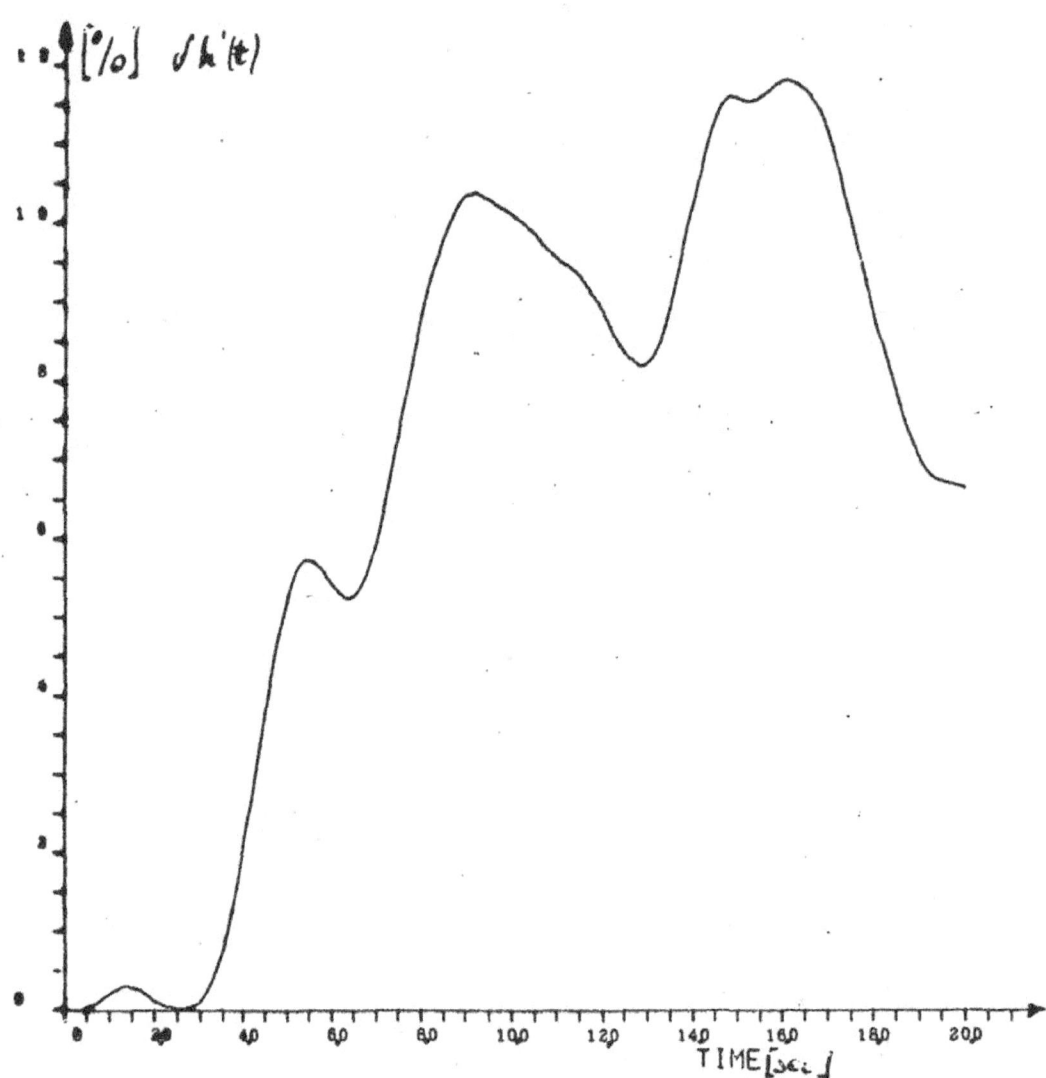

Fig. 60: Deviation of the Height δh' Above the Glide Path [Case II]

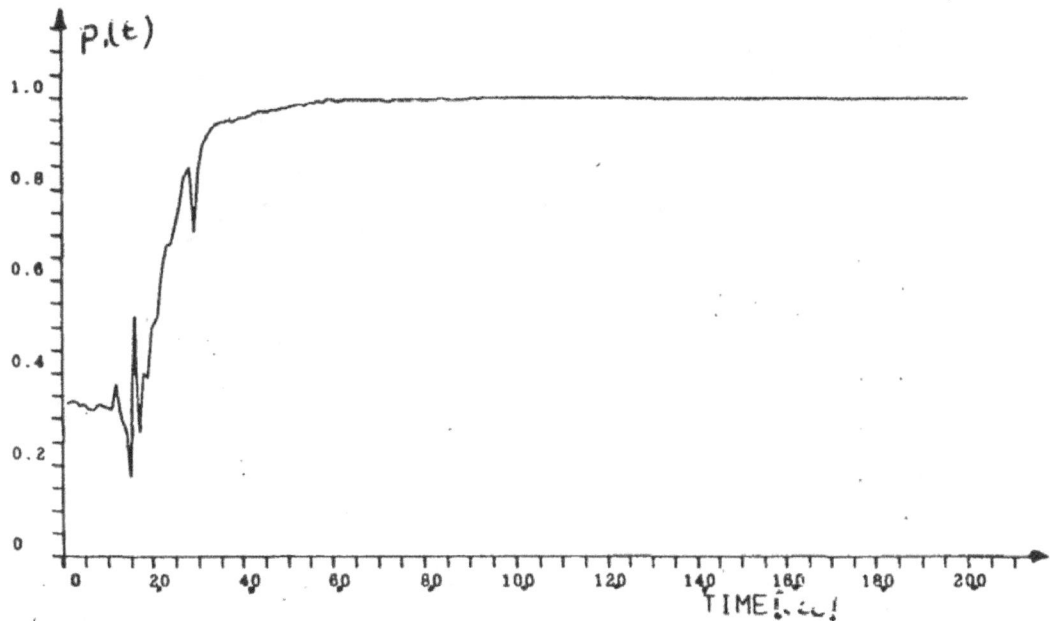

Fig. 6d: Probability $p_1(t)$ of Observing System 1, [Case II]

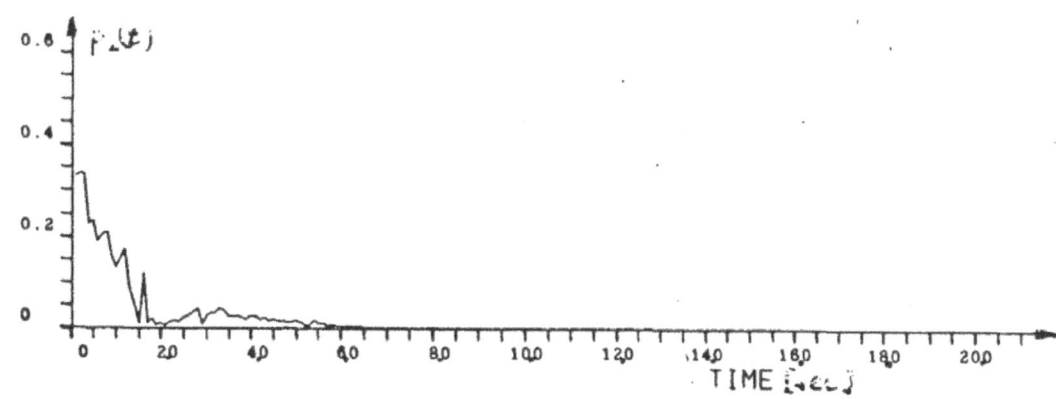

Fig. 6e: Probability $p_2(t)$ of Observing System 2, [Case II]

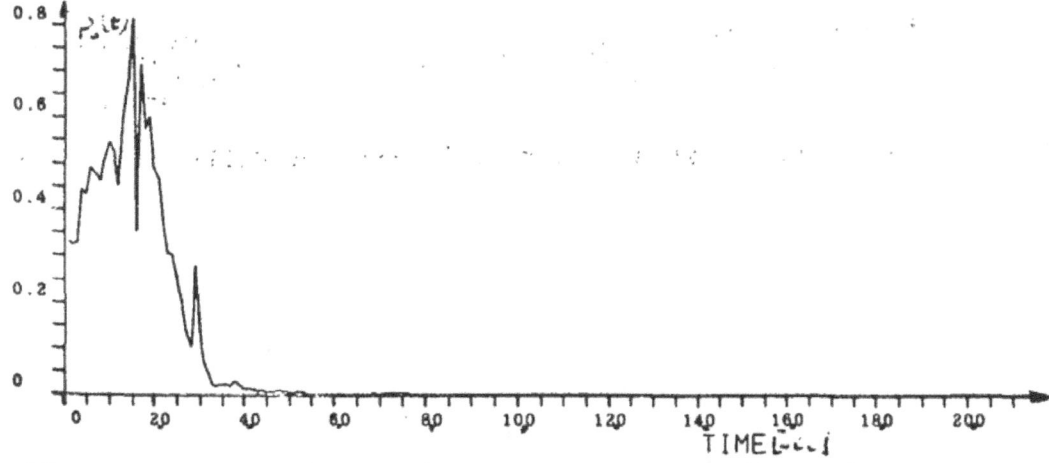

Fig. 6f: Probability $p_3(t)$ of Observing System 3, [Case II]

333

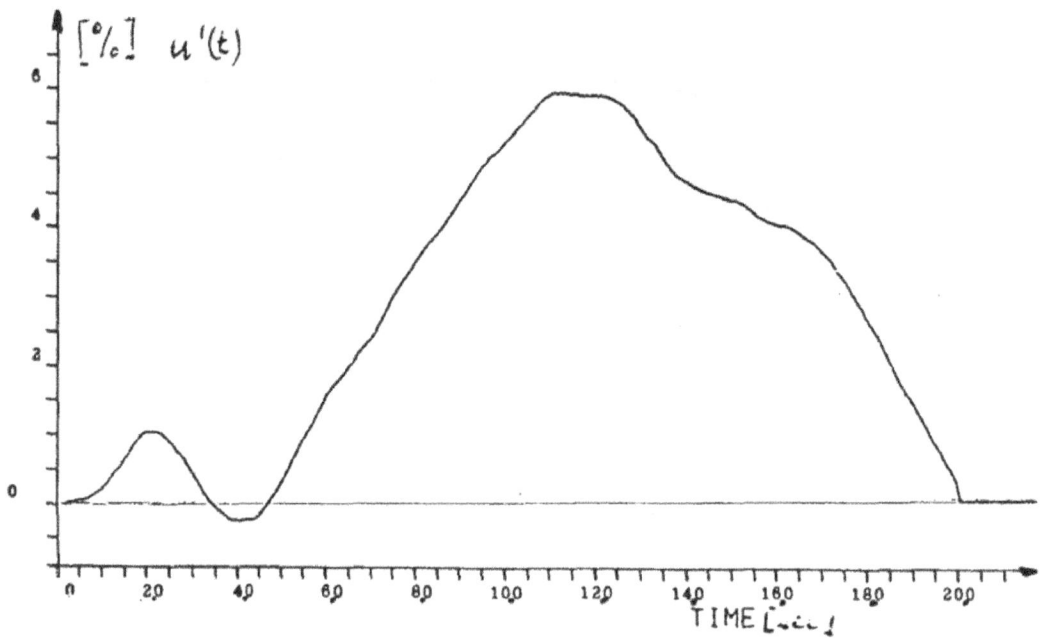

Fig. 7a: Deviation of the Forward Velocity u' Due to Disturbances
[Case III]

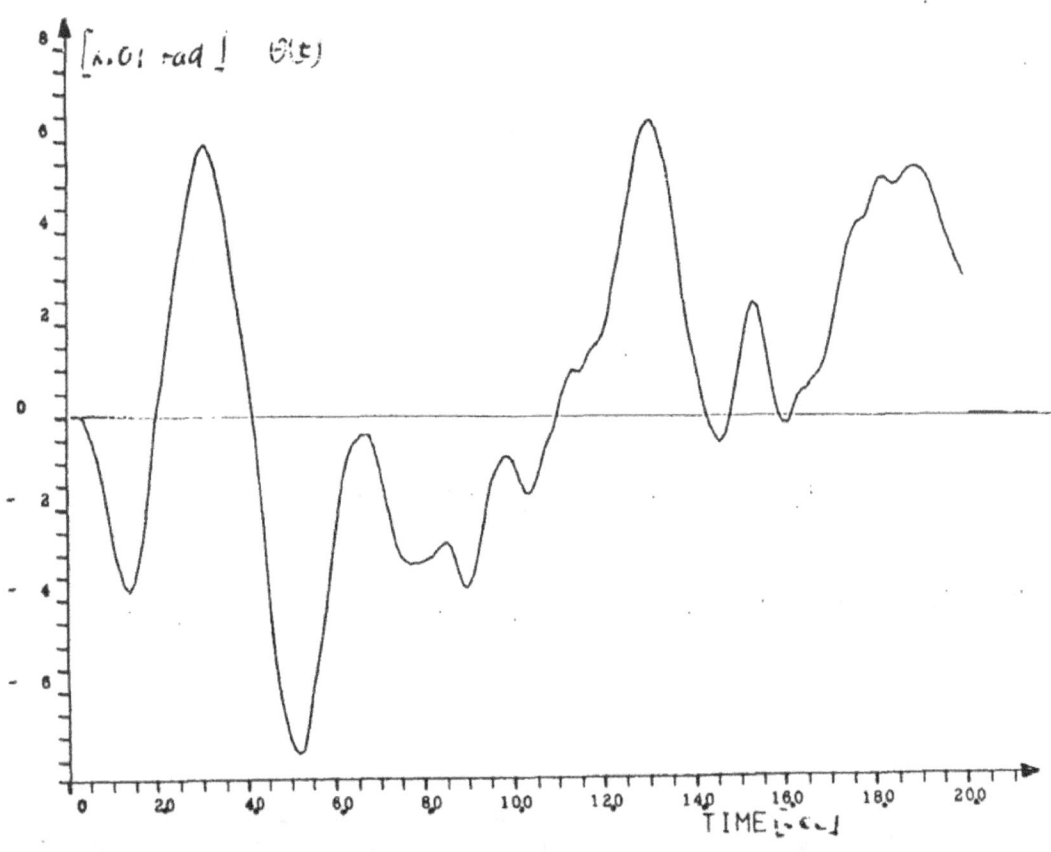

Fig. 7c: Deviation of the Pitch θ, [Case III]

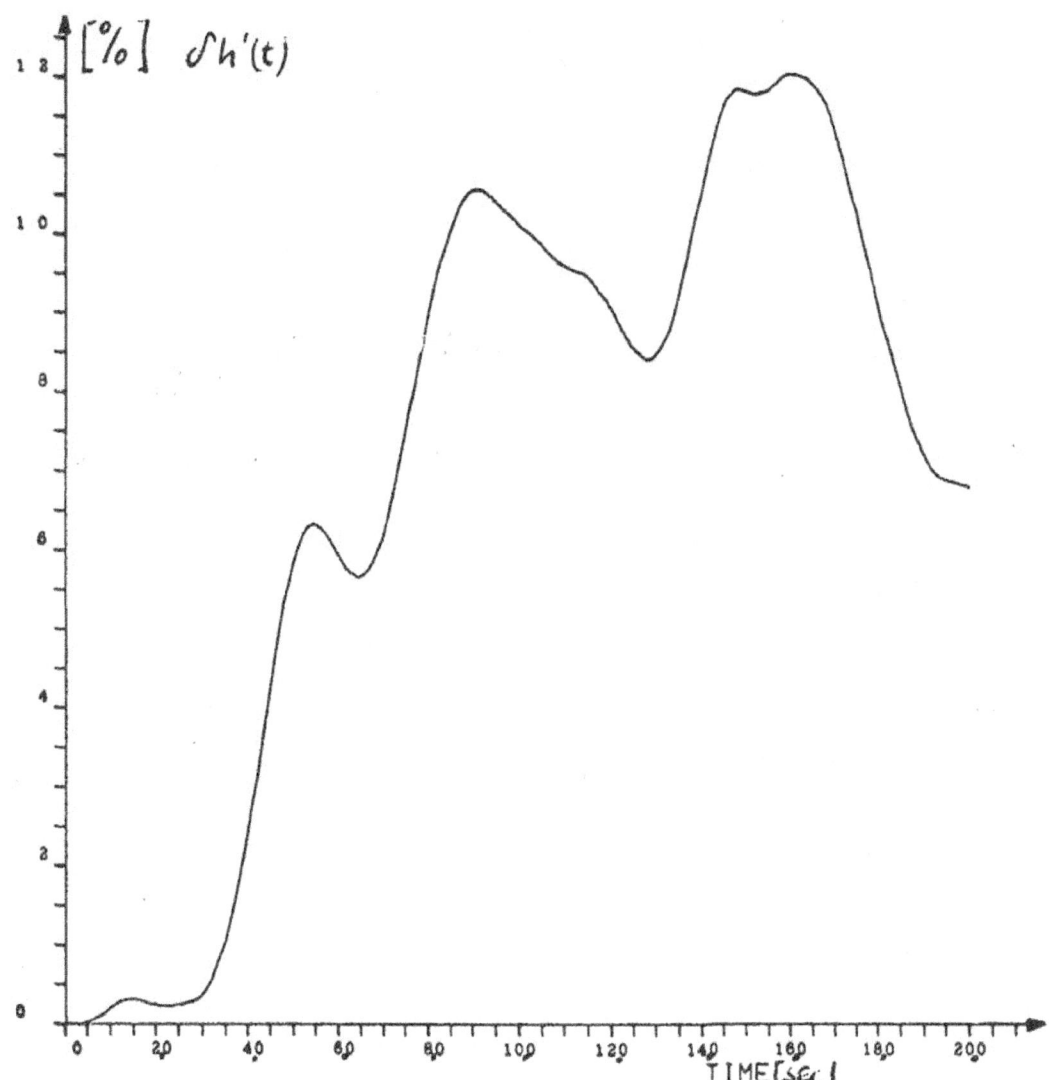

Fig. 7b: Deviation of the Height δh' Above the Glide Path [Case III]

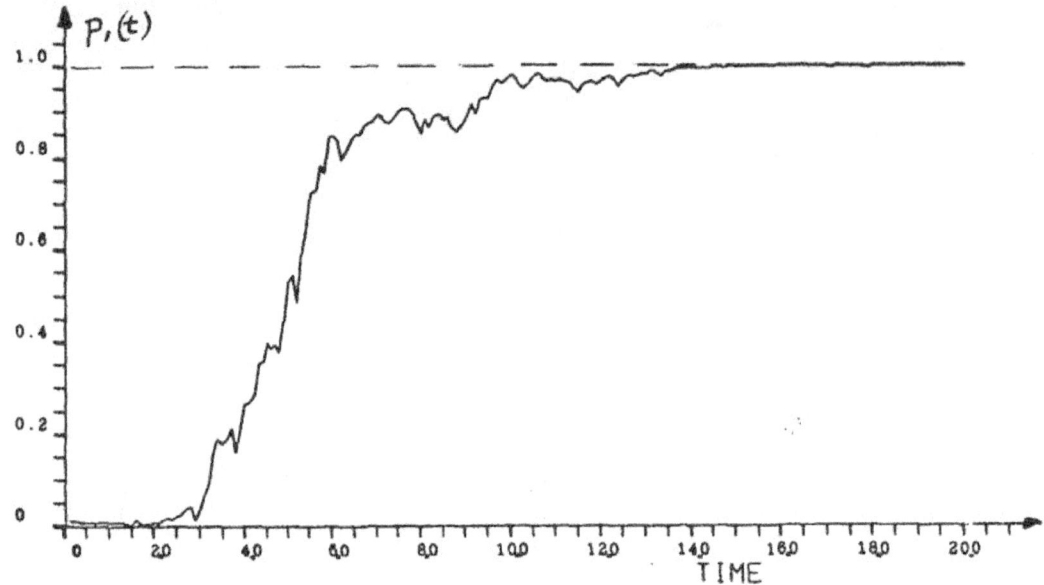

Fig. 7d: Probability $p_1(t)$ of Observing System 1 [Case III]

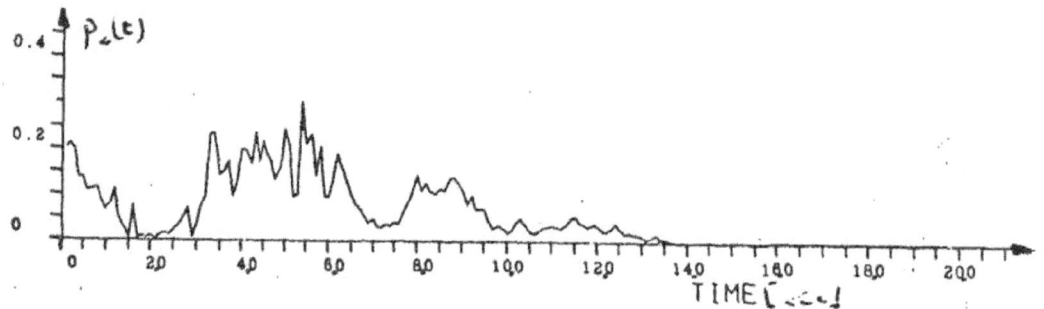

Fig. 7e: Probability $p_2(t)$ of Observing System 2 [Case III]

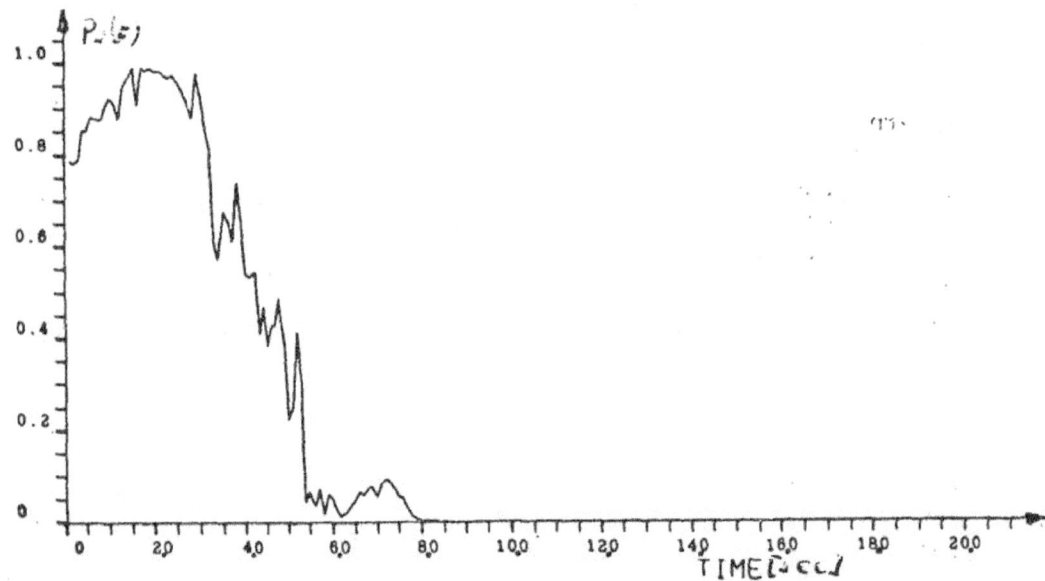

Fig. 7f: Probability $p_3(t)$ of Observing System 3 [Case III]

ESTIMATION OF ELASTIC AIRCRAFT PARAMETERS
USING THE MAXIMUM LIKELIHOOD METHOD
By R. C. Schwanz and Dr. W. R. Wells

Control Criteria Branch
Flight Control Division
Air Force Flight Dynamics Laboratory
Wright-Patterson Air Force Base, Ohio

ABSTRACT

This paper discusses the application of the maximum likelihood method to estimate the aerodynamic parameters of elastic flight vehicles in a symmetric flight condition. In this application, particular attention is directed toward the center of mass, elastic deformation, and sensor equations of motion. It is shown that the two major computational problems to be overcome are the inversion of large-sized matrices and the time-wise integration of a large number of linear, ordinary, differential equations. This method will be verified using the B-52 CCV aircraft flight data.

INTRODUCTION

On modern aircraft, the practical necessity of removing all excessive structural weight, whether thru conventional design practices or active aerodynamic control systems, has resulted in vehicles that are more aeroelastic then previous vehicles with similar operational missions. To a degree, all flight vehicles, including fighter aircraft, are aeroelastic. The degree of aeroelasticity depends upon the particular flight condition-Mach number, dynamic pressure, and mass distribution-at which measurements or observations are made.

In order to minimize the technical risks involved in the design of flexible vehicles, a prototype vehicle is often constructed prior to committing a large amount of resources to a production vehicle. The SST program is an obvious example. The intent of the prototype vehicle is to demonstrate that the design meets all the mission objectives. This demonstration entails flight tests of the prototype to verify the basic design decisions and to isolate any configuration problems that would be objectionable in the production vehicle.

337

An obvious first suggestion in optimizing the flight testing of the prototype is to apply modern parameter estimation methods to determine the important stability and control parameters that affect the handling qualities and flight control of the vehicle. However, most existing parameter estimation methods treat the flight vehicle as a "rigid" structure, thus eliminating the possibility of explicitly identifying important aeroelastic parameters that affect the flight data. At the very best, the existing methods model the vehicle as QUASI STATIC aeroelastic by assuming that the structural distortion occurs with infinite accelerations which produce an instantaneous response to motions of the center of mass of the vehicle. Thus, the structural motion is assumed to have the phase of the body-fixed, axis system motions such as $u, \alpha, \beta, p, q, r,$ et cetera.

If the flight tests of a vehicle indicate a QUASI STATIC behavior, modern estimation methods will determine vehicle parameters that are a product of the aerodynamic stability and control derivatives of the "rigid" vehicle multiplied by an aeroelastic correction factor, e.g.,

$$C_{L_\alpha}^{ELASTIC} = C_{L_\alpha}^{RIGID} \left(C_{L_\alpha}^{ELASTIC} / C_{L_\alpha}^{RIGID} \right)$$

where the correction factor $C_{L_\alpha}^{ELASTIC} / C_{L_\alpha}^{RIGID}$ is analytically determined. The inclusion of even the simplest aeroelastic correction factor complicates the desired design verifications, since the data determined from flight test must then be correlated with wind-tunnel-measured data that have been modified by analytically-determined, QUASI STATIC correction factors. Often the factors will change both the magnitude and the sign of important stability and control parameters.

With the advent of more highly elastic flight vehicles, characterized as Control Configured Vehicles (CCV), the determination of the aeroelastic stability and control parameters increase in importance. These vehicles are designed such that a flight control system (FCS) solves structural and dynamic stability design problems, e.g., handling quality, ride quality, reduced static stability, maneuver loadability, structural fatigue, and flutter. Of course, with the dependence of the vehicle on FCS, highly accurate design data sets are required such that the gains of the FCS are near their final value on the prototype [1,2]. It is difficult to envision the conventional approach of "gain twiddling" on a CCV FCS, since the primary structure is not designed to tolerate loads and dynamics with the FCS inoperative or improperly operative.

Thus, the purpose of this paper is to examine the problems involved in modelling complex aeroelastic flight vehicles such that modern parameter estimation methods may be applied. The commonly available mathematical formulations of the dynamic elastic vehicles as discussed by Schwanz [2] will be reviewed and the approximate formulation, MODAL TRUNCATION, will be selected for further consideration. Previously neglected structural motions measured by sensors will be modelled rather than "filtered" as in present day methods. Then the equations of motion and accelerometer equations will be combined to a state vector form amenable to the maximum likelihood method as discussed by Wells [3]. A brief discussion of the problems involved in mechanization of the computational algorithm ends the paper. More detailed discussions of the main points in the paper may be found in another article by the authors [4]. As the

338

computer implementation of the algorithm has just begun, numerical results are presently unavailable.

In this discussion of the maximum likelihood method applied to the elastic aircraft, numerous simplifications are assumed in order to illustrate the effects of aeroelasticity. The primary reason for these simplifications is to first solve the linearized problems and then to proceed to more complex non-linear problems as experience is accrued. Thus, this analysis herein serves as a test case for subsequent non-linear analyses.

The linearized equations discussed in this paper describe the so-called symmetric, small disturbance motions of an elastic flight vehicle. Further, the problem is restricted to the initial conditions of steady, non-rotating flight, with the wings level, and with the relative velocity vector of the center of mass parallel to the flat earth. In addition, the following simplifications and assumptions are employed:

The thrust is assumed to be constant during the perturbation motion and its magnitude determined solely by the initial conditions of flight.

All motions, body-fixed axis system and elastic deformation , are of small perturbation magnitude and are of the same order.

The sensor locations are assumed to be precisely prescribed for some aircraft shape and the signals of the sensors are assumed to be free of bias.

The generalized stiffness, mass, and damping as well as the total mass and inertia of the aircraft are assumed to be known (within some tolerance) by previous measurements or calculations.

The elastic flight vehicle is adequately represented by lumped masses related structurally and aerodynamically by finite element theory (Figure 1).

Process noise or turbulence is assumed negligible.

The extension of this analysis to the coupled, non-linear ordinary differential equations of motion, to arbitrary initial conditions, and to the inclusion of generalized mass, stiffness and damping as unknown parameters is analytically possible. However, practical computer considerations preclude this extension at the present time.

STATE EQUATION

The state equation to be used in the maximum likelihood method for elastic aircraft describes the body-fixed axis system motions, e.g., $u, w,$ $\theta, \dot{\theta}$, the elastic deformations relative to the body-fixed axis, e.g., $\phi_i u_i$ and $\phi_i \dot{u}_i$, and the motions of accelerometers, e.g., a . In the particular case discussed, the axis system motions are expressed in terms of the non-inertial body-fixed axis base vectors[5], ϕ_i are the retained invacuum mode shapes, and u_i are the generalized motions. In most analyses, i ranges from 1 to r and, usually, $3 \leq r \leq 20$. The final state equation, equation (8), provides the relationship between the parameters that are to be identified using the maximum likelihood method.

Equations of Motion

The equations that describe the motion of the elastic flight vehicle are complex and their derivation is lengthy. Several publications that are of most interest are those by Bisplinghoff and Ashley[6], Milne[7], Schwendler, McNeal[8] and Dusto, et.al.[9]. The notation and conventions to be followed in this paper are those of Schwanz[2].

As discussed in reference 2, the many formulations of the linearized equations of motion of elastic aircraft commonly found in the literature and employed in aeroelastic stability and control analysis may be grouped into six categories:

QUASI STATIC - The motions of the structure are assumed to be in phase with the rigid body motions: elastic motion acceleration is infinite. The method is used primarily for handling quality and reduced static stability FCS design for elastic aircraft with wide frequency separation between the axis system and elastic motions.

EXACT - The motion of the structure is determined by the eigenvalue (root) and eigenvector (mode shape) solutions of the equations of motion for the elastic aircraft. The mode shape coordinates contain complex numbers. The accuracy of the solution is limited by the existing computerized routines that calculate the complex number eigenvalues and eigenvectors.

MODAL SUBSTITUTION - The motions of the structure are assumed to be related to the orthogonal, invacuum eigenvectors (mode shapes). All eigenvectors contain only real numbers.

RESIDUAL STIFFNESS - The mode shapes representing the elastic motion in the MODAL SUBSTITUTION formulation are separated into "retained" and "deleted" modes. The deleted modes are represented in the dynamic stability analysis as quasi static aeroelastic corrections, using a correction factor related to the deleted modes and the stiffness of the "free-free" structure.

RESIDUAL FLEXIBILITY - Similar to the RESIDUAL STIFFNESS formulation, except the quasi static aeroelastic correction is related to the retained modes and the flexibility of the free-free structure.

MODAL TRUNCATION - The deleted modes of the RESIDUAL FLEXIBILITY formulation are not represented by any correction factor. This is the most common dynamic aeroelastic formulation reported in the literature.

340

A close examination of the six approaches indicates that the MODAL TRUNCATION formulation is the easiest to implement when determining the aerodynamic parameters of an elastic aircraft. The reasons are that the MODAL TRUNCATION formulation:

  Includes a significant amount of the structural dynamics that affect the FCS design.

  Explicitly employs "rigid" airplane stability and control derivatives that may be estimated and directly compared to wind tunnel measurements.

  Is the most commonly employed dynamic aeroelastic formulation presently used in design.

  Is the least complex of all the dynamic aeroelastic formulations that are identified.

The other five formulations each have an undesirable complication. As mentioned previously, the QUASI STATIC formulation does not properly represent all the structural dynamics important to the FCS and, in addition, requires precise analytical method to calculate the aeroelastic stability and control correction factors applied to the "rigid" airplane data. The EXACT and MODAL SUBSTITUTION formulations are more precise than the MODAL TRUNCATION formulation, but they consist of an enormous number of equations in most analyses and, thus, their selection would lead to an estimation procedure that would require a larger computer than presently available. The RESIDUAL STIFFNESS and RESIDUAL FLEXIBILITY formulations have the same number of equations as the MODAL TRUNCATION formulation, but they have more complex aerodynamic terms due to their representation of the static effects of the modal degrees of freedom that have been deleted dynamically.

The MODAL TRUNCATION formulation of the equations of symmetric and antisymmetric motion discussed in reference 2, has the following form:

$$
\begin{bmatrix} M & 0 & 0 \\ 0 & I_n & 0 \\ 0 & 0 & m \end{bmatrix}
\begin{Bmatrix} \dot{v} \\ \ddot{r}_{op} \\ \ddot{u} \end{Bmatrix}
+
\begin{bmatrix} M\,M^{(1)}_{12} & 0 \\ 0 & 0 \\ 0 & d \end{bmatrix}
\begin{Bmatrix} \dot{r}_{op} \\ \dot{u} \end{Bmatrix}
+
\begin{bmatrix} M\,M^{(2)}_{12} & 0 \\ 0 & 0 \\ 0 & k \end{bmatrix}
\begin{Bmatrix} v_{op} \\ u \end{Bmatrix}
=
\begin{Bmatrix} \delta_v^T f \\ \delta_r^T f \\ \phi^T f_i \end{Bmatrix} \quad (1)
$$

The definition of the matrices in equation (1) are:

$v^T = \lfloor u \ v \ w \rfloor$, the axis system perturbation velocities.

$r_{op}^T = \lfloor \phi \ \theta \ \psi \rfloor$, the axis system perturbation rotations.

$M = \begin{bmatrix} M & & \\ & M & \\ & & M \end{bmatrix}$, the mass of the aircraft.

$$I_n = \begin{bmatrix} I_{xx} & 0 & -I_{xz} \\ 0 & I_{yy} & 0 \\ -I_{xz} & 0 & I_{zz} \end{bmatrix}$$, the inertia of the aircraft.

$$M_{12}^{(1)} = \begin{bmatrix} 0 & V_3 & -V_2 \\ -V_3 & 0 & V_1 \\ V_2 & -V_1 & 0 \end{bmatrix}$$, the initial velocity of the center of mass of the aircraft.

$$M_{12}^{(2)} = \begin{bmatrix} 0 & g & 0 \\ -g & 0 & 0 \\ 0 & 0 & 0 \end{bmatrix}$$, the perturbation gravitational forces.

$$M = \begin{bmatrix} m_1 & & \\ & \ddots & \\ & & m_j \end{bmatrix}$$, the generalized mass.

$$d = \begin{bmatrix} d_1 & & \\ & \ddots & \\ & & d_j \end{bmatrix}$$, the generalized damping.

$$R = \begin{bmatrix} R_1 & & \\ & \ddots & \\ & & R_j \end{bmatrix}$$, the generalized stiffness.

$$\Phi = \begin{bmatrix} \Phi_{1,1} & \cdots\cdots & \Phi_{1,j} \\ \vdots & & \\ \Phi_{6n,1} & \cdots\cdots & \Phi_{6n,j} \end{bmatrix}$$, the retained invacuum modes.

$$\{\bar{\Phi}_v^T f\}^T = \lfloor f_x \; f_y \; f_z \rfloor$$, the total aerodynamic forces acting on the aircraft.

$$\left\{\bar{\Phi}_r^T f\right\}^T = \lfloor M_x \; M_y \; M_z \rfloor, \text{ the total aerodynamic moments acting on the}$$
aircraft.

$$\left\{\phi^T f_i\right\}^T = \lfloor f_1 \; f_2 \cdots f_j \rfloor, \text{ the generalized aerodynamic forces acting}$$
on each structural mode.

$$u^T = \lfloor u_1 \; u_2 \cdots u_j \rfloor, \text{ the time-dependent amplitude of each}$$
generalized structural motion.

The symmetric degrees of freedom of interest in this analysis may be
separated from equations (1) by rewritting the first, third and fifth equations
and the equations for the symmetric modal degrees of freedom (defined to be
number of the j number of retained invacuum modes). The specific forms of
these equations in terms of Etkin's stability axis definitions, including the
expansion of f_x, f_y, f_z and $f_1, f_2, \cdots f_r$ in terms of the non-dimensional
force coefficients, are:

Axis System Motion

$$M\dot{u} + Mg\theta = \tfrac{1}{2}\rho V_1 (2C_{x_0} + C_{x_u})u + \tfrac{1}{2}\rho S V_1 C_{x_\alpha} w$$

$$+ \tfrac{1}{4}\rho S V_1 \bar{c}\, C_{x_q} \dot{\theta} + \tfrac{1}{2}\rho S \bar{c}\, C_{x_{\dot{u}}} \dot{u} + \tfrac{1}{2}\rho S \bar{c}\, C_{x_{\dot{\alpha}}} \dot{w} + \tfrac{1}{4}\rho \bar{c}^2 S C_{x_{\ddot{q}}} \ddot{\theta}$$

$$+ \sum_{i=1}^{r} \left(\frac{\rho V_1^2 S}{2\bar{c}} C_{x_{u_i}} u_i + \tfrac{1}{2}\rho V_1 S C_{x_{\dot{u}_i}} \dot{u}_i + \tfrac{1}{2}\rho S \bar{c}\, C_{x_{\ddot{u}_i}} \ddot{u}_i \right)$$

$$+ \sum_{i=1}^{c} \left(\tfrac{1}{2}\rho V_1^2 S C_{x_{\delta_i}} \delta_i + \tfrac{1}{2}\rho V_1 S \bar{c}\, C_{x_{\dot{\delta}_i}} \dot{\delta}_i + \tfrac{1}{2}\rho S \bar{c}^2 C_{x_{\ddot{\delta}_i}} \ddot{\delta}_i \right) \quad (2a)$$

343

$$-MV_1\dot{\theta} + M\dot{w} = \tfrac{1}{2}\rho S V_1 (2C_{z_0} + C_{z_u})u + \tfrac{1}{2}\rho S V_1 C_{z_\alpha}w$$

$$+ \tfrac{1}{4}\rho S V_1 \bar{c}\, C_{z_q}\dot{\theta} + \tfrac{1}{2}\rho S \bar{c}\, C_{z\dot{u}}\dot{u} + \tfrac{1}{2}\rho S \bar{c}\, C_{z\dot{\alpha}}\dot{w} + \tfrac{1}{4}\rho\bar{c}^2 C_{z\dot{q}}\ddot{\theta}$$

$$+ \sum_{i=1}^{r}\left(\tfrac{1}{2}\rho\frac{V_1^2 S}{\bar{c}}C_{z u_i}u_i + \tfrac{1}{2}\rho V_1 S C_{z\dot{u}_i}\dot{u}_i + \tfrac{1}{2}\rho S \bar{c}\, C_{z\ddot{u}_i}\ddot{u}_i\right)$$

$$+ \sum_{i=1}^{c}\left(\tfrac{1}{2}\rho V_1^2 S \bar{c}\, C_{m_{\delta_i}}\delta_i + \tfrac{1}{2}\rho V_1 S \bar{c}^2 C_{m\dot{\delta}_i}\dot{\delta}_i + \tfrac{1}{2}\rho S \bar{c}^3 C_{m\ddot{\delta}_i}\ddot{\delta}_i\right) \tag{2b}$$

$$I_{xx}\ddot{\theta} = \tfrac{1}{4}\rho\bar{c}^3 S C_{m\dot{q}}\ddot{\theta} + \tfrac{1}{4}\rho S V_1 \bar{c}^2 C_{m_q}\dot{\theta} + \sum_{i=1}^{r}\tfrac{1}{2}\rho S \bar{c}^2 C_{m\ddot{u}_i}\ddot{u}_i$$

$$+ \tfrac{1}{2}\rho S \bar{c}^2 C_{m\dot{u}}\dot{u} + \tfrac{1}{2}\rho S \bar{c}^2 C_{m\dot{\alpha}}\dot{w} + \sum_{i=1}^{r}\tfrac{1}{2}\rho V_1 S \bar{c}\, C_{m\dot{u}_i}\dot{u}_i$$

$$+ \tfrac{1}{2}\rho V_1 S \bar{c}\, C_{m_u}u + \tfrac{1}{2}\rho S V_1 \bar{c}\, C_{m_\alpha}w + \sum_{i=1}^{r}\tfrac{1}{2}\rho V_1^2 S C_{m u_i}u_i$$

$$+ \sum_{i=1}^{c}\left(\tfrac{1}{2}\rho V_1^2 S \bar{c}\, C_{m_{\delta_i}}\delta_i + \tfrac{1}{2}\rho V_1 S \bar{c}^2 C_{m\dot{\delta}_i}\dot{\delta}_i + \tfrac{1}{2}\rho S \bar{c}^3 C_{m\ddot{\delta}_i}\ddot{\delta}_i\right) \tag{2c}$$

Elastic Deformation

$$K u_i + d_i \dot{u}_i + m_i \ddot{u}_i = \tfrac{1}{2}\rho V_1 S C_{u_i u}u + \tfrac{1}{2}\rho V_1 S C_{u_i \alpha}w$$

$$+ \tfrac{1}{4}\rho V_1 S \bar{c}\, C_{u_i q}\dot{\theta} + \tfrac{1}{2}\rho S \bar{c}\, C_{u_i \dot{u}}\dot{u} + \tfrac{1}{2}\rho S \bar{c}\, C_{u_i \dot{\alpha}}\dot{w} + \tfrac{1}{4}\rho\bar{c}^2 S C_{u_i \dot{q}}\ddot{\theta}$$

$$+ \sum_{j=1}^{r}\left(\rho\frac{V_1^2 S}{2\bar{c}}C_{u_i u_j}u_j + \tfrac{1}{2}\rho V_1 S C_{u_i \dot{u}_j}\dot{u}_j + \tfrac{1}{2}\rho S \bar{c}\, C_{u_i \ddot{u}_j}\ddot{u}_j\right)$$

$$+ \sum_{j=1}^{c}\left(\tfrac{1}{2}\rho V_1^2 S C_{u_i \delta_j}\delta_j + \tfrac{1}{2}\rho V_1 S \bar{c}\, C_{u_i \dot{\delta}_j}\dot{\delta}_j + \tfrac{1}{2}\rho S \bar{c}^2 C_{u_i \ddot{\delta}_j}\ddot{\delta}_j\right) \tag{2d}$$

Equations (2) are characteristic of those employed in the B-52 LAMS[10], the B-52 CCV[11], the F-4 Flutter FCS[12], and the B-1 Ride Quality FCS[13] studies. It should be noted at this point, that equations (2) differ from the usual structural dynamic equations used by flutter specialists due to the frequency-independent, non-dimensional, aerodynamic force and moment coefficients. The reason for the uncomplicated, frequency-independent formulation is that the development of FCS, particularly those characterized as a CCV-type, requires an order of magnitude greater number of analyses of linear and non-linear equations of motion than presently analyzed during flutter calculations. Thus, specialized formulations of the equations of motion and the steady and unsteady aerodynamics are used that are uniquely suited to stability and control and flight control design work and that are generally much faster to solve than those employed in the flutter design.

Next, equations (2) are rewritten such that all derivatives of the motions are to the left of the equals sign. In addition, a vector of motions, ζ, and a control vector, δ, are defined :

$$\zeta^T = \lfloor u \ w \ \theta \ \dot{\theta} \ u_1 \ u_2 \cdots u_r \ \dot{u}_1 \ \dot{u}_2 \cdots \dot{u}_r \rfloor$$

$$\delta^T = \lfloor \delta_1 \ \dot{\delta}_1 \ \ddot{\delta}_1 \cdots \cdots \delta_c \ \dot{\delta}_c \ \ddot{\delta}_c \rfloor$$

In terms of ζ and δ, the equations of motion then become:

$$F\dot{\zeta} = B\zeta + G\delta \tag{3}$$

or alternately

$$\dot{\zeta} = F^{-1}B\zeta + F^{-1}G\delta \tag{4}$$

where F is a square matrix of aerodynamic coefficients and specified inertia terms, B is a square matrix of aerodynamic coefficients, specified inertia, damping and stiffness terms, and G is a rectangular matrix of aerodynamic coefficients. For instance:

$$F_{11} = M - \tfrac{1}{2}\rho S \bar{c} \, C_{x\dot{u}}$$

$$B_{11} = \tfrac{1}{2}\rho S V_1 (2C_{x_0} + C_{xu})$$

$$G_{11} = \tfrac{1}{2}\rho S V_1^2 C_{x\delta_1}$$

The objective is to determine the unknown aerodynamic coefficients in F, B, and G by analyzing time histories from the flight test of the vehicle. No process noise, e.g., turbulence induced forces and moments, are included in this analysis, although, at this point in the study, the removal of the noise via the Kalman filter as in "rigid" airplane analyses[14] seems plausible.

Sensor Equations

There are four types of sensors commonly employed in the design of the flight control systems of aircraft. These are accelerometers, rate gyros, inertial platforms, and air data sensors. Several authors have discussed the appropriate equations of motion for those sensors that are used on "rigid" aircraft[3, 15]. Very little work in sensor representation has been done for elastic aircraft; most notable is the work by Dornfeld and Schaeffer[16] on the FLEXSTAB/CCVMOD Computer Program System[17].

It is the specific intent of this section to develop the equations that describe the motions of accelerometers placed on the aircraft. In this process, equations for the four major types of sensors are developed, prior to specializing only to accelerometers. The more general equations are reserved for future analyses, should the accelerometers prove to inadequately describe the elastic aircraft in the maximum likelihood method.

Prior to developing the equations of the sensors, a discussion of the effects of aeroelasticity on the sensor signals is required, since most previous analyses of flight test data assume that the effects of elastic deformation on sensor signals are small. In practice, this assumption is valid when the sensors have been placed on the fuselage where structural deformations are negligible. In the cases in which elastic deformations are measured by the sensors, a notch filter or washout filter is employed to "purify" the signal such that it contains only "rigid" airplane motions.

Since some of the sensors to be used in the maximum likelihood method discussed herein, are placed at the extremes of wings, tails, and fuselages to measure elastic distortions, the reorientation of the sensor axis from "jig shape" or "runway shape" to the "reference shape" may be important, viz., the change in the wing dihedral angle that occurs on the B-52 and U-2. The accepted definition of these elastic flight vehicle shapes should be recalled:

Jig: the shape of the vehicle supported by the construction jigs.

Runway: the shape of the vehicle on the runway at zero dyanmic pressure and at some prescribed fuel and payload mass distribution.

Reference: the shape of the vehicle during the reference flight condition about which perturbation elastic deformation occurs.

Cruise: the shape of the vehicle at its design mission operating condition; trim drag should be zero. Thus, the "reference shape" at the cruise design point.

A sketch of the jig, runway, and reference shapes in Figure 2 indicates the importance of precisely determining the initial orientation of the sensor measurement axis and the reorientation of the sensor axis for the reference flight condition.

In the most general cases, the instantaneous reorientation of the sensor axis may be described by a time dependent, direction cosine transformation between the body-fixed mean axis and the sensor axis. For the case considered herein, the reorientation is divided into (1) an initial time-independent reorientation from the installation shape to the reference shape and (2) a time-dependent reorientation from the reference shape to the instantaneous shape defined by the vectors γ and δ in equations (4). The time-dependent reorientation is assumed small compared to the initial reorientation. Thus, the direction cosine transform is time-independent and is written as:

$$\{y_i\} = [T_i]\{\Delta_i\} \tag{5}$$

where $\{\lambda_i\}$ is a column vector of the ith sensor signal, expressed in terms of the reference shape, sensor axis base vectors.

$[T_i]$ is a matrix of the direction cosine of the angles between the base vectors of the mean axis and the sensor axis system.

$\{y_i\}$ is a column vector of the ith sensor signals, expressed in terms of the mean axis base vectors consistent with equations (4).

The equations of motion for $\{y\}$, the air data system, the gyros, and the accelerometers, that are consistent with the initial conditions of the motion, follow from reference 4:

$$\{y\} = \begin{bmatrix} 0 & \bar{\Phi}_v & 0 & \bar{\Phi}_r & 0 \\ 0 & 0 & 0 & 0 & \bar{\Phi}_v \\ 0 & 0 & 0 & \bar{\Phi}_v & 0 \\ \bar{\Phi}_v & 0 & \bar{\Phi}_r & \bar{\Phi}_v M_{12}^{(1)} & \bar{\Phi}_v M_{12}^{(2)} \end{bmatrix} \begin{Bmatrix} \dot{v} \\ v \\ \ddot{r}_{op} \\ \dot{r}_{op} \\ r_{op} \end{Bmatrix} + \begin{Bmatrix} \dot{P}_i \\ \theta_{e_i} + \theta_{P_i} \\ \dot{\theta}_{e_i} + \dot{\theta}_{P_i} \\ \ddot{P}_i \end{Bmatrix} \quad (6)$$

where $\{y\}^T = \lfloor y^T \text{air data} \; y^T \text{Euler angles} \; y^T \text{Rate Gyro} \; y^T \text{accelerometer} \rfloor$

θ_{P_i} is a vector of rotations due to the translational structural distortions.

P_i is a vector of the translational elastic distortions.

θ_{e_i} is a vector of rotational elastic distortions.

$$\bar{\Phi}_v = \begin{bmatrix} \begin{bmatrix} 1 & 0 & 0 \\ 0 & 1 & 0 \\ 0 & 0 & 1 \end{bmatrix}_{j=1} \\ \vdots \\ \begin{bmatrix} 1 & 0 & 0 \\ 0 & 1 & 0 \\ 0 & 0 & 1 \end{bmatrix}_{\substack{Number \\ Sensors}} \end{bmatrix} \qquad \bar{\Phi}_r = \begin{bmatrix} \begin{bmatrix} 0 & P_{3j} & -P_{2j} \\ -P_{3j} & 0 & P_{1j} \\ P_{2j} & -P_{1j} & 0 \end{bmatrix}_{j=1} \\ \vdots \\ \begin{bmatrix} 0 & P_{3j} & -P_{2j} \\ -P_{3j} & 0 & P_{1j} \\ P_{2j} & -P_{1j} & 0 \end{bmatrix}_{\substack{Number \\ Sensors}} \end{bmatrix}$$

In the expression for $\bar{\Phi}_r$, P_{1j}, P_{2j}, and P_{3j} are the x, y, and z distances of the j th sensor from the center of rotation (assumed to be the center of mass).

As mentioned previously, only the accelerometer equations will be selected to augment the vector, z, and to supplement the measurements of the state. For this reason it is convenient to define and augment state vector, x, as follows:

$$x^T = \lfloor z^T \; a^T \rfloor$$

where the vector, z, is defined by equation (4) and the vector, a, can be developed from the fourth row of equations (6) by specializing to symmetric flight:

$$a^T = \lfloor a_{x_{cm}} \; a_{z_{cm}} \; a_{x_1} \; a_{z_1} \quad a_{x_{\ell-1}} \; a_{z_{\ell-1}} \rfloor$$

347

$$a_{xcm} = \dot{u} + P_{3j}\ddot{\theta} + g\theta \tag{7a}$$

$$a_{zcm} = \dot{w} - P_{ij}\ddot{\theta} - V_i\dot{\theta} \tag{7b}$$

In terms of invacuum modes, ϕ, and the generalized coordinates, u, the a_{xj} and a_{zj} may be expressed as:

$$a_{xj} = \dot{u} + P_{3j}\ddot{\theta} + g\theta + \sum_{i=1}^{r} \phi_{xji}\ddot{u}_i \tag{7c}$$

$$a_{zj} = \dot{w} - P_{ij}\ddot{\theta} - V_i\dot{\theta} + \sum_{i=1}^{r} \phi_{zji}\ddot{u}_i \tag{7d}$$

Here, ϕ_{xji} and ϕ_{zji} are directly related to $\phi(x,y,z)$ in equation (1), provided the accelerometers are "placed" upon a lumped mass. In the event the accelerometer are not placed upon a lumped mass, the individual elements in ϕ may be interpolated to give ϕ_{xji} and ϕ_{zji}. The assumption in using only accelerometers to augment $\underline{3}$ is that the structure experiences structural rotational degrees of freedom, θ_{ei}, much smaller in magnitude than the translational degrees of freedom, P_i. In the event this assumption is incorrect, rate gyro equations may be required as additional equations to augment $\underline{3}$.

Finally, the state equation for x may be written by combining equations (4) and (7):

$$\dot{x}_i = C_{ij}x_j + D_{ik}\delta_k \qquad i,j = 1, 4+2r, \quad k = 1, 3c$$

$$x_i = e_i \qquad\qquad i = 4+2r+1, \ 4+2r+2\ell$$

Here, x is a vector $[(4+2r) + (2\ell)]$ X1 in size, δ is defined as a vector $3c \times 1$ in size and C and D are:

$$C = \begin{bmatrix} F^{-1}B & | & 0 \end{bmatrix} \qquad D = \begin{bmatrix} F^{-1}G \end{bmatrix}$$

where e_i is defined in equations (7).

MAXIMUM LIKELIHOOD METHOD

The methods used to estimate the stability and control parameters of rigid aircraft may be characterized [18] as "Equation Error", "Output Error", and "Advanced Non-linear". The maximum likelihood method falls into the latter characterization along with the extended Kalman filter [19]. The advantages of the advanced methods are that they can be applied to problems which contain both process noise (turbulence-induced aircraft motions) and instrument noise.

The maximum likelihood method has been selected for the analysis presented herein due to its prior success on "rigid" aircraft. In addition, the computer demands of the method serve to identify the areas of major flight testing and computational difficulties prior to an expansion of the work effort to other estimator methods. And finally, the maximum likelihood formulation reduces to a form of the Output Error method in the case of negligible process noise.

The computational algorithm for the maximum likelihood method is developed along the lines presented by Wells[3]. Other discussions of the method may be found in the works of Mehra, Stepner, and Tyler[14], Steinmetz and Parrish[20], and Suit[21].

As a first step in this method, the parameters to be estimated in Equations (4) are identified. These aerodynamic parameters, first defined in equations (2), are ordered to form the column vector of parameters, P:

$$P^T = \lfloor C_{x_0},\ C_{x_u},\ C_{x_\alpha},\ C_{x_q},\ C_{x u_i}\,(i=1,\cdots r),\ C_{x \delta_i}\,(i=1,\cdots c),\ C_{x\dot{u}},$$

$$C_{x\dot{\alpha}},\ C_{x\dot{q}},\ C_{x\dot{u}_i}\,(i=1,\cdots r),\ C_{x\dot{\delta}_i}\,(i=1,\cdots c),\ C_{x\ddot{u}_i}\,(i=1,\cdots r),\ C_{x\ddot{\delta}_i}\,(i=1,\cdots c),$$

$$C_{z_0},\ C_{z_u},\ C_{z_\alpha},\ C_{z_q},\ C_{z u_i}\,(i=1,\cdots r),\ C_{z\delta_i}\,(i=1,\cdots c),\ C_{z\dot{u}},\ C_{z\dot{\alpha}},\ C_{z\dot{q}},$$

$$C_{z\dot{u}_i}\,(i=1,\cdots r),\ C_{z\dot{\delta}_i}\,(i=1,\cdots c),\ C_{z\ddot{u}_i}\,(i=1,\cdots r),\ C_{z\ddot{\delta}_i}\,(i=1,\cdots c),\ C_{m_\alpha},\ C_{m_\alpha},\ C_{m_q},$$

$$C_{m u_i}\,(i=1,\cdots r),\ C_{m\delta_i}\,(i=1,\cdots c),\ C_{m\dot{\alpha}},\ C_{m\dot{\alpha}},\ C_{m\dot{q}},\ C_{m\dot{u}_i}\,(i=1,\cdots r),\ C_{m\dot{\delta}_i}\,(i=1,\cdots c),$$

$$C_{m\ddot{u}_i}\,(i=1,\cdots r),\ C_{m\ddot{\delta}_i}\,(i=1,\cdots c),\ C_{u_i \alpha}\,(i=1,\cdots r),\ C_{u_i \alpha}\,(i=1,\cdots r),\ C_{u_i q}\,(i=1,\cdots r),$$

$$C_{u_i u_j}\,(i,j=1,\cdots r),\ C_{u_i \delta_j}\,(i=1,\cdots r, j=1,\cdots c),\ C_{u_i \dot{\alpha}}\,(i=1,\cdots r),\ C_{\alpha_i \dot{\alpha}}\,(i=1,\cdots r),\qquad (9)$$

$$C_{\alpha_i \dot{q}}\,(i=1,\cdots r),\ C_{\alpha_i \dot{\delta}_i}\,(i=1,\cdots r, j=1,\cdots c),\ C_{u_i \ddot{u}_j}\,(i,j=1,\cdots r),\ C_{u_i \ddot{\delta}_j}\,(i=1,\cdots r, j=1,\cdots c) \rfloor$$

m is the number of parameters, $3r^2 + 3rc + 9c + 15r + 20$.

r is the number of retained invacuum modes.
c is the number of control surfaces.

The elements of P are the familiar stability and control derivatives, plus some unconventional aeroelastic derivatives that experience has shown to be important in aeroelastic stability and control analyses.

Once the parameter vector is defined, the remainder of the algorithm follows in six steps. Each of the steps is the result of extensive analytical analysis that will only be summarized herein:

1. An initial estimate of P, defined to be P°, is formed using the data from experiment and analytical analyses. The values assigned P° are used to calculate $J(P^\circ)$, the performance function:

$$J(P^\circ) = \det\left\{ \frac{1}{N} \sum_{i=1}^{N} \nu(t_i)\,\nu(t_i)^T \right\} \qquad (10)$$

Here, $\nu(t_i)$ is the innovation sequence:

$$\nu(t_i) = y_s(t_i) - H\,x(P^\circ, t_i) \qquad (11)$$

and $y_s(t_i)$ are the measurements:

$$y_s(t_i) = T\Delta(t_i) = T H x(t_i) + T \eta_2(t_i) \qquad (12)$$

where T is the general cosine transformation matrix for the sensors.

$$H = \begin{bmatrix} H_{11} & H_{12} \\ H_{21} & H_{22} \end{bmatrix}$$

$H_{11} = I, \quad 4 \times 4$ $\qquad\qquad H_{21} = 0, \quad 2\ell \times (4+2r)$

$H_{12} = 0, \quad 4 \times (2r+2\ell)$ $\qquad H_{22} = I, \quad 2\ell \times 2\ell$

η_2 is instrument measurement noise having the expectancy properties

$$E\{\eta_2(t_i)\} = 0$$

$$E\{\eta_2(t_i)\eta_2(t_j)^T\} = R\delta_{ij}$$

where R is the covariance matrix of measurement noise and δ_{ij} is the familiar Kronecker delta function:

$$R \doteq \frac{1}{N} \sum_{i=1}^{N} \nu(t_i)\nu(t_i)^T \tag{13}$$

$x(p^o, t_i)$ is the value of x obtained from integrating equations (8) for
$$p = p^o$$

2. The value of $x(p^o, t_i)$ is compared to available measurements, $y_x(t_i)$. Usually, the agreement is "poor".

3. Assuming the first fit is poor, the initial estimate of p is updated to :

$$\hat{p} = p^o + \left[\sum_{i=1}^{N} A^T(t_i) R^{-1} A(t_i) \right]^{-1} \left[\sum_{i=1}^{N} A(t_i) R^{-1} \nu(t_i) \right] \tag{14a}$$

The rectangular matrix A is the sensitivity matrix:

$$A^T = \left[\frac{\partial \delta}{\partial p}^T \quad \frac{\partial a}{\partial p}^T \right] \tag{14b}$$

whose elements are determined by the sensitivity equations, formed from a solution of differential equations obtained by taking the partial of equations (8) with respect to p:

$$\frac{d}{dt}\left(\frac{\partial \delta}{\partial p}\right) = F^{-1}\left[B\left(\frac{\partial \delta}{\partial p}\right) + \left(\frac{\partial B}{\partial p}\right)\delta + \left(\frac{\partial G}{\partial p}\right)\delta \right] - \left[F^{-1}\left(\frac{\partial F}{\partial p}\right)F^{-1} \right](B\delta + G\delta) \tag{14c}$$

$$\frac{\partial a}{\partial p} = \frac{\partial e}{\partial p} \tag{14d}$$

with the initial conditions

$$\frac{\partial \delta_i}{\partial p_j} = 0, \quad \frac{\partial a_i}{\partial p_j} = 0$$

4. The new estimate of P is used to determine an updated value for $x(p, t_i)$ by again integrating equations (8) with p defined by $\hat{p}$.

5. Again, a value of $J(\hat{p})$ is determined from equation (10) and compared to the previous estimate, $J(\overset{o}{p})$.

6. If the values of $J(\hat{p})$ and $J(\overset{o}{p})$ do not agree with some criteria, the process in steps 1. thru 5. is repeated until convergence (or divergence) is indicated.

PROBLEMS OF COMPUTATION AND APPLICATION

The development of a parameter estimation method for elastic flight vehicles is recognized to be a high payoff venture. However, it is also recognized to be a high risk venture due to the lack of previous experience with the computation and application problems involved in elastic aircraft parameter estimation. For this reason, the solution of the problems will first be attempted as a Flight Dynamics Laboratory in-house analytical effort. Once satisfactory results are obtained from analytical test cases created by the FLEXSTAB/CCVMOD computer programs, the work effort will be expanded. The computer program being developed has been given the acronym FLEXFLT.

To date, six problems of computation and application have been identified. Three of these problems would be experienced by any parameter estimation procedure:

- Availability of the direction cosine transformation matrix, T, in equation (10).

- Excitation of all the states of x in equation (8).

- Determination of the optimum number, type, and location of sensors on the aircraft.

The other three problems are somewhat unique to the maximum likelihood method:

- The absence of realistic start-up data for $\overset{o}{p}$ in equation (14a).

- The inversion of large and possibly ill-conditioned matrices in equation (14a).

- The integration of a large number of state and sensitivity equations in equations (8) and (14c) and (14d).

Subsequent paragraphs in this section of the paper describe the approach to be followed during the in-house study. As will be noted, the study makes maximum usage of the experiences gained during past studies on the estimation of parameters of "rigid" aircraft.

The first problem, the calculation of T in equation (10), can be solved using the combined output data of the NASTRAN and the FLEXSTAB/CCVMOD Computer Program Systems. Other less precise calculations of T are possible from FLEXSTAB/CCVMOD alone, provided only dihedral angle changes and aerodynamically significant rotations are of interest. Since T is in part a function of P, an iterative cycle between FLEXFLT and FLEXSTAB/CCVMOD may be required for some applications. For expendiency in this restricted in-house effort, T will be assumed to be a diagonal unity matrix.

The excitation of the body-fixed axis motions and important elastic deformations is essential if the signal to noise ratio is to be large enough for optimum parameter extraction from the flight test data. This can be assured by careful selection of precision instrumentation and by well-planned flight tests. In the event a particular motion is not excited, its parameters will be deleted from P and from equations (7) and (14) by the FLEXFLT user since the aerodynamics of an unexcitable mode would be of little interest in the FCS design.

The third problem, the determination of the optimum number, type, and location of sensors, has plagued the methods developed for "rigid" aircraft. The inclusion of aeroelasticity effects could conceivably either complicate or alleviate the problem. The complication introduced us the requirement for a larger number of sensors. The alleviation introduced is a more precise representation of the sensor signals in equations (6). Figure 3 presents a fraction of the total number of sensors on the B-52E. The type and location of these sensors is nearly adequate for the first flight test applications of FLEXFLT. With these thoughts in mind, the in-house study assumes that proper air data sensors, rate gyros, and accelerometer measurements are now available. These available sensor signals will first be approximated with similar-type sensor signals which have been analytically created by the FLEXSTAB/CCVMOD programs. Prior to applying the FLEXFLT program to actual flight data, the analytical test cases will be corrupted with instrumentation noise and bias to provide insight into the difficult flight test problems.

The absence of realistic start-up data for $P°$ in equations (10) and (14a) poses a difficult computational problem. Many of the parameters defined in equation (9) may be of small magnitude, difficult to estimate analytically, and impossible to measure experimentally during wind tunnel tests of "rigid" and elastic models of the flight vehicle. Fortunately most of the parameters of importance are calculated analytically by the FLEXSTAB/CCVMOD programs and other advanced stability and control programs. Also, the MODAL TRUNCATION formulation of the dynamics allows direct inclusions of the available wind tunnel measurements of "rigid" airplane stability and control derivatives. The in-house study will begin by using altered values of the parameters that will be used by FLEXSTAB/CCVMOD to generate the analytical sensor signals mentioned in the previous paragraph. This approach may also prove to be feasible during actual flight test applications of FLEXFLT, although simple estimation methods may also prove of value in estimating $P°$.

The fifth problem, the inversion of large and possibly ill-conditioned matrices in equation (14a), is a major computational difficulty. The size of

R in equation (14a) is $(4+2r+2\ell)$ square. For the case of 3 retained invacuum modes $(r=3)$ and 12 accelerometers $(2\ell=12)$, the size of R is 22x22. A matrix this size may be easily inverted, provided it is properly conditioned. In the event it is nearly singular, the in-house study will employ the approximate inverse methods used by Callahan (22). In inversion of the matrix $[\sum_{i=1}^{k} A(t_i) R^{-1} A(t_i)]$ is more difficult since this matrix has the dimensions of $m\times m$. The magnitude of m is $3r^2+3rc+9c+15r+20$, where c is the number control surfaces. For the previous example, and for 5 control surfaces, m is 182. The inversion of a 182x182 matrix is routinely accomplished in aeroelastic stability and control computations, provided the matrix is well-conditioned. The in-house study will begin the checkout of FLEXFLT with restricted analytical test cases in which the derivatives associated with $\ddot{\delta}_c$ and $\ddot{u}_i$, are eliminated. These two simplifications reduce m to a value of 134 in the illustrative example. If necessary, further reductions will be employed to bring m to the order of 80 as in the analyses described in reference 22.

The time-wise integration of a large number of state equations and sensitivity equations is another major computational problem of FLEXFLT. The number of state equations is $(4+2r)$. Using the previous illustrative example, the state equations number only 10. However, the number of sensitivity equations to be integrated is immense. Again, for the example cited, the number is $(m)(4+2r)$ or 1820. Reducing m to the order of 80 reduces the number of sensitivity equations to 800, still a significant computer programming task to say the least! Fortunately these are linear, first order equations and the programming work is only tedious. Currently, the in-house study described in reference 22 integrates 545 sensitivity equations for $\partial 3/\partial p$, using a first order prediction method characterized as the "Adams-Bachford backwards difference method". This will be the method used in the initial FLEXFLT studies.

CLOSING COMMENTS

The dependence of the USAF on flight vehicle prototypes requires an accurate and timely evaluation of the flight test data from the prototype prior to a committment of resources to a production vehicle. The trend toward higher-performance, lighter-weight vehicles introduces complex aeroelastic phenomena that can contribute significant problems to vehicle design. At present, the stability and control parameter estimation methods available treat only the simplest of aeroelastic phenomena (QUASI STATIC) and even then, require an accurate analytical method to estimate aeroelastic corrections to the "rigid" airplane stability and control derivatives. Thus, a high payoff in terms of a realistic evaluation of the prototype vehicles, and in terms of a reliable method to identify the sources of troublesome stability and control problems of the prototype vehicles, can be gained with an elastic parameter estimation method.

This paper presents a maximum likelihood algorithm that will estimate the aerodynamic stability and control parameters of a flexible flight vehicle experiencing small perturbation longitudinal motions from a symmetric, steady, non-rotating, wings level, initial flight condition. The computer program being developed to use the algorithm is named FLEXFLT. The initial feasibility study reported herein indicates the two major computational problems to be overcome are the inversion of a large sized matrix (possibly ill-conditioned) and the time-wise integration of a large number of sensitivity equations.

It is usually the case, high payoff in research and development also implies a risk. The risk is that computational problems involved in mechanizing the maximum likelihood parameter estimation method on a digital computer may be too difficult to overcome. To date this does not appear to be the case, although analytical test cases from FLEXSTAB/CCVMOD and experimental test cases from the B-52 CCV program have not been analyzed. The dollar cost of the risk is minimized thru the emphasis placed upon an in-house effort by the Flight Dynamics Laboratory prior to an extensive contractor effort. The key element in minimizing the risk to achieve the high payoff is the FLEXSTAB/CCVMOD computer programs that will be used to create sophisticated analytical test cases for FLEXFLT.

REFERENCES

1. Schoeman, R., "Fly-by-Wire and Artificial Stabilization Design," Paper Submitted to AGARD Flight Mechanics Panel Symposium, 10-13 April 1972, Brounschweig, Federal Republic of Germany.

2. Schwanz, R., "Formulation of the Equations of Motion of an Elastic Aircraft for Stability and Control and Flight Control Applications," AFFDL-FGC-TM-72-14, August 1972.

3. Wells, W., "A Maximum Likelihood Method for the Extraction of Stability Derivatives from High-Angle-of-Attack Flight Data," AFFDL-FGC-TM-72-16, September 1972.

4. Schwanz, R., Wells, W., Rough Draft of an AFFDL-FGC-TM to be published in June, 1973.

5. Etkin, B., Dynamics of Flight, 1959.

6. Bisplinghoff, R., Ashley, H., Principals of Aeroelasticity, 1962.

7. Milne, R., "Dynamics of the Deformable Airplane, Part I and II," Her Majesty's Stationery Office, London, 1964.

8. Schwendler, R., McNeal, R., "Optimum Structural Representation in Aeroelastic Analysis," ASD-TR-61-680, March 1962.

9. The Boeing Company (Seattle), "A Method for Predicting the Stability Characteristics of an Elastic Airplane," NASA/ARC Contract NAS 2-5006, Rough Draft Report, February 1972.

10. The Boeing Company (Wichita), "Aircraft Load Alleviation and Mode Stabilization (LAMS)," AFFDL-TR-68-161, September 1968.

11. The Boeing Company (Wichita), "Detailed Program Plan for Control Configured Vehicles Program (B-52 Phase) - Revision B," Doc. No. D6-8652, June 1972.

12. Triplett, W., Kappus, H., Landy, R., "Active Flutter Suppression Systems for Military Aircraft, A Feasibility Study," AFFDL-TR-72-116, October 1972.

13. Wykes, J., "B-1 Flexible Vehicle Equations of Motion for Ride Quality, Terrain Following, and Handling Qualities Studies", North American Aviation/Los Angeles Report, TFD-71-430-1, 29 January 1973.

14. Mehra, R., Stepner, D., Tyler, J., "A Generalized Method for the Identification of Aircraft Stability and Control Derivatives from Flight Test Data," Proceedings of the 1972 Joint Automatic Control Council, Stanford, California, 16-18 August 1972.

15. Sorensen, J. A., Tyler, J. S., Jr., and Powell, J. David, "Evaluation of Flight Instrumentation for the Identification of Stability and Control Derivatives," AIAA Paper No. 72-963, 11-13 September 1972.

16. Dornfeld, G., Schaeffer, D., "Linear Systems Analysis Program," Attachment to C/S AERO-490, Aerodynamics Research Group, The Boeing Company (Seattle), 21 June 1972.

17. USAF Contract F33615-72-C-1172, to be completed 1 December 1973.

18. Mehra, R., Taylor, J., "Lecture Notes on Parameter Estimation Methods," Presented to AFFDL on 9-12 May 1972.

19. Jazwinski, A., Stochastic Processes and Filtering Theory, 1970.

20. Steinmetz, George G., Parrish, Russell V., and Bowles, Ronald L., "Longitudinal Stability and Control Derivatives of a Jet Fighter Airplane Extracted from Flight Test Data by Utilizing Maximum Likelihood Estimmation," NASA TN D-6532, March 1972.

21. Suit, William T., "Aerodynamic Parameters of the Navion Airplane Extracted from Flight Data," NASA TN D-6643, March 1972.

22. Callahan, J., "Maximum Likelihood Estimation of Aircraft Stability and Control Coefficients for Low to Near Stall/Spin Angle of Attack Flight Regimes," Master's Thesis, GAM/MC/73-5, March 1973.

355

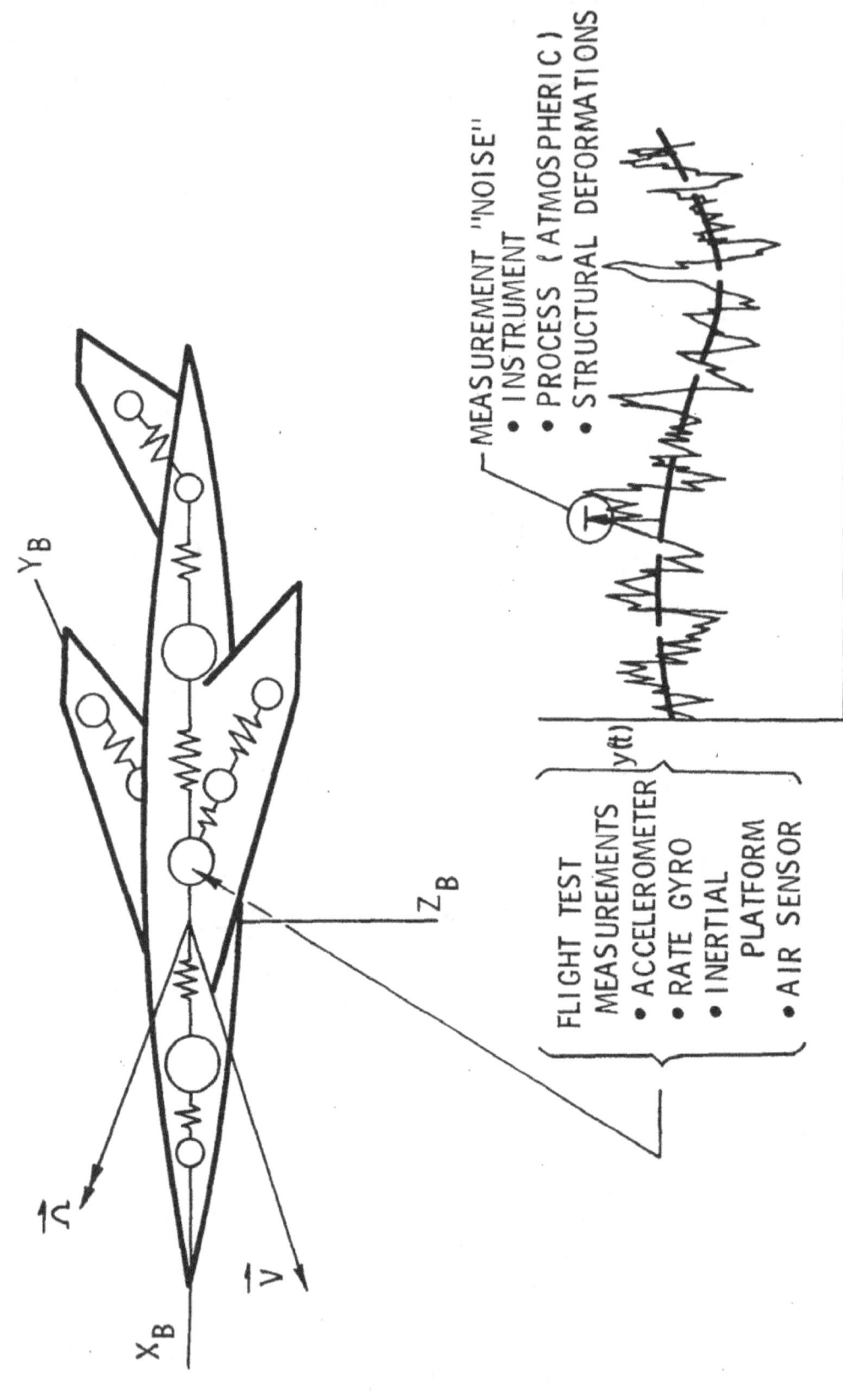

Figure 1. Lumped Mass Representation Of An Elastic Flight Vehicle

356

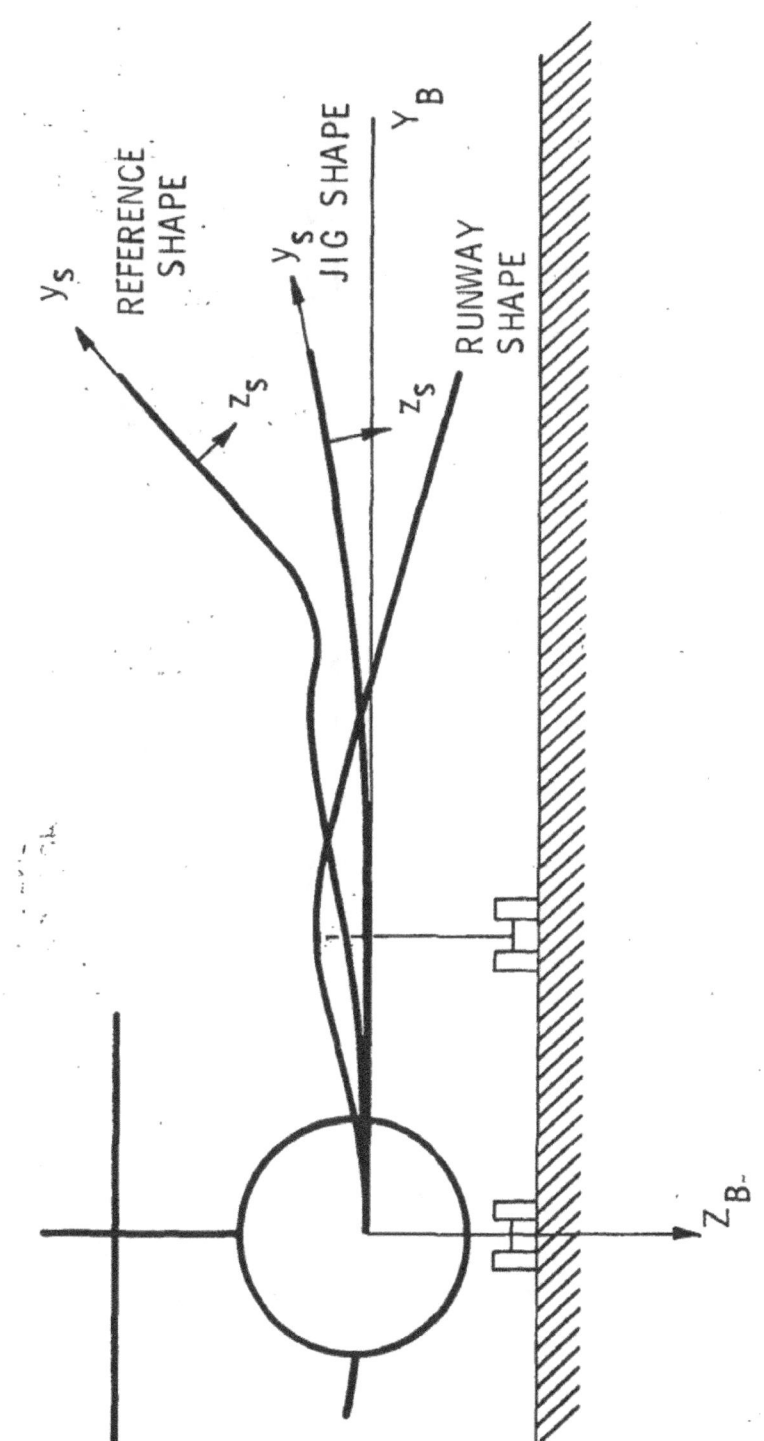

Figure 2. Effect Of Vehicle Elasticity On Sensor Measurement Axis Reorientation

357

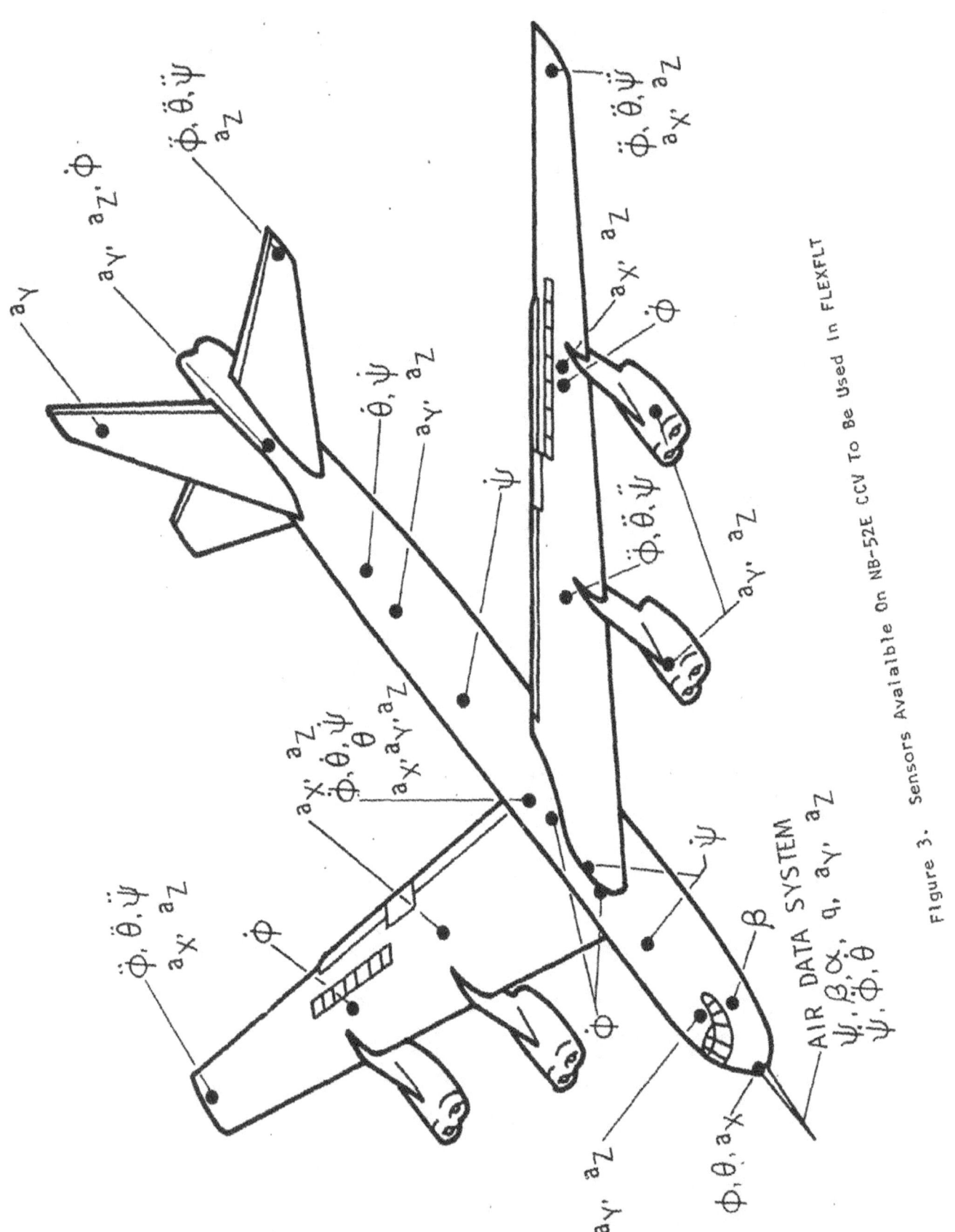

Figure 3. Sensors Availalble On NB-52E CCV To Be Used In FLEXFLT

358

ESTIMATION OF STRUCTURAL PARAMETERS
FROM DYNAMIC TEST DATA

B. M. Hall
Branch Chief, Flight Control Systems

and

M. S. Sholar
Senior Staff Engineer, Flight Control Systems
McDonnell Douglas Astronautics Company — West
McDonnell Douglas Corporation

ABSTRACT

Parameters for a lumped linear model approximating a distributed elastic structure are determined from dynamic test data comprised of several mode shapes and frequencies. Measurement errors, nonlinear response, and nonmeasurable quantities such as mode slope components are accommodated. Some mass and stiffness parameters may be known accurately, whereas the remainder are to be estimated. The method entails minimizing a quadratic function of the difference between corresponding modes and frequencies of the theoretical model and the test specimen. This technique has been applied to some actual vibration test data, and the special techniques that are required to overcome convergence problems are described.

INTRODUCTION

Ground vibration tests are performed on nearly every aerospace vehicle which has a control system or which may tend to exhibit aeroelastic instabilities such as flutter. The process of performing these tests has presented formidable experimental problems for many years. Among the problems which present themselves to the test engineer are those of exciting normal modes of the structure and dealing with the presence of ground restraints which create boundary conditions not representative of flight. Another inherent problem which prevents the measurement of the normal modes of an undamped structure is the fact that the structure, indeed, has distributed damping. This causes bothersome phase shifts in the experimental data and makes the excitation and measurement task that much more difficult. Since the ultimate use of the data from the vibration test is to serve as inputs to a dynamic or static modal analysis, it is necessary to examine the manner in which the data are used, and to see if better analytical techniques can be devised to use the data which come from the vibration test. Of particular importance when considering this problem is the fact that the resulting analysis will be performed on the vehicle in some flight condition which is considerably removed from the boundary conditions which exist during the vibration test. This is true for the airplane which is tested on its landing gear; the missile which is tested horizontally in a sling; and particularly true for the spacecraft which is tested in the atmosphere at 1g. In all of these cases, the vehicle must be free of sloshing fluids in order to excite the orthogonal structural modes.

In view of these restrictive test conditions, one may then conclude that the basic reasons for performing the test are (1) to verify the analytical predictions of the behavior of the structure in the test environment, and (2) refine the math model of the vehicle. This problem, which is essentially that of estimating the vehicle structural parameters from test data, was addressed by the authors several years ago (Reference 1). The method suggested in this original work has since been applied by the authors as well as others at different aerospace companies. These subsequent applications have met with varying degrees of success. The purpose of this paper is to present some additional techniques for the parameter estimation method and to show an application to a typical missile vibration test.

DEFINITION OF THE VEHICLE MATH MODEL

A detailed description of the general lumped parameter model of the vehicle is given in Reference 1. Essentially, the model consists of a linear structure, with lumped mass and stiffness characteristics described by mass and spring matrices. It is assumed that the mass data are known without error, and that structural stiffness only needs to be estimated.

It is assumed that the structure has undergone a vibration test and that the mode shapes and frequencies, which characterize the unknown stiffnesses, have been measured. These data contain measurement errors as well as noise. It should be pointed out that it is practical to measure only a few of the infinite number of modes characteristic of a continuous structure.

The problem then is to choose a lumped mass model for the system and estimate the stiffness for this model which yields modes giving the best agreement between the model modes and the experimental modes.

360

THE PARAMETER ESTIMATION TECHNIQUE

The following method finds the linear model having certain modes and frequencies which are as close as possible in a weighted least squares sense to the corresponding experimental modes and frequencies. This is achieved by minimizing a quadratic cost function of the following form:

$$F = 1/2 \sum_{i=1}^{N_1} (X_{ie} - X_{it})^T W_i (X_{ie} - X_{it}) + 1/2 \sum_{i=1}^{N_2} (\omega_{ie}^2 - \omega_{it}^2)^2 \tag{1}$$

Where X_{ie} is the ith experimental mode, an nxl vector normalized such that

$$X_{ie}^T M X_{ie} = \frac{1}{\omega_{ie}^2}$$

X_{it} is the ith theoretical mode (eigenvector), an nxl vector normalized with respect to spring matrix, K

W_i is an nxn symmetric positive semidefinite weighting matrix of the ith mode error

ω_{ie}^2 is the ith experimental eigenvalue

ω_{it}^2 is the theoretical eigenvalue

N_1 is the number of eigenvectors to be considered

N_2 is the number of eigenvalues to be considered

In this formulation, it is assumed that all elements of the mass matrix (M) are known and the elements of the stiffness matrix (K) are to be determined. The K matrix is taken to be comprised of unknown parameters which occur as linear factors. An expression for K after Rubin (Reference 2) is

$$K = \overline{K} + \sum_{j=1}^{n} k_j K^j \tag{2}$$

where the k_j are the scalar parameters of the system which are to be determined, $\overline{K}$ is the known portion of the matrix, and the K^j are matrices locating the k_j in correct positions according to the model and its boundary conditions.

The eigenproblem defining the theoretical modes and frequencies is

$$\left[K - \omega_{it}^2 M \right] X_{it} = 0 \tag{3}$$

361

The normalization of the theoretical modes is chosen with respect to the K matrix such that

$$X_{it}^T K X_{it} = \delta_{ij}$$

$$X_{it}^T M X_{it} = \frac{1}{\omega_{it}^2} \delta_{ij} \tag{4}$$

$$\delta_{ij} = \begin{cases} 1 & i = j \\ 0 & i \neq j \end{cases}$$

The experimental modes will not be orthogonal with respect to the mass matrix in general due to errors. The major sources of error are (1) lumping the mass distribution of a continuous structure, (2) errors in the estimate of the mass data, (3) distributed damping in the structure, and (4) errors in mode measurement and data reduction. However, in order to compare them in the cost function with the theoretical modes, they are normalized by requiring that

$$X_{ie}^T M X_{ie} = \frac{1}{\omega_{ie}^2} \tag{5}$$

The global minimization of Equation 1 is effected over the parameter space of the k_j by a modified Newton Raphson technique which makes use of analytically derived first derivatives and numerically determined second derivatives. The minimization procedure is constrained to positive values for physical realizability. The first derivatives of F with respect to k_j are

$$\frac{\partial F}{\partial k_j} = \sum_{i=1}^{N_1} (X_{it} - X_{ie})^T W_i \frac{dX_{it}}{dk_j} + \sum_{i=1}^{N_2} (\omega_{it}^2 - \omega_{ie}^2) \frac{d\omega_{it}^2}{dk_j} \tag{6}$$

The derivatives of the eigenvectors and eigenvalues in Equation 6 for M fixed and normalization with respect to K are

$$\frac{\partial \omega_{it}^2}{\partial k_j} = \frac{X_{it}^T K^j X_{it}}{X_{it}^T M X_{it}} \tag{7}$$

$$\frac{\partial X_{it}}{\partial k_j} = \sum_{\substack{k=1 \\ k \neq i}}^{N} \frac{X_{kt}^T K^j X_{kt}}{\omega_{it}^2 X_{kt}^T M X_{kt} - 1} X_{kt} - 1/2 (X_{it}^T K^j X_{it}) X_{it} \tag{8}$$

which corresponds to the second formulation in Fox and Kapoor (Reference 3).

362

The Hessian matrix of second derivatives is approximated numerically at a point k_{j_0} in the parameter space from finite differences of the gradient vectors evaluated at $k_j = k_{j_0} \pm \delta$ using Equations 6, 7, and 8. This procedure requires solving for both eigenvalues and eigenvectors at each incremented point since they are required for evaluation of the first derivatives.

The modified Newton Raphson procedure for iterating the design vector $\underline{k}^{(n)}$ is given in Equation 9 where $\underline{k}^{(n)}$ is a column vector of the k_j parameters at the nth iteration.

$$\underline{k}^{(n+1)} = \underline{k}^{(n)} - c\left[H^{-1}\right]^{(n)} \left(\nabla_{\underline{k}}F\right)^{(n)} \tag{9}$$

$\left[H^{-1}\right]^{(n)}$ is the inverse of the Hessian matrix evaluated at the nth iteration.

$\left(\nabla_{\underline{k}}F\right)^{(n)}$ is the gradient vector at the nth iteration

c is a step-size control parameter chosen for convergence and satisfaction of constraints

The set of equations contained in this section completes the formal mathematical description of the minimization process. Given enough computer time, patience, and, hopefully, good experimental data, one could presumably crank through the repetitive operations and find the design vector of stiffnesses which would describe that linear lumped model which best describes the real physical system. As with all processes of this type, there are many formidable problems which lie in the path of achieving a converged solution to a problem with a rather simple concept. Some of the roadblocks encountered and procedures for dealing with them follow.

SYSTEM RESTRAINTS

The basic restraint to be considered is that of guaranteeing that the $\underline{k}$ design vector will, when subtracted from the basic value of the appropriate term of the spring matrix, yield a realizable physical stiffness (Equation 2). For example, if one is working with a bending beam, the elements of K are linear functions of the beam bending stiffness EI. The restraint in this case is that the EI cannot go negative.

CONVERGENCE

Assuring convergence to the desired answer is a more difficult problem, and one that requires cut and try methods. One of the most valuable methods for speeding convergence was developed while working on a particular problem. This method can be compared to the multilevel optimization technique used in trajectory optimization. In this case the multilevel technique consists of lumping some of the elements of the design vector together, and restraining these to vary from their initial values as a unit. For example, if one is working with a bending beam of 30 sections, he might lump together sections over which the stiffness is relatively constant. If he does this for several groupings of, say, 4 sections each, he has reduced the degrees of freedom from 30 to maybe 7. Convergence is speeded and, after achieving a minimum for the course lumping, one can now release one or more of the restrained sections and again solve the minimization problem.

There is an important point to be noted when applying this technique. When subdividing the structure into a lumped model, the number of segments is chosen on the basis of that subdivision which is required to compute accurate mode shapes and frequencies independently of the parameter estimation process. This may require that a segment of a missile booster or airplane fuselage will be divided into several sections even though the stiffness is known to be constant over the segment. For this case, one would constrain these segments to move together throughout the entire minimization process; because, to allow one section to have a final value different from a neighboring section is physically not possible.

In examining the convergence process, one must necessarily consider the question of local minima, and the existence of a global minimum. For this type of problem, the search for the global minimum over the entire permissible parameter space is not so important. First of all, we are working with noisy and inaccurate vibration test data; hence, an error function equal to zero does not exist. In fact, the whole scale of the error function is shifted by the choice of weighting factors alone. Also, the global minimum, if one were fortunate enough to find it, might lead to an unrealistic stiffness distribution. We must be content then with locating the best minima we can find which yields a reasonable design vector. This search narrows to that of the structural analyst, who has a complete knowledge of the structure, interacting with the computer program in order to find the best answer.

THE WEIGHTING FUNCTION

The weighting function serves two basic purposes. The first is to provide dimensionality consistency between the eigenvalues and eigenvectors in the error function; the second is to weight the experimental data relative to the errors which are known to exist in the experimental data. For instance, the higher frequency modes or closely coupled modes are frequently difficult to excite. In these cases, one would weight the test data lower than the data from the more easily excited modes. For the purpose of providing dimensional consistency, it has been found that the following procedure suffices. Referring to Equation 1, each element of the theoretical eigenvector is divided by the maximum value element of the experimental eigenvector.

A similar procedure is adopted for the eigenvalues. For this case, the square of each theoretical eigenvalue is divided by the square of each corresponding experimental eigenvalue.

In addition to normalization, one must consider the weighting or emphasis to be given to each mode. As mentioned before, this will vary depending on the particular case under consideration. For the purpose of the example presented later on in this paper, the weighting matrix is chosen to give each mode equal weight. When both the normalization and equal mode weighting are combined, the final error function of Equation 1 becomes

$$F = \frac{1}{2}\left\{ \left[X_{1E} - X_{1T}\right]^T \left[\frac{1}{I3}\right]\left[X_{1E} - X_{1T}\right] + \left[X_{2E} - X_{2T}\right]^T \frac{M_1 \omega_1^2}{M_2 \omega_2^2 I3} \left[X_{2E} - X_{2T}\right] \right.$$

$$\left. + \left[X_{3E} - X_{3T}\right]^T \frac{M_1 \omega_1^2}{M_3 \omega_3^2 I3} \left[X_{3E} - X_{3T}\right] + \left[1 - \frac{\omega_{T1}^2}{\omega_{E1}^2}\right]^2 + \left[1 - \frac{\omega_{T2}^2}{\omega_{E2}^2}\right]^2 + \left[1 - \frac{\omega_{T3}^2}{\omega_{E3}^2}\right]^2 \right\}$$

(10)

TYPICAL MISSILE BENDING PROBLEM

In the previous paper (Reference 1), some simple examples were presented in order to show how the process works, and to make an assessment of the convergence problems which might arise when actual test data are used. These simple examples were based on a free-free beam of varying mass and stiffness. Computer runs were made in order to determine the effect on the process due to the lack of measured slope data, and to assess the convergence process when the experimental data contains measurement errors. The results of these computer runs showed that the process did converge, and a best linear model was generated from each case.

In the present example, a typical missile structure is investigated. This structure has vibration data associated with it and therefore constitutes a very real case that one might encounter in practice. The first three experimental missile mode shapes and frequencies are shown in Figure 1.

Using these mode shapes and frequencies in tabulated form, along with the mode shapes and frequencies derived from the initial estimate of the stiffness, one can form the error function and begin the minimization procedure using Equations 1 through 10. Thirteen elastic sections are chosen for the math model describing this actual continuous structure. The number of sections chosen for the elastic model is a compromise between the number of sections required to avoid errors due to lumping of the continuous structure and to avoid an excessive number of sections, which greatly increases computer time. For this typical structure, 13 elastic sections are chosen in order to give reasonable definition to the first 3 bending modes. Because of the constant cross

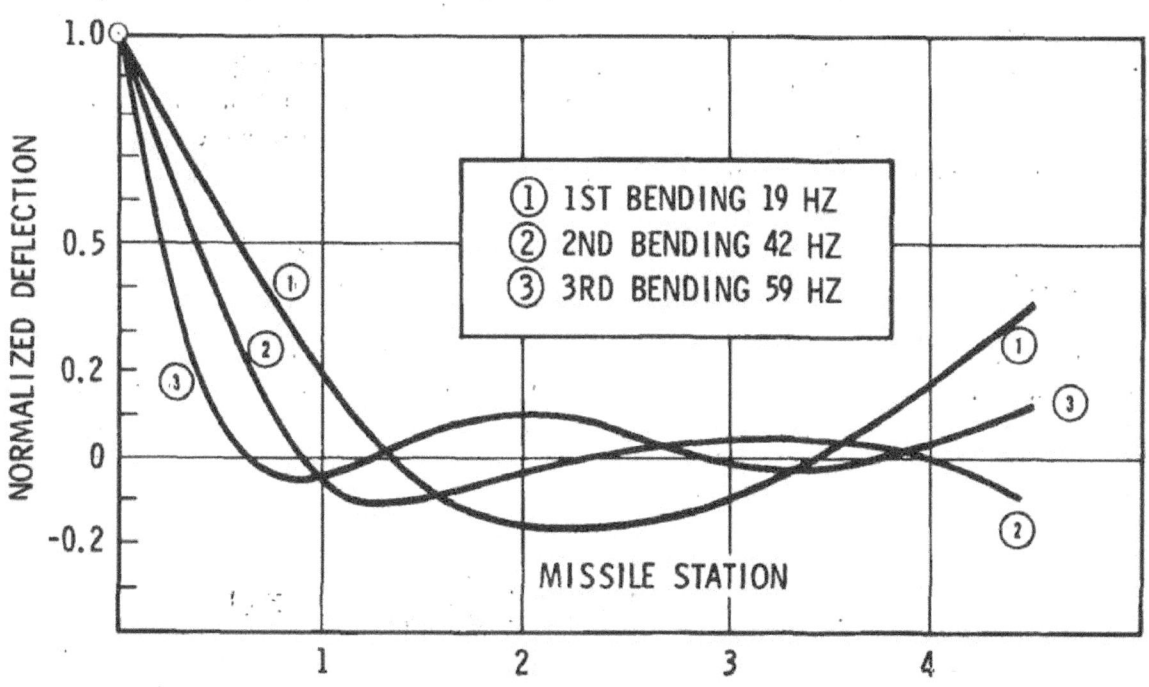

Figure 1. Missile Mode Shapes

section properties over portions of this structure, the following sections must have the same stiffness in the final 13-section math model: Sections 1 and 2; 3 and 4; 5 and 6; and 7, 8, and 9. Section 12 has elastic characteristics determined by other means and the k_j associated with this section will remain at unity.

The results of the multilevel minimization procedure are shown in Table 1. This table shows the k_j scalar parameters, along with the value of the error function F. Each level of minimization is listed vertically. Starting with the first grouping, it will be noted that all k_j are unity, which means we are assuming as our first estimation the computed bending stiffness of the missile. The actual value of F is not so important; what is important is how much one can reduce F as the stiffness parameters k_j are varied. This first grouping required 9 iterations to converge and the value of the error function was reduced from 0.0211228 to 0.00877676. Using the converged values of the k_j's to start the next grouping, and freeing section 11 to seek a new value, 14 iterations were required to further reduce the error function to 0.00518092. For the next grouping, section 13 was allowed to seek a new value. This final group converged in 4 iterations to an error function value of 0.00517841. As pointed out before, the absolute value of this error function is not important. What is important is that the k_j's have converged, and the stiffness of the lumped model has been determined.

Table 1

TABULATED RESULTS

SECTION		1	2	3	4	5	6	7	8	9	10	11	12	13	ERROR FUNCTION
STIFFNESS SCALAR		k_1	k_2	k_3	k_4	k_5	k_6	k_7	k_8	k_9	k_{10}	k_{11}	k_{12}	k_{13}	
GROUP													RESTRAINED TO UNITY		
1	INITIAL	1.0000	1.0000	1.0000	1.0000	1.0000	1.0000	1.0000	1.0000	1.0000	1.0000	1.0000	1.0000	1.0000	0.0211228
	FINAL	1.2963	1.2963	0.98292	0.98292	1.1484	1.1484	0.76012	0.76012	0.76012	0.10630	1.0000	1.0000	1.0000	0.00877676
2	INITIAL	1.2963	1.2963	0.98292	0.98292	1.1484	1.1484	0.76012	0.76012	0.76012	0.10630	1.0000	RESTRAINED 1.0000	1.0000	0.00877676
	FINAL	1.4557	1.4557	0.81303	0.81303	1.4698	1.4698	0.77703	0.77703	0.77703	0.78233	0.20021	1.0000	1.0000	0.00518092
3	INITIAL	1.4557	1.4557	0.81303	0.81303	1.4698	1.4698	0.77703	0.77703	0.77703	0.78233	0.20021	RESTRAINED 1.0000	1.0000	0.00518092
	FINAL	1.4555	1.4555	0.81415	0.81415	1.4716	1.4716	0.77307	0.77307	0.77307	0.77834	0.19900	1.0000	1.4087	0.00517841

CONCLUSION

A method for determining the parameters of a linear structural model from dynamic test data has been presented. A typical problem involving experimental data from a bending missile has been solved using the technique described in this paper. The solution shows that the technique is indeed a practical one. Because of the requirements for grouping the elastic sections during the minimization process, a good deal of judgment and foresight are required. Other problems, including that of large computer times, will undoubtedly be encountered as one applies this technique to more complex structures requiring many more modes for definition.

REFERENCES

1. B. M. Hall, E. D. Calkin, and M. S. Sholar. *Linear Estimation of Structural Parameters from Dynamic Test Data.* Presented to the AIAA/ASME 11th Structures, Structural Dynamics, and Materials Conference, Denver, Colorado, April 22-24, 1970.

2. C. P. Rubin. *Dynamic Optimization of Complex Structures.* Proceedings of the AIAA Structural Dynamics and Aeroelasticity Specialist Conference, New Orleans, Louisiana, April 16-17, 1969.

3. R. L. Fox and M. P. Kapoor. *Rates of Change of Eigenvalues and Eigenvectors.* AIAA Journal, Vol. 6, No. 12, December 1968, p. 2426.

DETERMINATION OF PROPULSION-SYSTEM-INDUCED FORCES

AND MOMENTS OF A MACH 3 CRUISE AIRCRAFT

Glenn B. Gilyard
NASA Flight Research Center

ABSTRACT

The YF-12 airplane (fig. 1) is an advanced, twin-engined, delta-wing inter-ceptor designed for long-range cruise at greater than Mach 3 and at altitudes above 21,336 meters (70,000 feet).

During the joint NASA/USAF flight research program with the YF-12 airplane, the Dutch roll damping was found to be much less during automatic inlet operation than during fixed inlet operation at Mach numbers greater than 2.5 and with the yaw stability augmentation system off. This significant reduction in Dutch roll damping is due to the forces and moments induced by the variable-geometry features of the inlet. Two stability-derivative extraction techniques were applied to the flight data—the recently developed modified Newton-Raphson technique (ref. 1) and the time vector method (ref. 2). These techniques made it possible to determine the forces and moments generated by spike and bypass door movement.

For efficient supersonic propulsive operation, the terminal shock (fig. 2) must be maintained in the inlet aft of the throat at an optimum position that is primarily a function of sideslip in a short-duration lateral-directional maneuver. A fast-acting fine control of normal shock position is provided by the forward bypass doors, which are positioned around the circumference of the front part of each nacelle. These doors regulate the amount of air that is expelled from the nacelle before it reaches the engines.

Two rudder-pulse maneuvers (fig. 3) were analyzed to determine airframe/ propulsion system interactions. The first maneuver was performed with automatic operation of the inlets, which resulted in a divergent Dutch roll oscillation, whereas the second maneuver was performed with the inlets fixed and was stable.

In both the Newton-Raphson and time vector analyses, the forces and moments produced by inlet spike and forward bypass door position variations were deter-mined as derivative coefficients and hence were considered as additional control variables in the equations of motion. It was assumed that the effect of bypass door position in the inlets-automatic maneuver was the same for both inlets, thus the right- and left-hand bypass door deflections could be combined. This combined parameter was denoted by η, and defined as the right-hand bypass door deflection minus the left-hand bypass door deflection.

369

The three variables used in the analysis were roll rate, yaw rate, and lateral acceleration. The sideslip parameter had a significant pneumatic lag.

The basic principle of the Newton-Raphson method is to minimize deviations between flight and calculated time histories of airplane responses to control inputs. The method can be used to determine all stability and control derivatives from a minimum of aircraft response variables.

The maneuvers with the inlets in the fixed and the automatic modes were matched simultaneously, thereby yielding one set of stability and control derivatives, including those of the forward bypass doors, which satisfied both maneuvers.

Figure 4 compares flight time histories and time histories calculated from the derivatives determined with the Newton-Raphson technique for inlets-automatic and inlets-fixed operation. The derivative results provide a good match with the flight data.

The time vector technique defines the derivatives explicitly in terms of measured frequency, damping, amplitude, and phase relations. The measurements of the two maneuvers are used simultaneously in the analysis to compute one set of stability and control derivatives. The time vector method as used is described in detail in reference 2.

Derivatives determined by using the two techniques are compared in figure 5. The bypass door derivatives, L_η, N_η, and Y_η, show good agreement. The remaining major stability derivatives also agree well.

The net effect of bypass doors with the inlets in the automatic mode increases the directional static stability by 40 percent and changes the effective dihedral from positive to negative, as shown in figure 6. Furthermore, the control effectiveness of the bypass doors is approximately the same as the control derivatives N_{δ_r} and L_{δ_a} when compared in terms of percent of control available (fig. 7).

Comparison of the flight-measured and calculated sideslip angles of figure 4 reveals a sideslip system lag of approximately 0.5 second. A closed-loop stability analysis (ref. 3) shows that the lagged sideslip input to the inlet computer is responsible for the divergent Dutch roll oscillation.

A more complete analysis of the airframe/propulsion system interaction phenomenon is presented in reference 3.

REFERENCES

1. Iliff, Kenneth W.; and Taylor, Lawrence W.: Determination of Stability Derivatives From Flight Data Using a Newton-Raphson Minimization Technique. NASA TN D-6579, 1972.

2. Gilyard, Glenn B.: Explicit Determination of Lateral-Directional Stability and Control Derivatives by Simultaneous Time Vector Analysis of Two Maneuvers. NASA TM X-2722, 1973.

3. Gilyard, Glenn B.; Berry, Donald T.; and Belte, Daumants: Analysis of a Lateral-Directional Airframe/Propulsion System Interaction of a Mach 3 Cruise Aircraft. AIAA Paper No. 72-961, Sept. 1972.

TEST AIRPLANE

Figure 1

INLET CONTROL SYSTEM

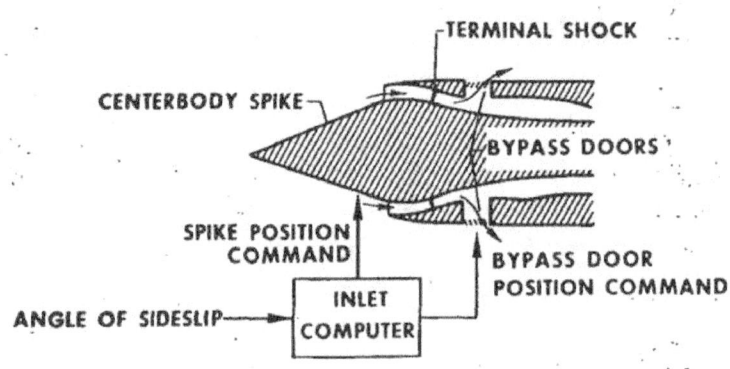

Figure 2

DUTCH ROLL RESPONSE TO RUDDER PULSE

YAW SAS OFF, M ≈ 3.0

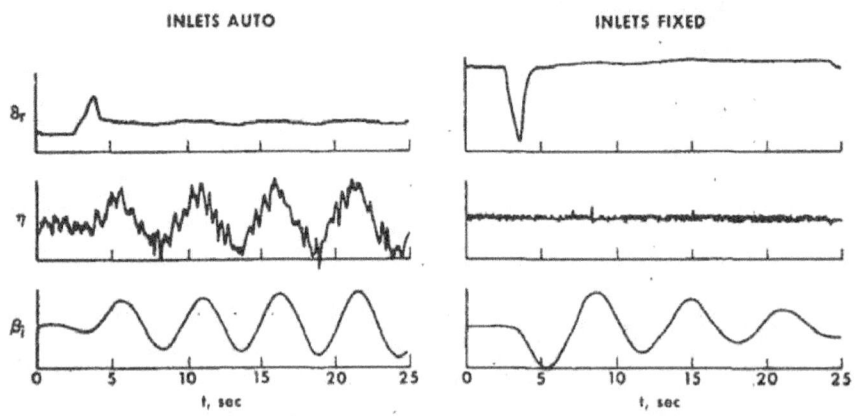

Figure 3

COMPARISON OF NEWTON-RAPHSON MATCHES
WITH FLIGHT DATA

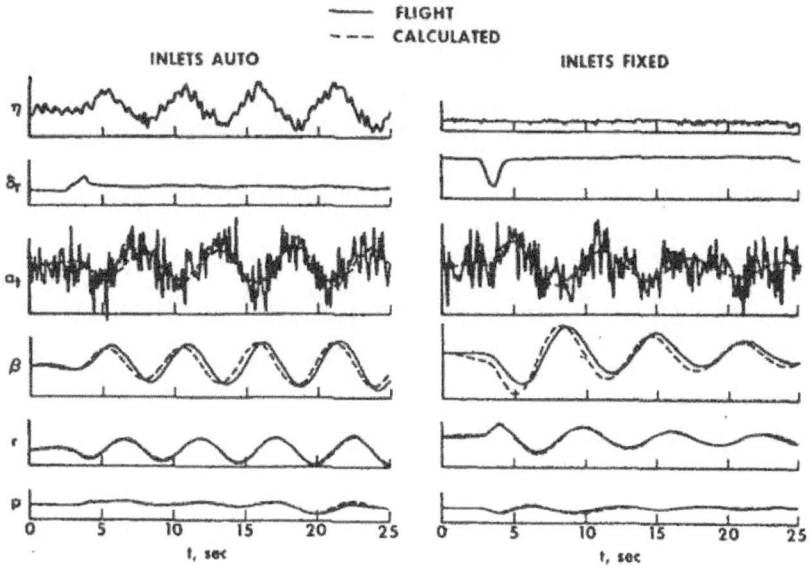

Figure 4

FLIGHT-DETERMINED DERIVATIVES

	TIME VECTOR	NEWTON-RAPHSON
L_β, $\frac{1}{sec^2}$	−0.97	−0.91
L_{δ_a}, $\frac{1}{sec^2}$	-------	2.95
L_η, deg/sec²-percent	.41	.35
N_β, $\frac{1}{sec^2}$	.86	.86
N_{δ_r}, $\frac{1}{sec^2}$	-------	−.73
N_η, deg/sec²-percent	.12	.11
Y_β, $\frac{1}{sec-rad}$	−.042	−.038
Y_η, $\frac{1}{sec-percent}$	−.0033	−.0027

Figure 5

INFLUENCE OF INLET OPERATION ON EFFECTIVE STATIC STABILITY

	INLETS FIXED	INLETS AUTO
N_β	0.86	1.23
L_β	-.91	.24

Figure 6

COMPARISON OF BYPASS DOOR AND CONTROL EFFECTIVENESS

$$L_\eta = 0.35 \ deg/sec^2\text{-percent}$$

$$L_{\delta_a} = 0.30 \ deg/sec^2\text{-percent}$$

$$N_\eta = 0.11 \ deg/sec^2\text{-percent}$$

$$N_{\delta_r} = -0.073 \ deg/sec^2\text{-percent}$$

Figure 7

374

SYSTEMS IDENTIFICATION - REPRISE AND PROJECTIONS

Lawrence W. Taylor, Jr.
NASA Langley Research Center

If systems identification is considered in the broadest sense to encompass all modeling activities, including those of describing observed phenomena by some "law," then systems identification is a fundamental part of obtaining knowledge of any physical system. If restricted to "laws" expressed in mathematical terms, perhaps the best modeler of dynamic systems was Sir Isaac Newton. Looking at figure 1, one can see that Newton not only extended Kepler's empirical laws of motion to the universal laws of motion which carry his name but also founded calculus. In addition, he devised the iterative method of obtaining roots, Newton's method, which is widely used today in systems identification. Another giant was Gauss who first solved the least squares problem and contributed the important Gauss or normal probability distribution. Next it was Fisher who contributed the maximum likelihood estimate, and Weiner, Komogorov, Swerling, and Kalman who have made important contributions to our understanding of stochastic estimation. An equally important contribution has been the development of automatic data processing. Von Neuman was instrumental in the early development of the electronic digital computer. This brief sketch is not in any way complete or fair to those making as important contributions to systems identification in other fields such as physics, chemistry, and electronics, but is meant only to give some perspective to the time over which contributions to systems identification have been made.

Another important aspect of systems identification is the rapidly increasing number of applications of the modeling techniques being made. Figure 2 depicts several of the areas of application: astronomy, maneuvering target identification, aircraft dynamics, gravitational fields, stock market, physiological systems, pilot tracking, adaptive control, process control, production testing, econometrics, psychological behavior, system checkout, and structural dynamics. There are, of course, more areas to which systems identification techniques are being applied. The wide use of modeling techniques points to their importance and usefulness beyond that of determining stability derivatives of airplanes.

Having considered the past contributors and current applications, let us turn to the future of systems identification. The reason for the question mark of figure 3 is that to a large degree the future of systems identification is up to us and others like us. In order to make the greatest use of any improvements of modeling techniques, it is important that we substantiate our claims and make full reference to related work as we report our work. We need to communicate more with analysts in other fields who are using systems identification techniques so that unnecessary duplication of effort is reduced and there is a free flow of ideas.

In trying to make a projection into the future of systems identification, I believe directions are clear. First, there will be a continuation of the increased number of considerations given in formulating the systems identification problems, that is, fit error, measurement noise, likelihood, a priori information, input design, and adaptive control. Another thrust that will be possibly even more productive is the greatly increased amount of data processed. This will give impetus to more efficient algorithms and data processing systems. Automatic editing of the data will be necessary because the current visual checks on data validity will be too costly. Other advances will be made, of course, which cannot be foreseen, but it is certain that the immediate future of systems identification is one of growing importance and growing interest.

In conclusion, it is up to each of us to strive to replace the question mark of the future with meaningful results, whether our job deals with theory or application or the funding of these endeavors.

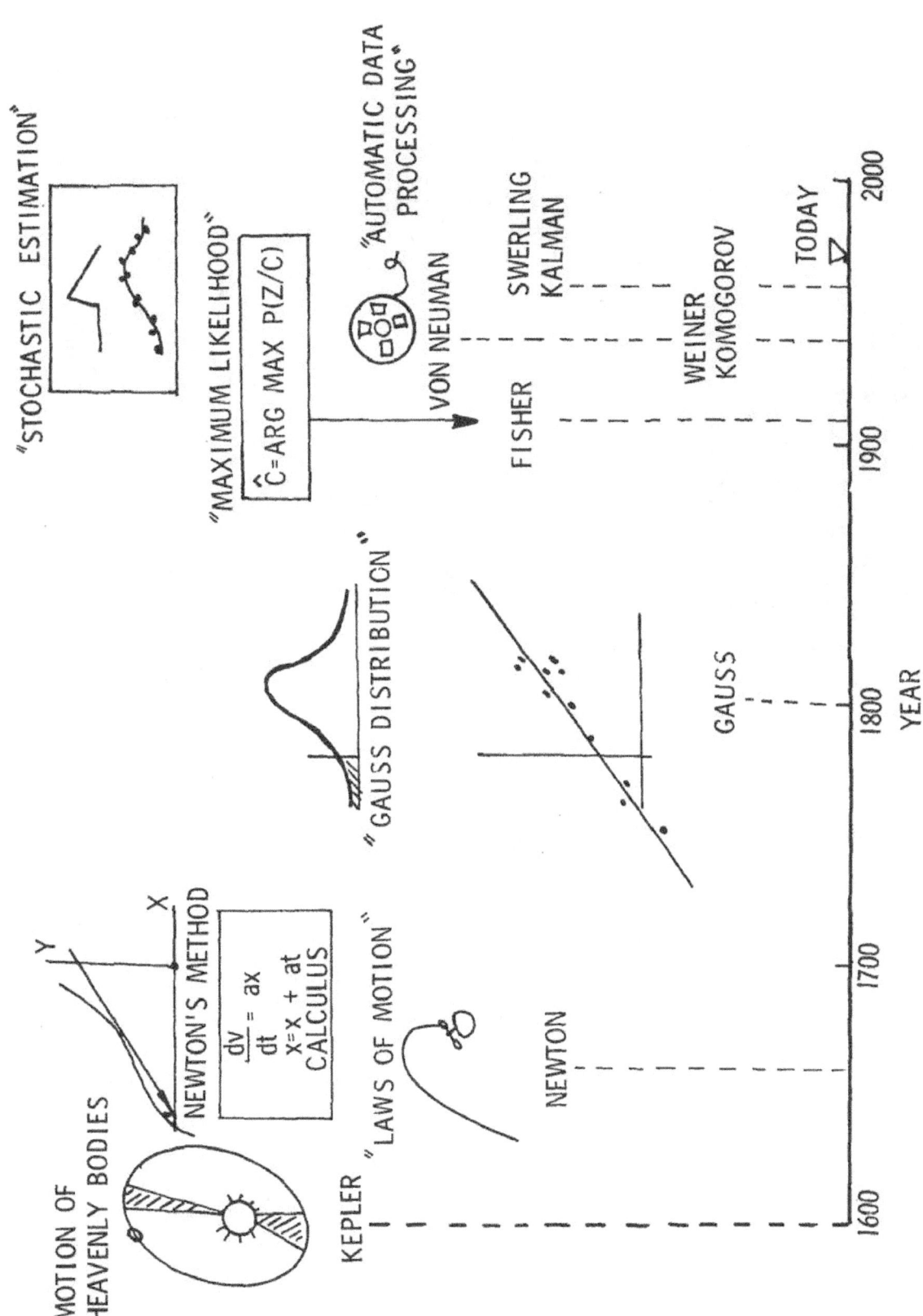

Figure 1. Our heritage in systems identification.

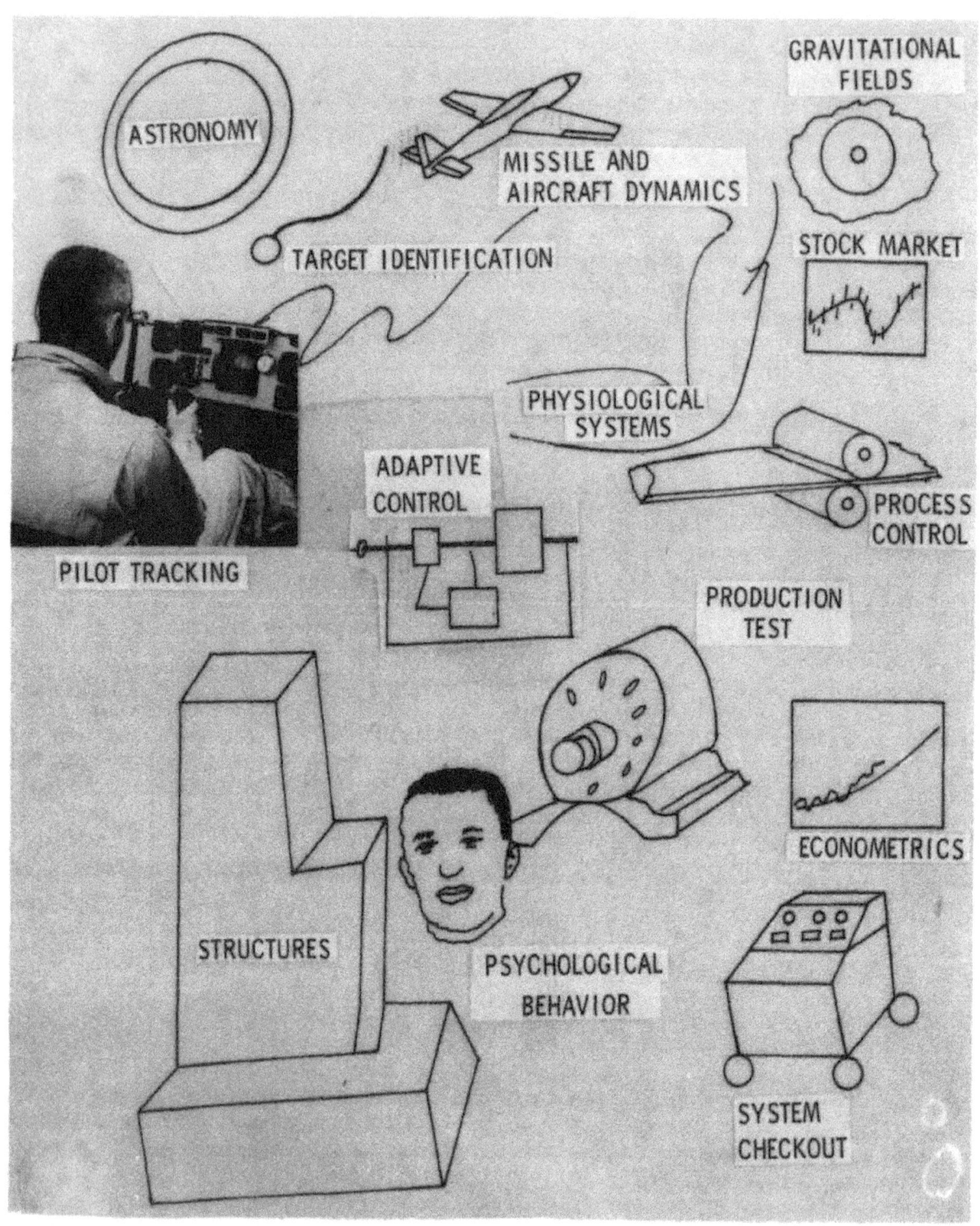

Figure 2. Systems identification applications.

Figure 3.   Future of systems identification.

SOME AFTER-DINNER REFLECTIONS ON SYSTEM IDENTIFICATION[+]

By A. V. Balakrishnan

Department of System Science
University of California
Los Angeles, California 90024

It's a great pleasure for me to be here today. I consider it a very great privilege to be able to participate in this symposium even in a ceremonial role as an after-dinner speaker. I am happy to be here because not only is this a very timely conference, as witness the large number of people here, but I have no doubt that this will be a watershed in the history of aircraft parameter identification, reflecting the first serious attempt to relate identification theory to practice. But, above all, I am happy to be here because it gives me a chance to acknowledge publicly the important, even crucial role played by the Flight Research Center in this whole area. I think this very symposium is evidence of the foresight and understanding by the Center and by Dr. Rediess because I know how hard it is to arrange a symposium in an area so totally unglamorous and dull as compared to the many other activities of the Center. In fact, I can see it is very difficult to explain to the general public just what it is that we do.

I owe my own personal start in System Identification to this Center. In 1963, I think it was, I had presented a paper at an obscure conference at Princeton University, on identifying a nonlinear system from input-output data using Volterra series - (as you know, this is only a fancy name for polynomials which you have already heard about so much today), and a gradient technique. By a happy coincidence it was noticed by a brilliant 'scientist' from the Center who happened by chance to take one of my classes. He saw in it an application to (what appears to me to be the perennial) problem of identifying human operators, and which I am sure is familiar to most of you and certainly still very much with us, and thus I was introduced to the world of aircraft identification problems. I am speaking of course of Mr. L. W. Taylor who is more famous and better known to you than I am. Larry's problems were always very challenging and since that time have guided almost all my research efforts in stochastic control. It was Larry also who saw the importance of second order methods - the modification of the Newton-Raphson method, which I had suggested in an obscure part of an unreadable tome on communication theory. In Larry's hand it became an extremely useful tool, so useful that I suggest that since the whole name is inapplicable anyway, be named Newton-Raphson-Taylor method. In this way, everybody will think you are talking about the Taylor of Taylor series and since Raphson didn't reportedly have anything to do with it either anyway, it would be very appropriate.

[+]Edited version of a tape transcription of the after-dinner talk.

Now by 1965 to 1967 the interest in identification problems had grown very large indeed; in fact, the International Federation of Automatic Control began a series of tri-annual symposia dedicated only to system identification. The first one was held in Prague, Czechoslovakia in '67, and the second also in Prague in 1970, and the latest one as you know, to be in Holland in 1973. The symposia were quite useful, especially the first one in 1967 that was held in Czechoslovakia because there was a large participation from East Europe and the USSR as well as the other smaller communist countries, and I can well remember a paper from Cuba on sugar refineries, identifying sugar refinery plants. I think that in looking over the symposia you could see a gradual progress from '64 to '67, and from '67 to '70 it is more or less steady state. In '67 there was only a trickling of papers on least squares and transfer functions, but by that time you could discern three or four groups across the world who were doing serious work. Of course I will omit references to the U.S. work because this work, in the first place is known to you and in any case, there are too many experts right here in the audience with you. Perhaps the most noticed work was the work of Karl J. Aström who had apparently worked on identification problems for a paper mill in Sweden with the aid of IBM, and he based his work for the first time on a linear difference equation model and standard statistical estimation theory including an asymptotic theory of consistency (his proof unfortunately turned out to be incorrect!). The British under Professor Hammond, with typical British thoroughness and emphasis on practicality were exploiting weighting patterns and using pseudo random inputs. More recently, the Imperial College in London under Professor Westcott has been applying identification techniques to steel mills. Professor Eykhoff in Holland was interested in application to biomedical engineering, an area which is truly enormous in scope and difficult to ascertain the impact. In some contrast to all of this, was the work of Dr. Rajbman from the USSR from the Central Institute of Control Problems in Moscow, based on an earlier theory of pugachev with considerable application to metallurgical problems. The work in Czechoslovakia under Dr. Peterka was on an even more practical turn, using difference equation models with application to steam boiler plants, etc. There was hardly any work at all reported on aircraft parameter identification from any country, and little or no work on continuous-time models.

I hope this gives you some idea of the mainstream of the activity across the world, since my intent here is far from any kind of survey. It was quite clear any way that nowhere else in the world was there any large-scale attempt at using computerized parameter identification techniques. So if we are behind, the rest of the world is farther behind. The literature on identification is quite large. At the 1973 symposium over 300 papers were submitted and 100 or so were eventually selected for presentation. If you think today's sessions were crowded, they are having sessions with all ten-minute presentations. Now it's quite understandable perusing the thick volumes of proceedings that the potential user is bewildered by the variety of techniques and claims. Most papers follow a canonical pattern roughly like this: they will do some mathematics with some techniques based on some model, often making ad hoc simplification because the total problem is too complex and then present a technique which they claim 'works' and by way of justification, sometimes a computer simulation of sorts is thrown in. There is a whole philosophical question here as to just what the author has proven by this exercise!

382

Nowhere is there at any time any use of actual data, and of course no information at all about what the customer or the actual data user thought of it or did with it. It is of course, in view of this, understandable when Dr. Queijo, for example, once remarked to me that it appeared to him that by properly massaging the data, you can produce any number you want. Now whereas in the usual engineering design problems, analysis usually need not be precise because the final design has to be experimental anyhow, and you can test whether the motor works or not by actually constructing it; the situaton is quite different in identification since you don't know what the answers are that you're looking for and as far as I can tell, there really is no means of verifying them except perhaps indirectly.

At this point, to make some of my points slightly more precise I would like to recall the main theoretical framework in System Identification. In the canonical identification problem you have a system model and you have some input and some noisy observations and then you are supposed to say what is inside (the black-box) and after a few years -- since '67- '68 -- we can also allow for load disturbance. The most important problem here of course before you do anything else, is the problem of modelling, and is quite evident to any of us here who work with these problems as indeed a crucial step. It is a readily conceded law that no model is good unless it originates with the user. Next I think it is quite pertinent to ask what do you mean by identifying, namely, what you want to do with it afterwards? This doesn't seem to bother many of the speakers today, but perhaps it should. I have been going around like Diogenes with a lamp trying to find out what you do with them. Nobody would really give me a straight answer, except to say, "Well, we're using it here and we're using it there." It seems to me that one can distinguish two possible uses: you may want to predict what the behavior of the system for inputs different from those used in the identification or verification; and the second one, which to some extent is implied in the first one, is that you want to somehow control the system afterwards, and since you know that the system parameters will change in different circumstances, and we don't know precisely how ahead of time, so you really have to have a controller which also has an identification phase in it, either implicit or explicit. If you do that, the problem of identification has slightly changed and it now goes into adaptive control and in terms of adaptive control you have one happy circumstance in that nobody really knows how to evaluate an adaptive control system that I know of (so that we are in the happy world where every adaptive system is optimal, or cannot be proved otherwise!). Adaptive control was the promised land of automatic control in 1960. There is a large tome by the Soviet author Feldbaum promising it to us, but I think it has died a natural death.

There may be a moral for us in identification, but I'll come to that later. By the way, I cannot help remarking at this point on the similarity between pattern recognition and between system identification, the modelling in the latter and the feature selection problem in the former. Once we select the right features and once we select the right model, then the rest of it is no longer much of a problem. Once we have formulated a model - including the disturbances and measurement errors - finding the parameters can be formulated as a statistical estimation problem. As far as estimation is concerned, there is of course quite a history of well established theory and techniques. In any identification or estimation problem you are looking for some functional that vanishes at the true number and only at that value, in a suitably restricted neighborhood. The gradient of the likelihood functional is one such. You look for a root of the gradient of the log-likelihood functional or in many cases you can take many variations of this functional to make it vanish at the right value. Then you have a matrix which figures in the Newton-Raphson-Taylor technique for finding the root, essentially, except for the averaging process, the one used by Fisher in 1922, and many people here I noticed did use the term 'Fisher Information Matrix', and the asymptotic variance of the estimate can be calculated from it. All of this is pretty much standard, so the main difficulty if you have a precisely parameterized model is then in the evaluation of the likelihood functional, and it turns out that we can make such a calculation provided the system is linear both to the input as well as the random disturbance. You could have a system described by partial differential equations - a distributed system with boundary data - so long as the response to the input is linear, the response to the external disturbance is linear and the observation error can be modelled as additive noise. If you have biases or scale changes or calibration errors, those are supposed to be known as part of your measurement system. The additive noise on the observation represents the unavoidable random error in any measurement. The calculation of the likelihood functional in the continuous case system as in the aircraft parameter problem, where the basic system dynamics are given in the continuous sense, can then be made using a ('prewhitening') procedure due to Krein (1958), or, in its recursive form, using Kalman filtering. See the paper by K. Iliff in the symposium which is the <u>only</u> one to use a continuous-time model! And here we also note that the measurement noise, although in theory is white, in practice the data that you observe is not, and allowance must be made for this. Even though such a theory was available, say since 1960 or so, it had to await the formulation by Rediess and Taylor in 1970 of the practical problem of aircraft parameter identification - stability augmentation to find relevance. It is unfortunate that the original paper by Rediess and Taylor is not generally available, at least in the aircraft area and, for example, I could not find any references to it. I would like to call that perhpas the Rediess-Iliff model because Taylor's name is used too many times already, and perhaps there is justice in that because without the formulation and work by Iliff, of course, it would simply remain a text book example.

Now, whether we use adaptive control or not, the importance of identificaton in modelling cannot be overemphasized. This is almost a pet peeve with me now, so I should like to go in a little bit deeper. As you know, modelling is now spreading and affecting every aspect of our lives. I am not talking about our airplane models or biomedical models, but the models now being used to make decisions of economic and political importance. Here one constructs the model without any attempt at all on identification - on verifying it on actual data. I would like to pick up now one example which has recently received lots of publicity. This is the work subsidized by the "Club of Rome", started by a group of businessmen in Italy, Germany, and the United States. The basic model was 'invented' by Professor Forester (interested readers will want to read his book, "Limits to Growth"). (Incidentally, his colleague Meadows claims that 'systems dynamics' was 'invented' by Forester of MIT

384

and that the model was the result of 30 years of work at MIT.) This model is now being used as justification for reducing the standard of living. There's absolutely no attempt at verification at all on this model (which is even conceptually impossible at the present time), and yet it is being used, to make decisions which can affect all of us here. I think in a larger sense this is the importance of the symposium because there are relatively few (very few) attempts at looking at the problem of identification in a real physical system hard as it is, and the progress we make will surely have a very great impact way beyond our own immediate problem. The sobering experience will be invaluable in models that affect our lives such as this one. Now, you can see what my fears are because all we have to do is make bids to companies to identify this model and of course translating for Dr. Queijo, they will produce any number you want, so depending upon which congressman you are, we'll furnish you with whatever numbers you want. This is why I think that the need for some consistent mathematical theory is quite essential, because a computer simulation proves little. I think that in taking at least one such problem, as the aircraft problem also points up the importance of the role played by the Center because one must take the first step in recognizing the need for improvement and more importantly judge the advances in the state of the art in theory and also in juding its value and relevance. If this is true in any application, I think it is even more true in identification for the reasons I've already mentioned to you but they're worth repeating, because without actual real live data identification with simulated data really does not prove anything. I don't know whether identification with real data proves anything or not, but at least we think it does beyond the simulated data. I am not saying that theory by itself has no value of course, on the contrary, the theoretical papers are of value in this area - they can shape even the basic modelling itself.

I think that I may end on this note: The need for a careful proven mathematical theory consistent within the model and not 'cluged; up, however complex, and tested on actual data. For once I would like to have the data-gathering planned as a statistical design problem with the definite purpose of identification in mind, and for once to broadcast our failures. Not merely successes but also the failures as a warning to those who would model the whole world- and emphasize the great need for caution in playing the identification game. I hope that this is but the first of a series of conferences which will allow the kind of interchange so essential before System Identification becomes an important valuable day-to-day activity, hopefully unlike adaptive control which was oversold and has faded away. And once again, I would like to express the indebtedness of my colleagues in the academic world as well as the engineering world in general for arranging this important conference.

☆   U.S. GOVERNMENT PRINTING OFFICE: 1974—739-160/111

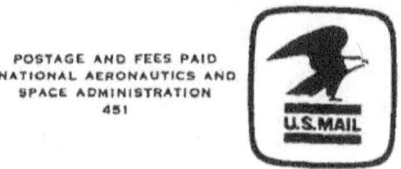
"The aeronautical and space activities of the United States shall be conducted so as to contribute . . . to the expansion of human knowledge of phenomena in the atmosphere and space. The Administration shall provide for the widest practicable and appropriate dissemination of information concerning its activities and the results thereof."
—NATIONAL AERONAUTICS AND SPACE ACT OF 1958

NASA SCIENTIFIC AND TECHNICAL PUBLICATIONS

TECHNICAL REPORTS: Scientific and technical information considered important, complete, and a lasting contribution to existing knowledge.

TECHNICAL NOTES: Information less broad in scope but nevertheless of importance as a contribution to existing knowledge.

TECHNICAL MEMORANDUMS: Information receiving limited distribution because of preliminary data, security classification, or other reasons. Also includes conference proceedings with either limited or unlimited distribution.

CONTRACTOR REPORTS: Scientific and technical information generated under a NASA contract or grant and considered an important contribution to existing knowledge.

TECHNICAL TRANSLATIONS: Information published in a foreign language considered to merit NASA distribution in English.

SPECIAL PUBLICATIONS: Information derived from or of value to NASA activities. Publications include final reports of major projects, monographs, data compilations, handbooks, sourcebooks, and special bibliographies.

TECHNOLOGY UTILIZATION PUBLICATIONS: Information on technology used by NASA that may be of particular interest in commercial and other non-aerospace applications. Publications include Tech Briefs, Technology Utilization Reports and Technology Surveys.

Details on the availability of these publications may be obtained from:

SCIENTIFIC AND TECHNICAL INFORMATION OFFICE

NATIONAL AERONAUTICS AND SPACE ADMINISTRATION
Washington, D.C. 20546

www.ingramcontent.com/pod-product-compliance
Lightning Source LLC
Chambersburg PA
CBHW080233180526
45167CB00006B/2257

* 9 7 8 1 4 9 9 1 6 1 7 6 2 *